BARRON'S

VISUAL LEARNING

Physics

Published by arrangement with UniPress Books Limited

Publisher: Nigel Browning

Page design: Mike LeBihan Studio

Illustrations: Sarah Skeate

Project manager: Viv Croot

Editorial consultant: Ken Rideout

First edition published in North America by Kaplan, Inc.,
d/b/a Barron's Educational Series

Published by Kaplan, Inc., d/b/a Barron's Educational Series

750 Third Avenue
New York, NY 10017

www.barronseduc.com

ISBN: 978-1-5062-6762-3

Kaplan, Inc., d/b/a Barron's Educational Series print books are
available at special quantity discounts to use for sales promotions,
employee premiums, or educational purposes. For more information
or to purchase books, please call the Simon & Schuster
special sales department at 866-506-1949.

Printed in China

10 9 8 7 6 5 4 3 2 1

Kurt Baker holds a degree in astrophysics from Cardiff University,
Wales, and a Ph.D. in astrophysics from Bristol University, UK, during which
time he authored and coauthored several astrophysics papers published
in NASA astrophysics journals.

BARRON'S

VISUAL LEARNING

Physics

AN ILLUSTRATED GUIDE FOR ALL AGES

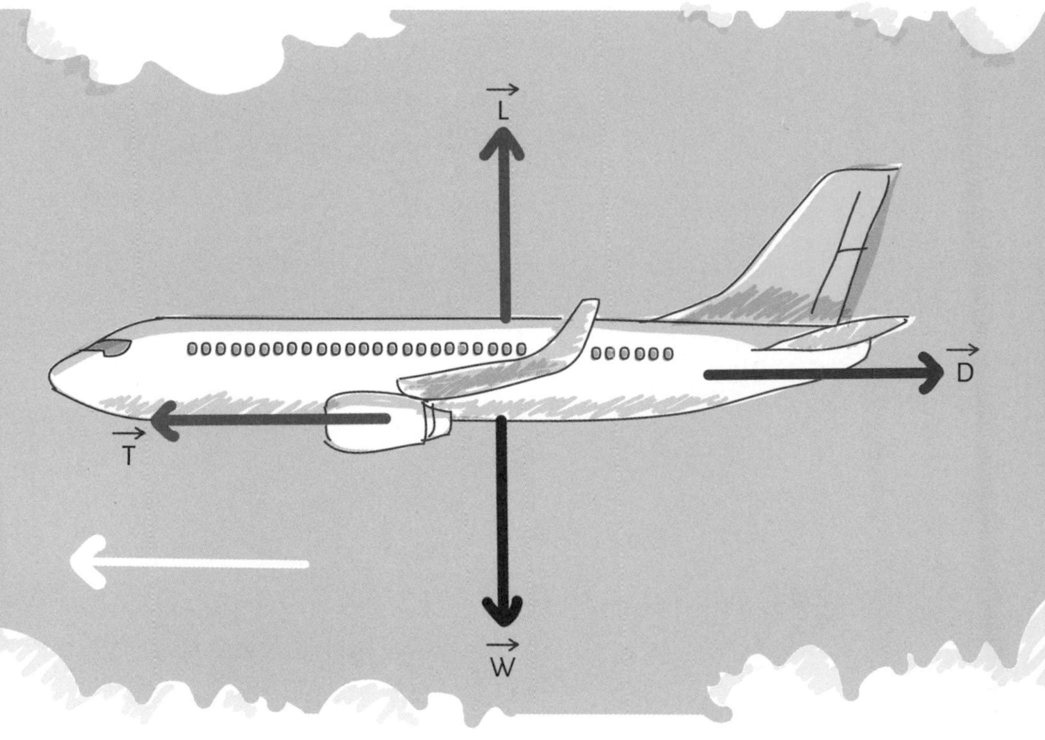

KURT BAKER

CONTENTS

Introduction 6

1 Forces 8
What Is a Force? 9
Contact Forces 10
Noncontact Forces 14
Newton's Laws 22
RECAP 24

2 Linear Motion 26
Particle Position 27
Particle Motion 28
Motion Graphs 30
Constant Acceleration 34
RECAP 36

3 Rotational Motion 38
Examples of Rotational Motion 39
Circular Motion 40
Orbital Motion 42
Rotational Kinematics and Dynamics 45
RECAP 48

4 Conservation Laws 50
Types of Conservation Laws 51
Closed Systems 52
Collisions 54
RECAP 60

5 Electricity 62
Electric Charge and Charge Transfer 63
Current, Voltage, and Resistance 64
Electrical Circuits 68
RECAP 74

6 Fields and Forces 76
Fields and Their Effects 77
Gravitational Fields 78
Magnetic and Electric Fields 80
RECAP 84

7 Electromagnetism 86
Faraday's Law of Induction 87
Electromagnetic Induction 88
Energy Loss and Energy Transfer 90
Electromagnetic Radiation and Spectrum 92
The Electromagnetic Spectrum 94
RECAP 96

8 Waves 98

Amplitude, Frequency, and Period 99
Simple Harmonic Oscillations 100
Traveling Waves 104
Wave Properties 106
Interference and Standing Waves 109
Doppler Shift 111
RECAP 112

9 Optics 114

The Laws of Reflection 115
Refraction, Snell's Law, and Total Internal Reflection 116
The Science of Optics 118
The Behavior of Light 120
Interference and Interferometry 123
RECAP 126

10 Thermodynamics 128

Temperature 129
Thermal Energy Transfer 130
Laws of Thermodynamics 134
RECAP 140

11 Fluids 142

Density and Pressure 143
Pressure Difference, Lift, and Buoyancy 144
Fluid Flow and Bernoulli's Principle 146
RECAP 148

12 Modern Physics 150

Special Relativity 151
General Relativity 152
Nuclear Physics 154
Nuclear Reactions 156
Quantum Physics 158
The Standard Model 160
Semiconductors 162
RECAP 164

13 Astrophysics 166

Stellar Evolution 167
The Hertzsprung-Russell Diagram 170
The Dynamics of Galaxies 172
Redshift and Recession Velocity 174
Hubble's Constant 176
The Beginning of the Universe 178
The End of the Universe 180
Gravitational Lensing and Gravitational Waves 182
Black Holes 184
RECAP 186

Index 188
Acknowledgments 192

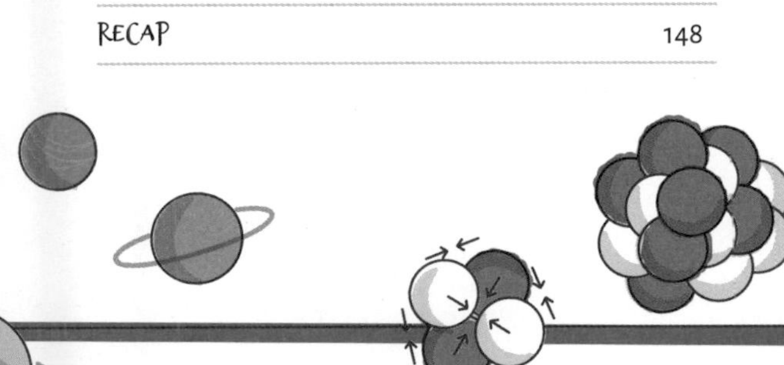

INTRODUCTION

Physics is the unifying science that defines and underpins everything around us. The universal and unequivocal laws of physics explain the mechanics of the world you live in, the atomic world beneath it, and the cosmic landscape of space above.

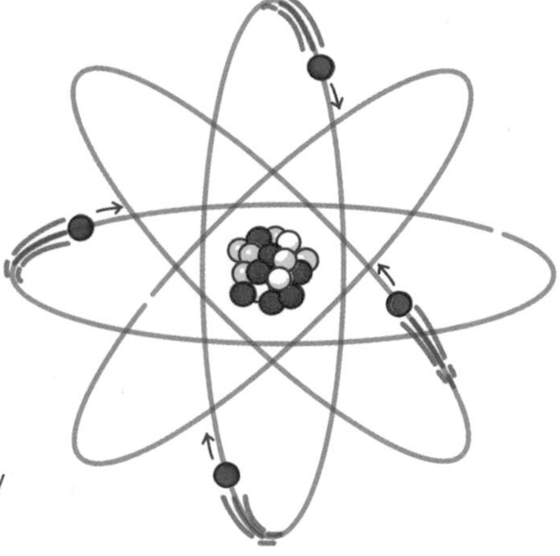

The secret nuclear heart of the atom, first revealed by Ernest Rutherford in 1909.

It was Sir Isaac Newton (1642–1727), English mathematician and physicist, who forged a path to understanding the movement of objects controlled by external forces. He bridged the gulf between physics and mathematics with the use of formulas to express complex physical relationships. He also began unraveling the mysteries of gravity and the dawn of classical mechanics and lit the forward path to unlock the mysteries of light, quantum physics, relativity, and cosmology. It was centuries before these enigmatic concepts gained clarity through the work of brilliant nineteenth- and twentieth-century minds, such as Albert Einstein, Max Planck, and Niels Bohr.

Each step and leap toward a comprehensive understanding of our universe was made from the shoulders of giants. Each discovery helped to propel humanity into a world of awe-inspiring ideas, providing the technologies for moon landings, long distance communication, and the development of ground and space-based telescopes that continue to probe the outer reaches of the universe, pushing the boundaries of our knowledge into new dimensions. Discoveries made in particle physics in the twenty-first century have pieced together the quantum realm and confirmed the existence of exotic particles predicted by theory. The acceleration of science has been exponential and continues to unveil new discoveries year on year.

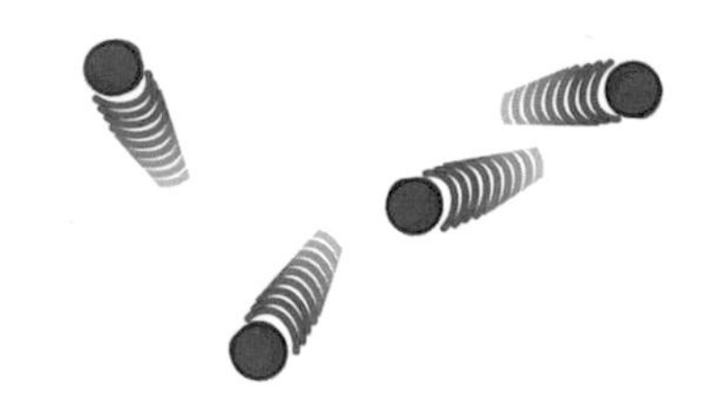

The unifying trait that binds all great physicists is curiosity. It's the "how does that happen?" factor that holds the scientific world together, and as more unknowns become known, more questions arise. If you share this curiosity, *Visual Learning Physics* is an invaluable resource that will help you grasp fundamental laws and concepts. Many of these ideas are complex and benefit from the visual learning approach in which concepts and ideas are presented with detailed images and diagrams, supported by clear and informative text.

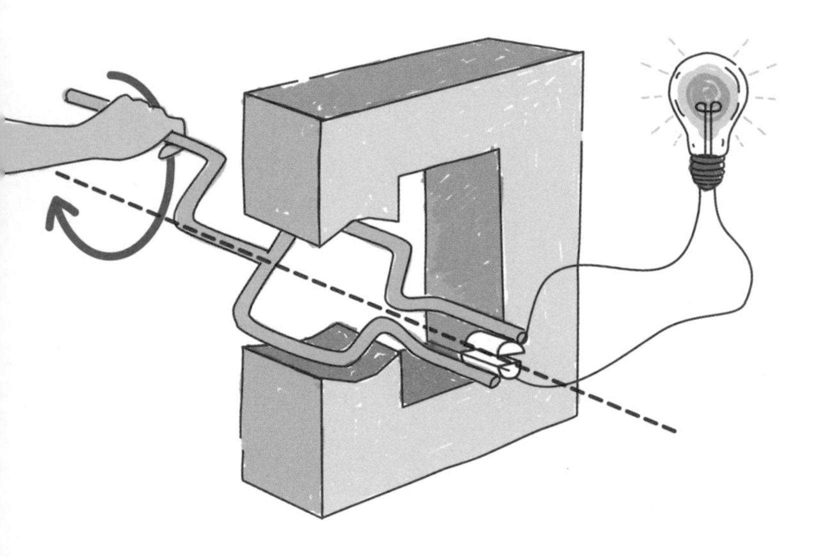

Power induced by passing a conductive wire through a magnetic field: the basis of electricity generation.

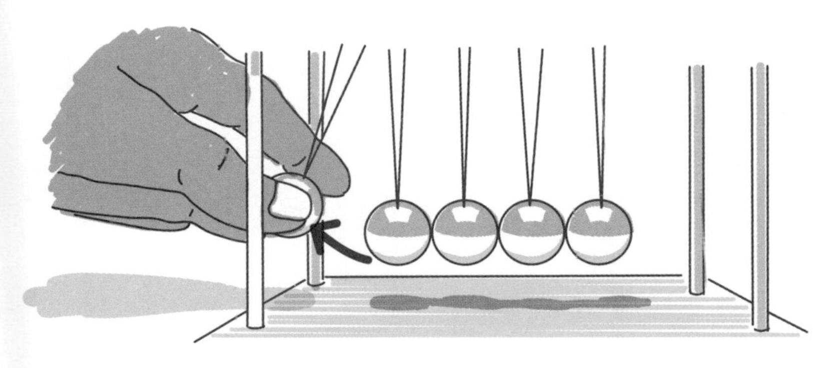

Newton's cradle perfectly demonstrates the conservation of linear momentum, a principle embodied in Newton's first law.

Broad in scope, *Visual Learning Physics* covers the subjects that are taught in physics courses and provides a strong foundation to venture further.

There are 13 chapters, each devoted to a specific area. You'll start by looking at the very basics of forces (gravity) and motion (linear and rotational), see how the universe obeys eternal laws that conserve every kind of energy, and examine the fields in which these laws operate. Moving on to specifics, you will learn about electricity, electromagnetism (without which the modern world would be nowhere), how waves behave in all mediums, what optics can show us (the World Wide Web, for example), the power of heat in thermodynamics, and how fluids are so much more than liquids.

When you have established the basics, it's time to follow Albert Einstein into the world of twentieth-century physics and confront the awesome might of nuclear power and the elegant enigma of quantum mechanics. Finally, in the chapter on astrophysics, you will reach the stars and discover, among other things, just how significant gravity is in the wider universe. Each chapter ends with a Recap feature, designed specifically with visual clues to help you recall what you have just learned.

Using careful and concise explanations, complemented by powerful and supportive imagery, this book is structured to gradually step upward from the basics of mechanics and forces into the more challenging ideas concealed in modern physics and astrophysics. It is the start of a journey that has no end but will stoke the fires of curiosity—the spearhead of discovery. Welcome to *Visual Learning Physics*!

Angular momentum is conserved as a skater changes her pose while she spins; the wider she makes herself, the slower she spins.

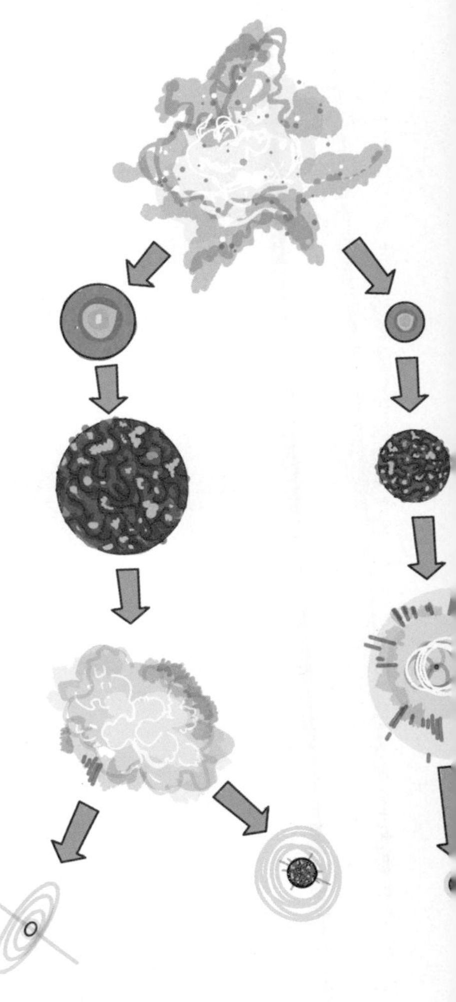

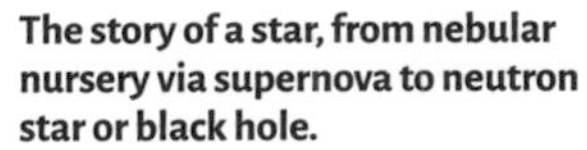

The story of a star, from nebular nursery via supernova to neutron star or black hole.

CHAPTER 1

FORCES

Forces are everywhere and yet you do not actually see them; instead, you experience the effects of forces. A force can transfer energy from one object to another, or a force can simply hold a body in position with no transfer of energy. Forces control the motion of planets and bind atoms together to form nuclei.

Forces can be classified into two categories: contact and noncontact forces. Contact forces are provided by physical mechanisms. Noncontact forces occur as the result of an influence at a distance.

WHAT IS A FORCE?

The internationally recognized unit of force is the newton (N), named for the English mathematician, physicist, and astronomer Sir Isaac Newton (1642–1727). Newton can be thanked for a very simple definition that is scientifically accurate.

If an object (referred to as a "body" by physicists) experiences an imbalance of forces, it will accelerate (speed up) or decelerate (slow down).

If two American football players collide with equal and opposite forces, there is no overall force and there is **no acceleration**.

An **imbalance of forces** means that there is an overall force in one particular direction: The body will accelerate or decelerate in that direction. By the same argument, if there is no overall force (forces are balanced), then a body's velocity (speed in a certain direction) will not change; it will remain motionless (at rest) or continue with constant velocity.

If a player tackles an opponent and the force of the tackle exceeds the opponent's force in the opposite direction, then there is an imbalance of forces and both players will either accelerate or decelerate in that direction.

If the same force is applied to both a large and a small player, the larger one will accelerate less rapidly because he has a larger mass; this is called **inertia**.

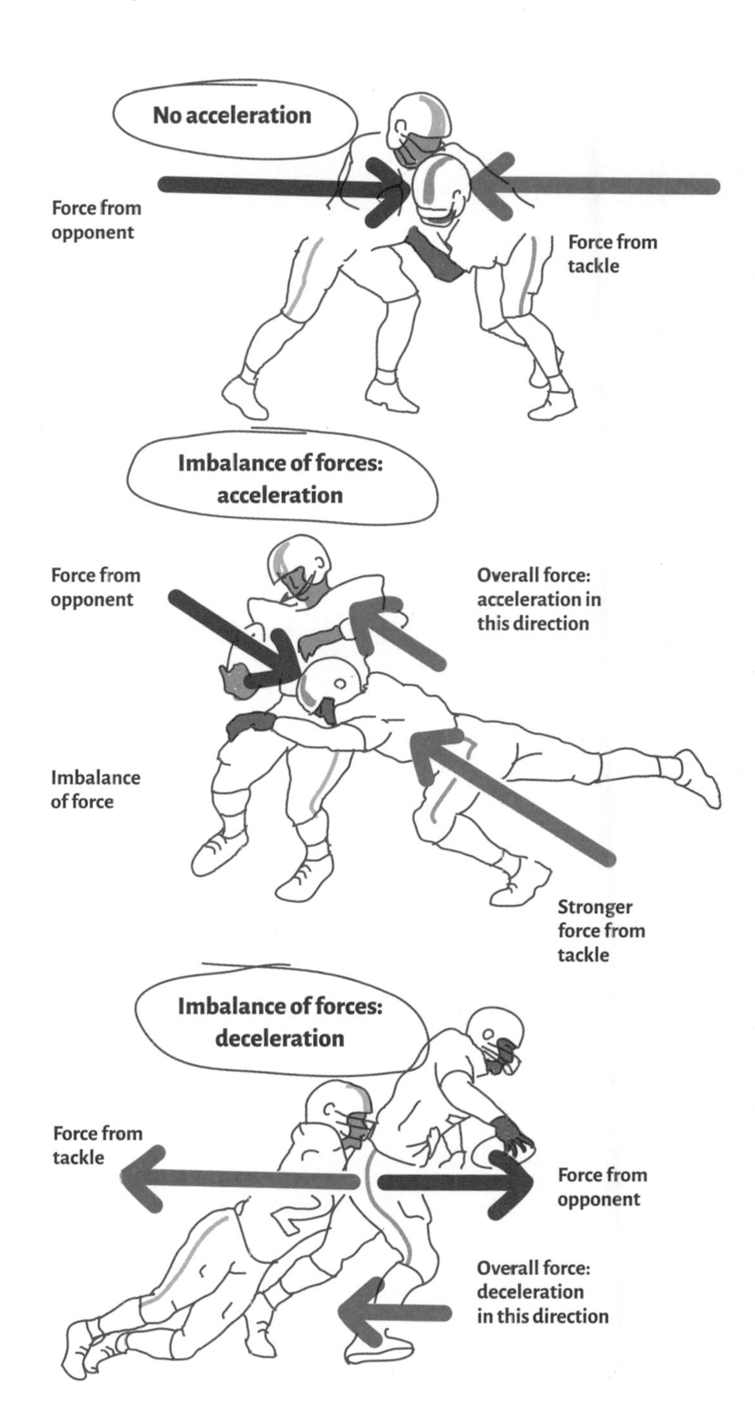

CONTACT FORCES

Contact forces are caused by different physical mechanisms in which the force provided is in contact with the body. These forces include applied (pushing), pulling (tension), frictional (air resistance), normal reaction (support), and spring.

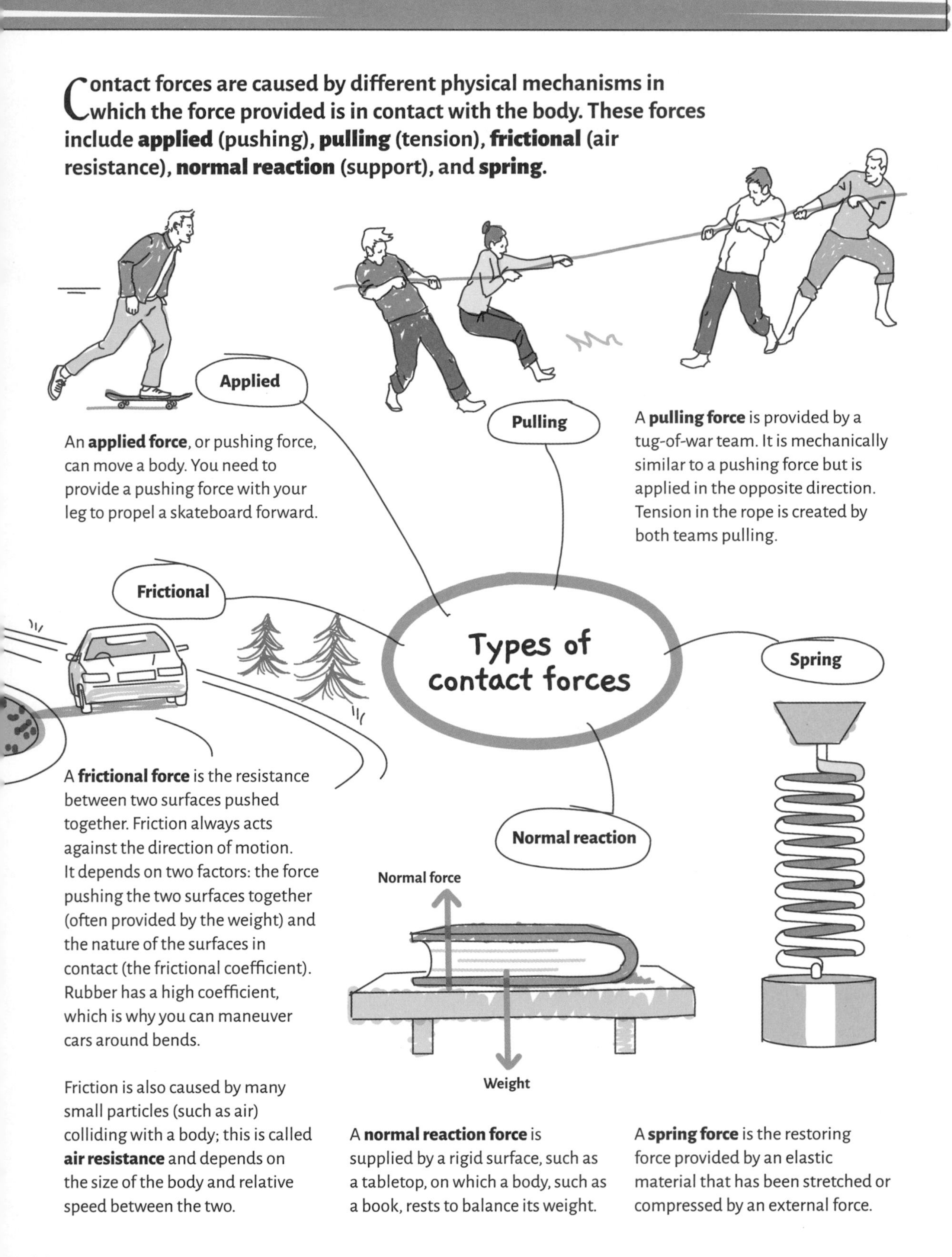

An **applied force**, or pushing force, can move a body. You need to provide a pushing force with your leg to propel a skateboard forward.

A **pulling force** is provided by a tug-of-war team. It is mechanically similar to a pushing force but is applied in the opposite direction. Tension in the rope is created by both teams pulling.

A **frictional force** is the resistance between two surfaces pushed together. Friction always acts against the direction of motion. It depends on two factors: the force pushing the two surfaces together (often provided by the weight) and the nature of the surfaces in contact (the frictional coefficient). Rubber has a high coefficient, which is why you can maneuver cars around bends.

Friction is also caused by many small particles (such as air) colliding with a body; this is called **air resistance** and depends on the size of the body and relative speed between the two.

A **normal reaction force** is supplied by a rigid surface, such as a tabletop, on which a body, such as a book, rests to balance its weight.

A **spring force** is the restoring force provided by an elastic material that has been stretched or compressed by an external force.

Normal reaction force

The normal reaction force prevents a body falling through a surface. Every object has **weight** (force due to gravity) on Earth and will accelerate downward. If a body is on any rigid surface that cannot move it will experience an equal force in opposition to its weight. As mass and therefore weight of a body increases, the normal reaction force will also increase to support it.

Imagine you are in an elevator, standing on a set of scales that monitors your weight. When the elevator is stationary, the weight registered by the scales is your "true weight."

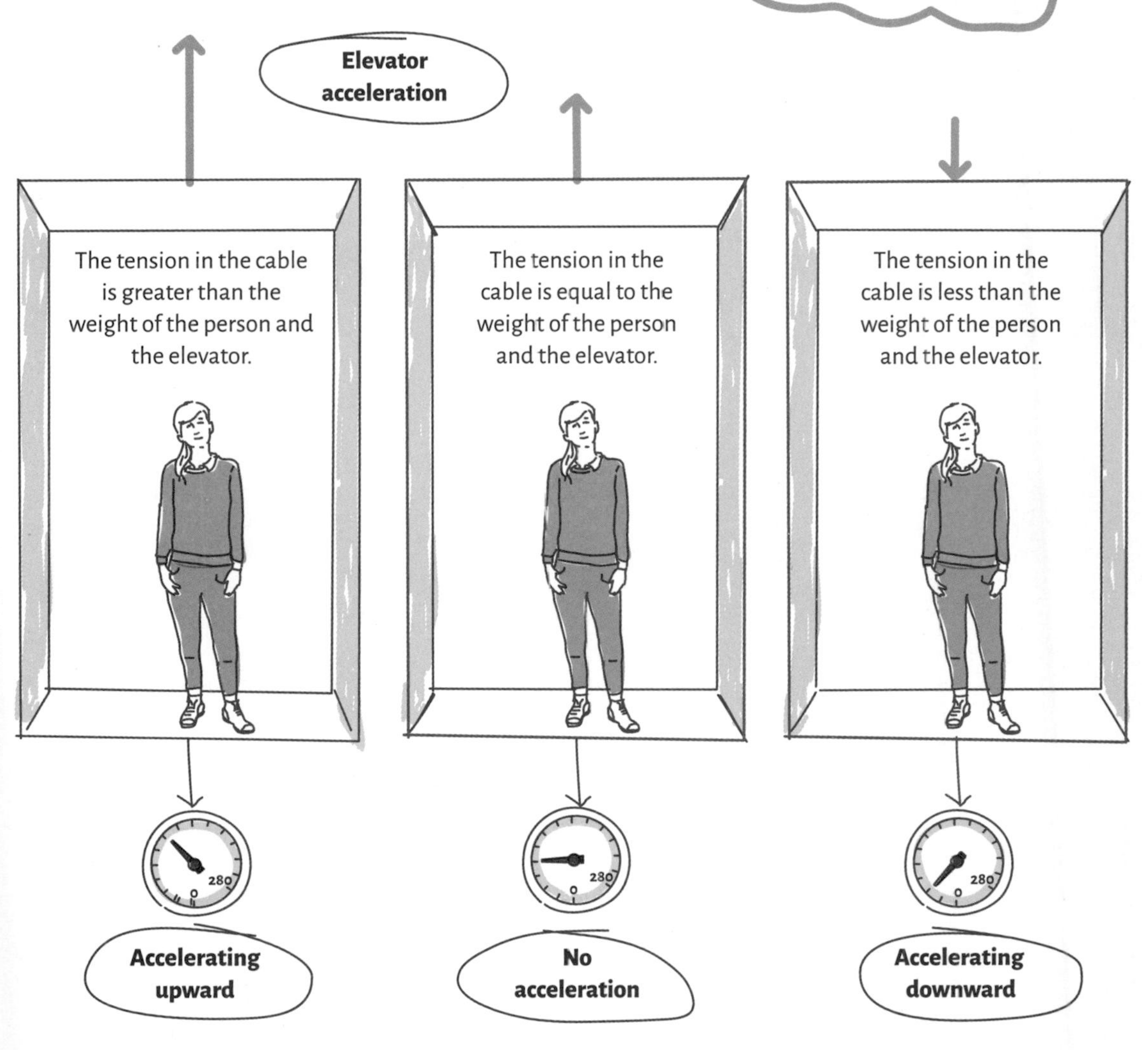

The motion of the elevator is controlled by the tension in the elevator cable. If the elevator is **accelerating** upward (getting faster), the floor of the elevator must support your weight and also provide the upward force to accelerate your mass. In this case, you feel heavier, and the scales will show an increase in weight.

When the elevator is traveling at a constant speed, there is no acceleration, and so the scales read your weight as it should be.

When the elevator starts to slow (**decelerate**), the opposite is true, and you feel lighter.

The reaction force you feel in your feet is not constant in an elevator throughout its motion. It will increase and decrease as the elevator changes velocity.

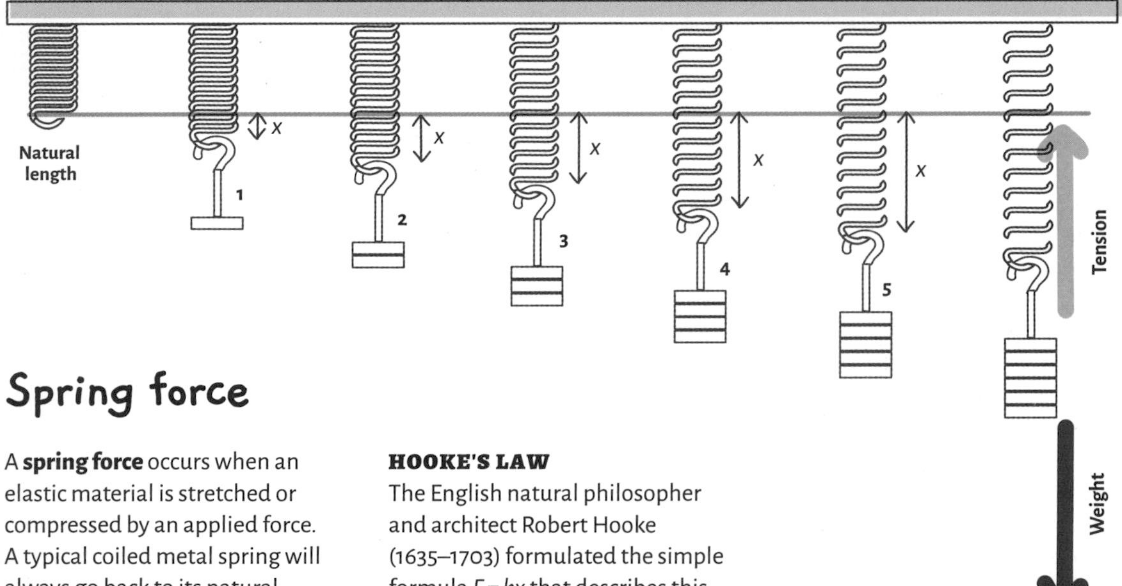

Spring force

A **spring force** occurs when an elastic material is stretched or compressed by an applied force. A typical coiled metal spring will always go back to its natural length when the force is removed. The tension or compression force in a spring is called the **restoring force**. The change in length of the stretched or compressed spring (x) depends on two factors: the magnitude of the force applied (F) and the stiffness of the spring (its spring constant, k).

HOOKE'S LAW

The English natural philosopher and architect Robert Hooke (1635–1703) formulated the simple formula $F = kx$ that describes this relationship. Expressed graphically, it is a straight line passing through the origin showing that an increase in applied force creates a uniform increase in length if the elastic limit is not reached.

This is the point at which the spring behaves differently under tension and depends upon the material. The **gradient** (steepness) of the graph is the spring constant.

The spring stiffness, k, is governed by a number of factors, such as material type and thickness as well as the diameter of the coils. In practice, it is unique for every spring and is measured in newtons per meter (N/m).

Hooke's law is not just confined to physical spring systems but is also a good model for the vibration mechanics of atoms in a material and the physics of waves. Both systems are governed by the magnitude of the restoring force from an equilibrium position and cause the system to oscillate.

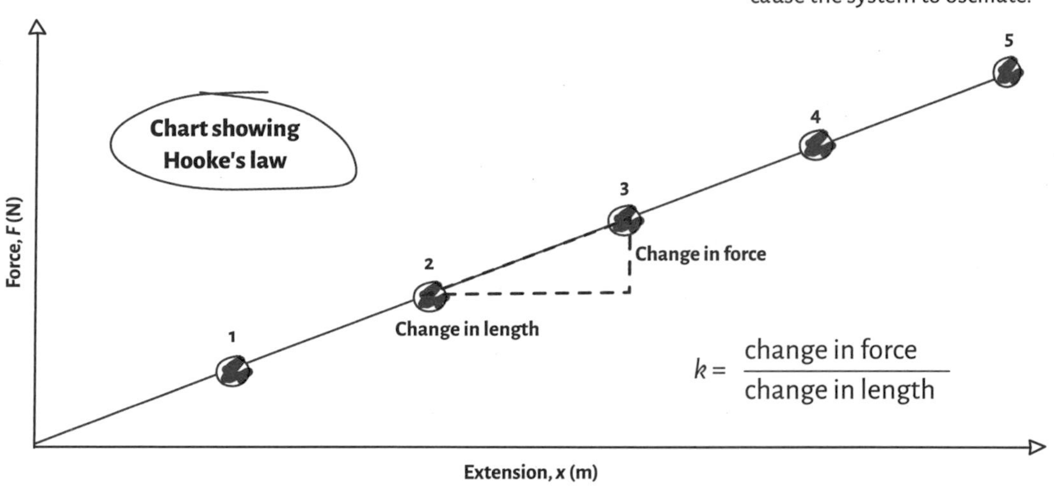

Vectors

Physicists deal with forces by quantifying their size (**magnitude**) and considering the direction in which each force is acting. The magnitude and direction are essential when dealing with forces, because the motion of the body as a result of the overall force is directly affected by both. The internationally adopted method for this is the use of vectors.

Vectors are an effective way of visualizing all the forces acting on a body and are a useful tool to understand what effect multiple forces have on the body's resultant motion. The overall force acting on a body can be calculated by the addition of vectors.

A **vector** is an arrow drawn from where the force is acting. The length of the arrow represents the magnitude of the force, and the direction of the arrow describes which way the force is acting. Each vector has a letter as a means to identify it in calculations—usually with a right-pointing arrow drawn above. For example, weight is usually represented by $\vec{W}$.

Often there is more than one force acting on a body, and these are represented by multiple arrows. An aircraft cruising at constant speed has various forces acting on it: **thrust** (T) from the engines, **drag** (D) from air resistance, the plane's weight (W), and the **lift** (L).

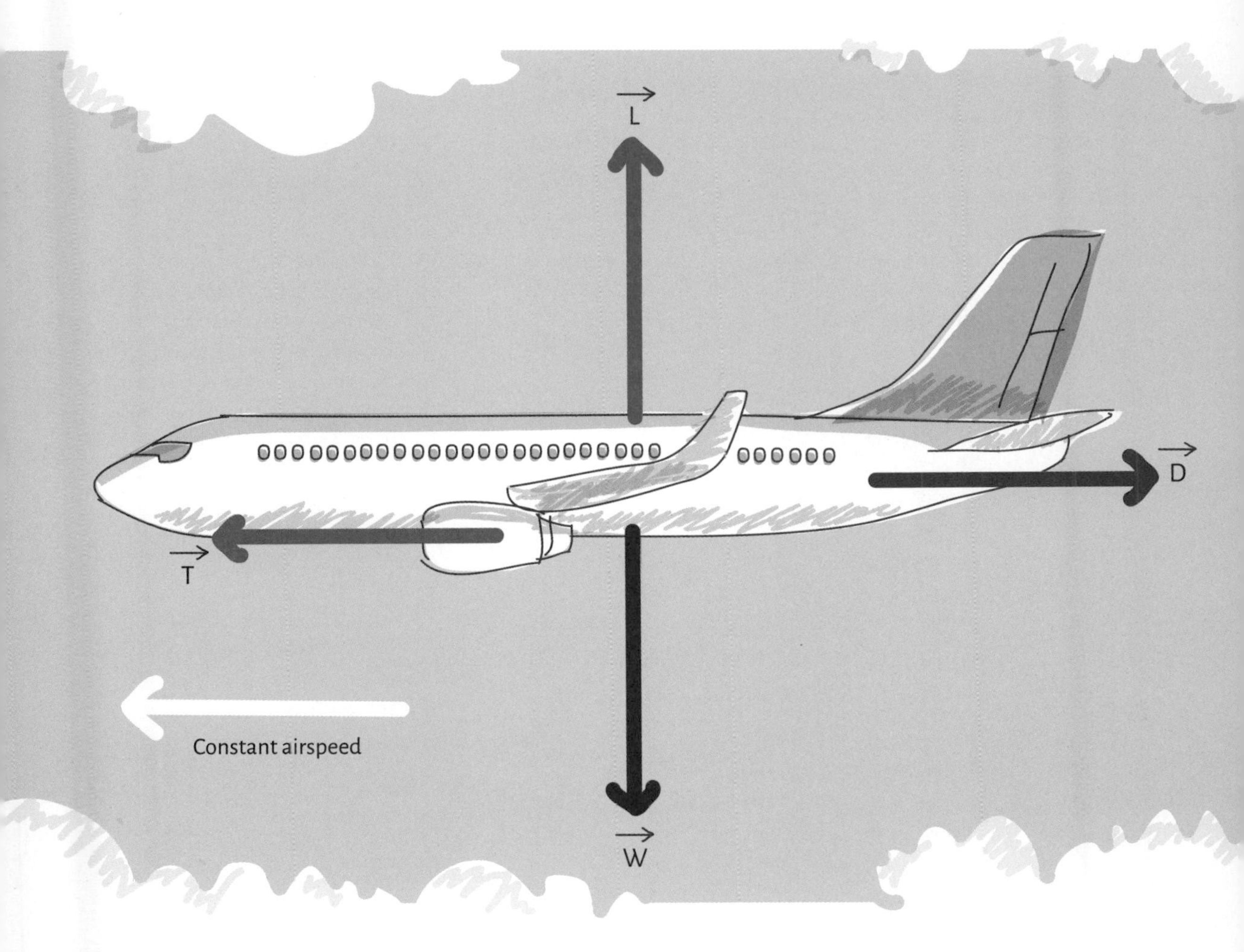

NONCONTACT FORCES

A noncontact (or action-at-a-distance) force has an effect on an object without physically touching it. All noncontact forces are affected dramatically by distance. These include gravitational (weight), electrostatic, magnetic, and nuclear forces.

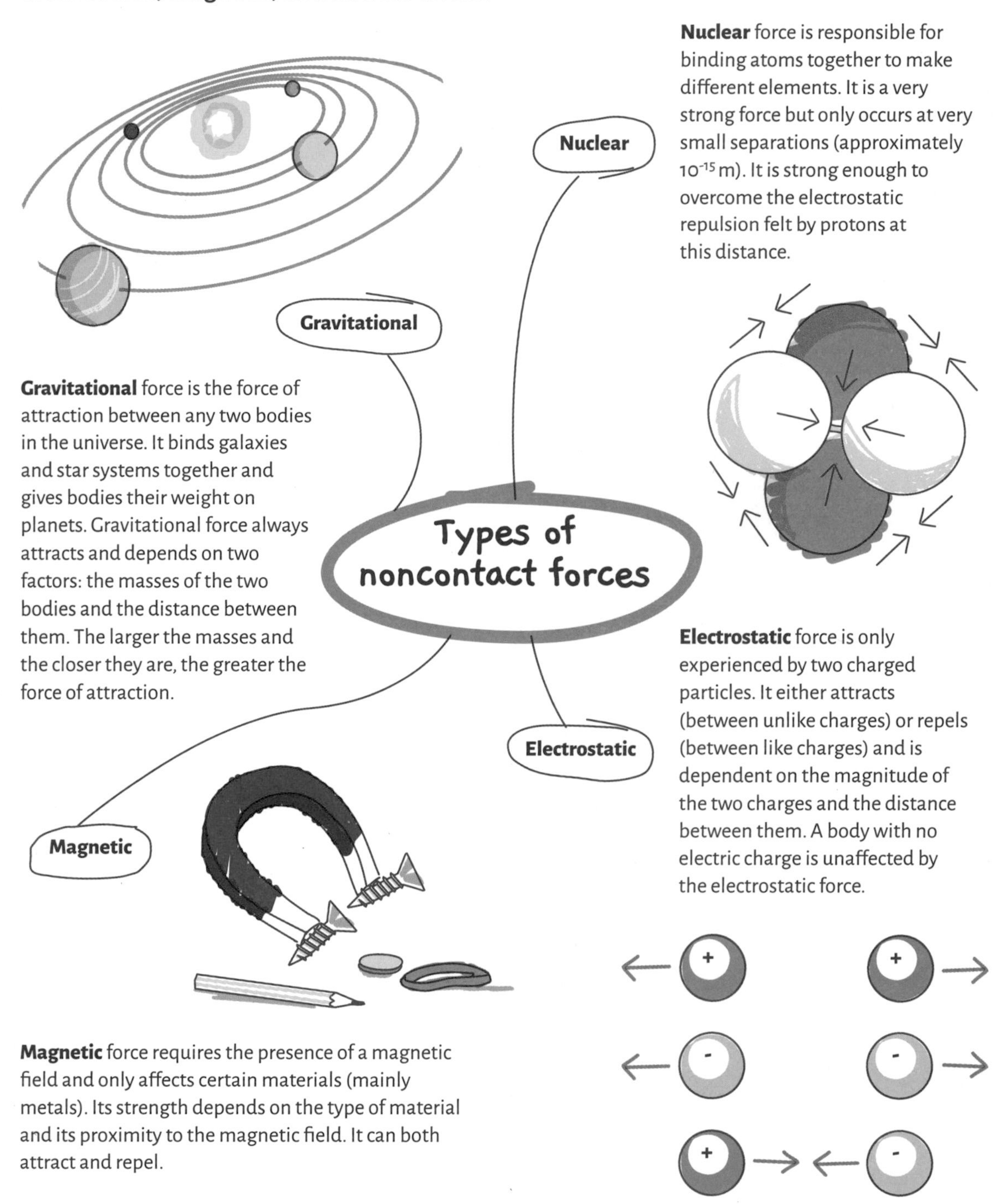

Nuclear force is responsible for binding atoms together to make different elements. It is a very strong force but only occurs at very small separations (approximately 10^{-15} m). It is strong enough to overcome the electrostatic repulsion felt by protons at this distance.

Gravitational force is the force of attraction between any two bodies in the universe. It binds galaxies and star systems together and gives bodies their weight on planets. Gravitational force always attracts and depends on two factors: the masses of the two bodies and the distance between them. The larger the masses and the closer they are, the greater the force of attraction.

Electrostatic force is only experienced by two charged particles. It either attracts (between unlike charges) or repels (between like charges) and is dependent on the magnitude of the two charges and the distance between them. A body with no electric charge is unaffected by the electrostatic force.

Magnetic force requires the presence of a magnetic field and only affects certain materials (mainly metals). Its strength depends on the type of material and its proximity to the magnetic field. It can both attract and repel.

Weight

Weight (the force due to gravity) and mass are not the same thing. Mass is measured in kilograms, and weight is a force, so it is measured in newtons (1 N ≈ 0.10197 kg). The **mass** of a body is constant—it's a measure of how much of something there is. Its weight, however, will depend on the magnitude and proximity of a mass to a large body that generates a **gravitational field**.

The strength of a gravitational field is denoted as g and gives the force in newtons experienced by each kilogram of mass. On (or close to) Earth, each kilogram has a weight of about 9.8 newtons (g = 9.8 N/kg). In fact, there is a force of equal magnitude acting on both the planet (mass, *M*) and the body (mass, *m*) toward each other, but the planet is so massive by comparison that this force has no observable effect on it.

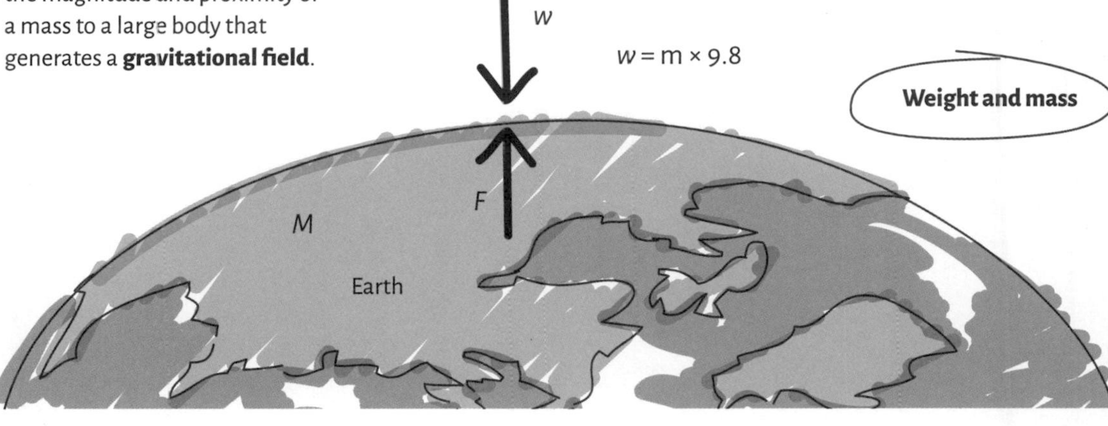

Which body falls faster?

If a body is dropped, it will accelerate toward the ground at a rate of 9.8 m/s each second (m/s^2). This is the same for any object regardless of mass. If the effect of air resistance is removed, a bowling ball and a feather dropped together will accelerate toward the ground at the same rate.

This idea is not intuitive, but it is a direct result of Newton's second law. The force experienced by the bowling ball is much greater than that experienced by the feather.

However, the bowling ball requires a larger force, proportionate to its mass, to accelerate it.

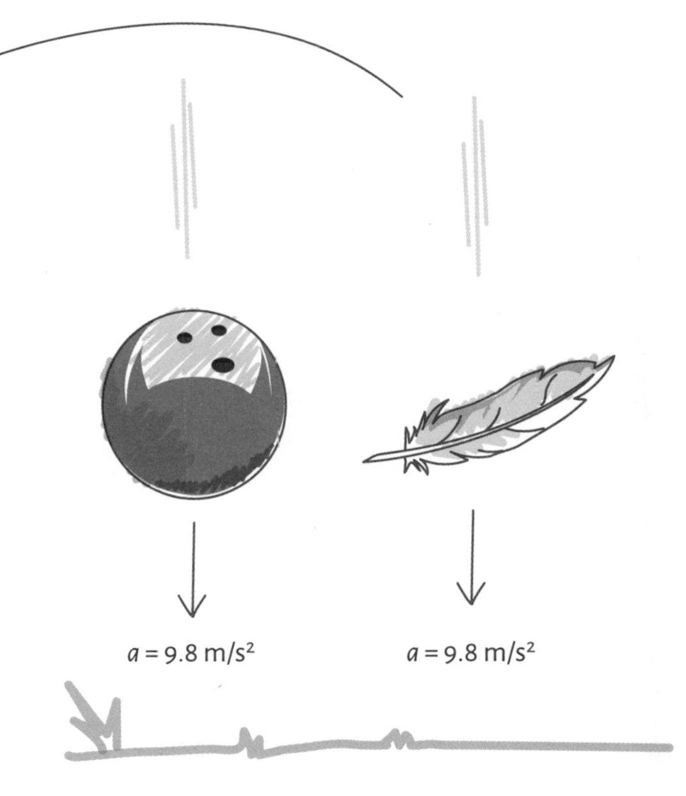

Gravitational force

Once again, it was Newton who defined the force due to **gravity**. He stated that the magnitude of the force between two masses is proportional to the product of the two masses and inversely proportional to the square of their separation. Simply put, the size of the force experienced between two bodies greatly increases the closer they are and scales directly according to their masses.

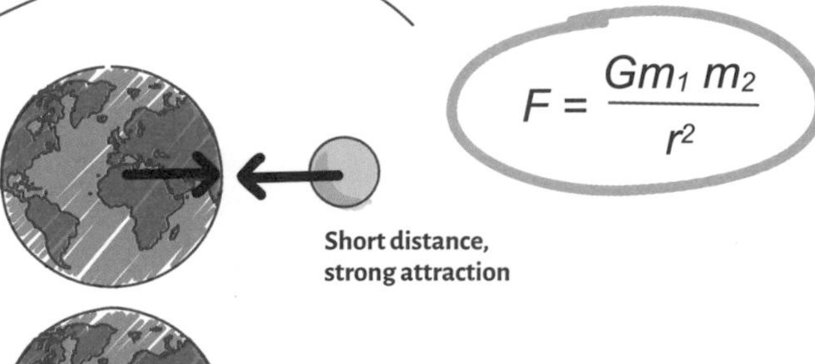

$$F = \frac{Gm_1 m_2}{r^2}$$

Short distance, strong attraction

Long distance, weak attraction

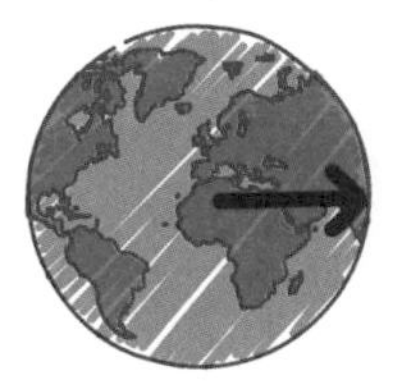

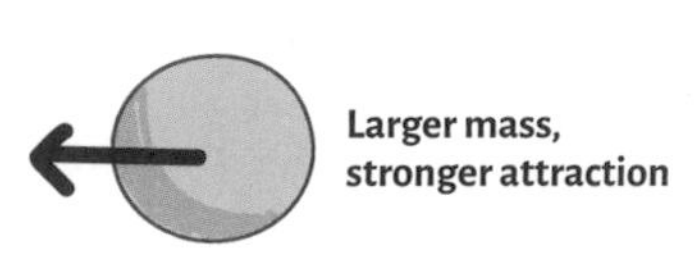

Larger mass, stronger attraction

F is the magnitude of the force in newtons, m_1 and m_2 are the masses of each body in kg, r is the separation distance in meters, and G is the **universal gravitational constant**. This is very small (numerically, it is approximately 6.67×10^{-11}), illustrating that forces between masses are only significant when either or both of the masses are very large, such as a planet or star. The interaction between small masses is so tiny, gravitational effects are almost undetectable.

Large masses distort the fabric of space-time, causing localized effects around a massive body, even redirecting the path of light beams.

Jumping on the moon

The moon's gravity is one-sixth of that on Earth. If you weighed 45 kg (100 lb) on Earth, your mass would be the same on the moon, but you would weigh 7.7 kg (17 lbs). If you could jump 1 meter (3.3 ft) on Earth, you could jump 5.5 meters (18 ft) on the moon.

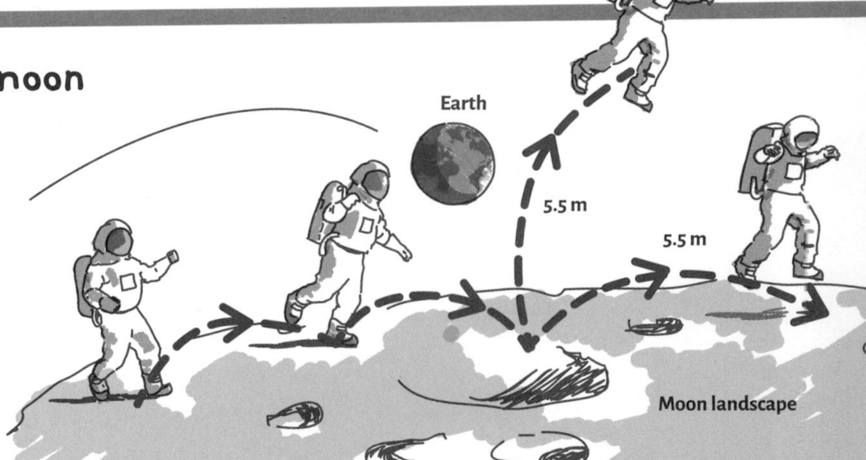

Moon and tides

Tidal bulge due to inertia

Tidal bulge due to gravity

Moon

Moon's gravitational pull

The moon and Earth exert equal and opposite forces upon one another that causes distortion of any fluids (such as water).

Over large areas of water in Earth's oceans, the force of the moon (called a **tidal force**) causes a bulge in its surface closest to the moon. On the opposite side of Earth, the moon's gravitational effects are much lower, due to the extra distance from it and the inertia of the water in **orbital motion**. These two factors work against the gravitational pull of the moon, creating an outward bulge on the opposite side of Earth. This causes the differences in sea levels.

What holds stars together?

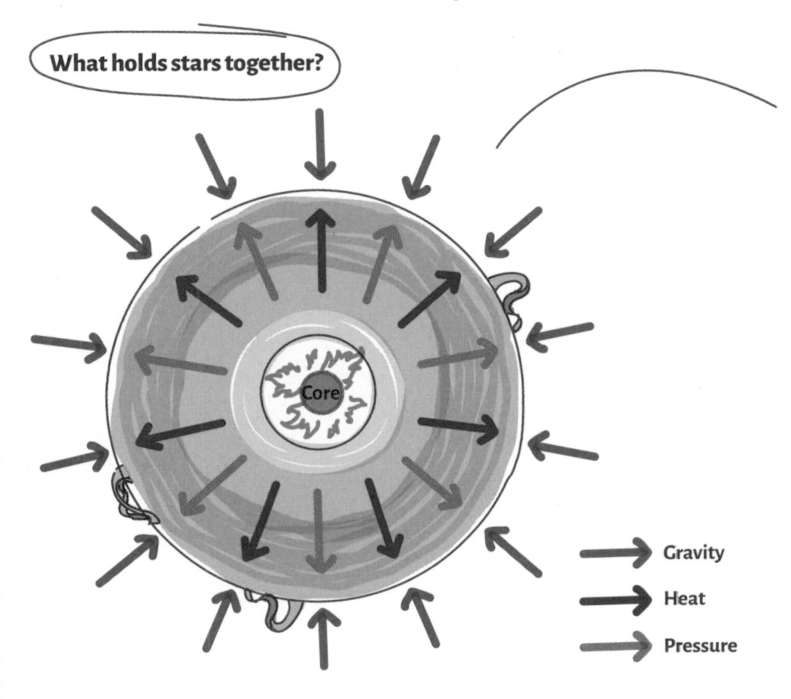

As a star fuses hydrogen in its center, a huge quantity of energy is produced, resulting in an outward pressure due to the radiation field. Coupled with **gas pressure**, the force of gravity due to the mass of the star is balanced, creating a stable sphere that for medium-to-small stars lasts for billions of years—this is called **hydrostatic equilibrium**. When the central region depletes its reserves of hydrogen, energy is no longer produced and the radiation pressure stops. The star becomes unstable, and gravity becomes the dominant force, causing it to collapse.

The force of gravity is what keeps objects on the ground on Earth (and other planets). It allows satellites and moons to orbit planets, and planets to orbit stars. It also binds together entire galaxies, from the combined mass of their stars, and will eventually dictate the fate of the entire universe.

Gravity is essential to our existence, but its origins are still a mystery. The *graviton* is a theoretical particle that is currently thought to be responsible for the transmission of gravity but is yet to be detected.

Electrostatic force

Like and unlike

Attract

Repel

Repel

Electrostatic force is felt between two (or more) charged particles. It is highly dependent on the proximity of the bodies to one another and scales with the magnitude of the charge (measured in **coulombs**) on each body. Unlike gravity, however, it can both attract and repel. Two like charges (either both positive, or both negative) will repel each other, whereas two unlike charges (one positive, one negative) will attract.

Within all matter ,there are protons, neutrons, and electrons. **Protons** carry a positive charge, **electrons** a negative charge, and **neutrons** are neutral (carry no charge). Two protons will strongly repel one another, but a proton and a neutron experience no electrostatic force due to the charge-neutral nature of the neutron. It is the opposite charges of protons and electrons that binds them together—all elements contain equal numbers of protons and electrons in their atoms, which maintains **charge neutrality**.

A helium atom

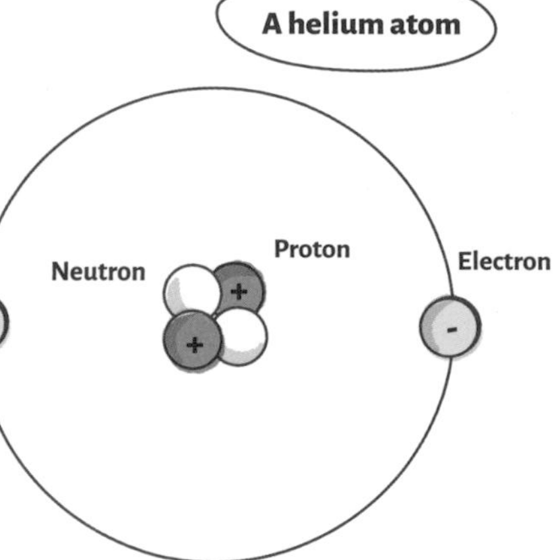

CREATING STATIC

It is possible to positively charge a surface by removing some of the electrons, leaving an overall positive charge. Simply rubbing a nylon cloth over a surface (such as a balloon) will transfer electrons to the cloth, creating an electrostatically charged balloon. When held near your hair, the positive charge of the balloon will attract the electrons in your hair, pulling it outward.

This simple demonstration shows that the electrostatic force is very strong compared to gravity, being able to overcome the weight of the hair.

Nuclear forces

Nuclear forces are extremely strong when acting at very short distances. They are encountered within the nucleus of all atoms and act between all **nucleons** (neutrons and protons).

The nuclear force binds nucleons together to form elements and is strong enough to overcome the electrostatic repulsion felt by neighboring protons when nucleons are separated by approximately 1×10^{-15} m (1 femtometer). If the separation of the nucleons exceeds roughly double this (approximately 2.5×10^{-15} m), the strong nuclear force no longer has an effect.

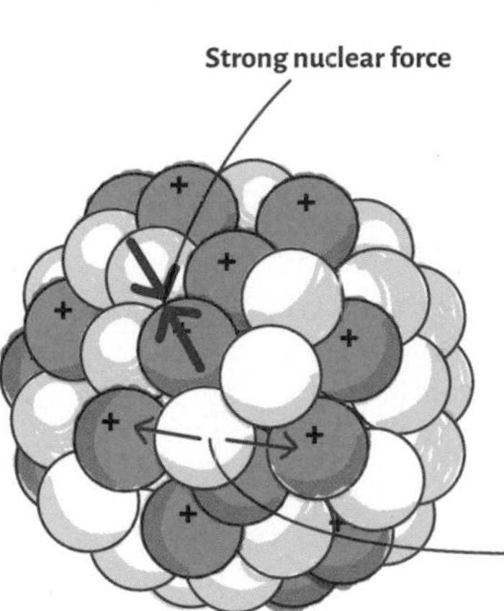

Nuclear fission

Most elements, such as oxygen or carbon, are stable, as the nuclear strong force is sufficient to overcome the electrostatic repulsion of the protons in the nucleus. However, **radioactive** elements have unstable nuclei and will decay into smaller fragments (see page 156) while producing energy and radiation. This is called **nuclear fission**.

Fuel elements or **rods** contain naturally occurring uranium-235. A **thermal** (moving) neutron is fired into the fuel rods and temporarily binds with the nucleus, creating highly unstable uranium-236. This undergoes nuclear fission (breaks apart), releasing three (fissile) neutrons while producing energy, a gamma ray, and radioactive waste.

The three neutrons are moving very quickly and are slowed by the **moderator fluid** in the reactor core. Two of the three fissile neutrons are then absorbed by the **control rods**, thus preventing a catastrophic **chain reaction** where one reaction would lead to three more, and so on. Unstable radioactive elements are unable to remain intact as the nuclear strong force that binds them is insufficient to hold so many nucleons together, so they break apart to create more stable elements with fewer nucleons, and in doing so, release energy.

A nuclear fission reactor uses uranium-235 as a fuel source.

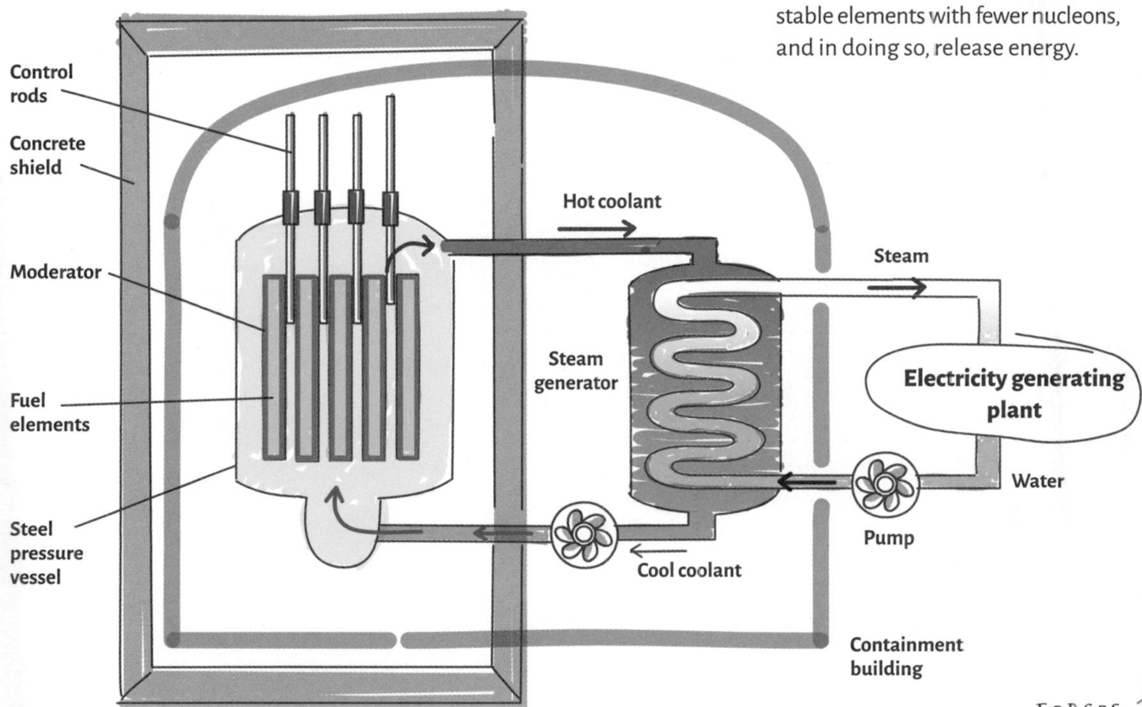

Magnetic force

A **magnetic** force is experienced either between two **magnetized bodies** (which can attract or repel) or between a magnetized body and a magnetic body, such as a metal. Magnetized materials are surrounded by a **magnetic field**, the strength of which depends on the material itself and its distance from the object.

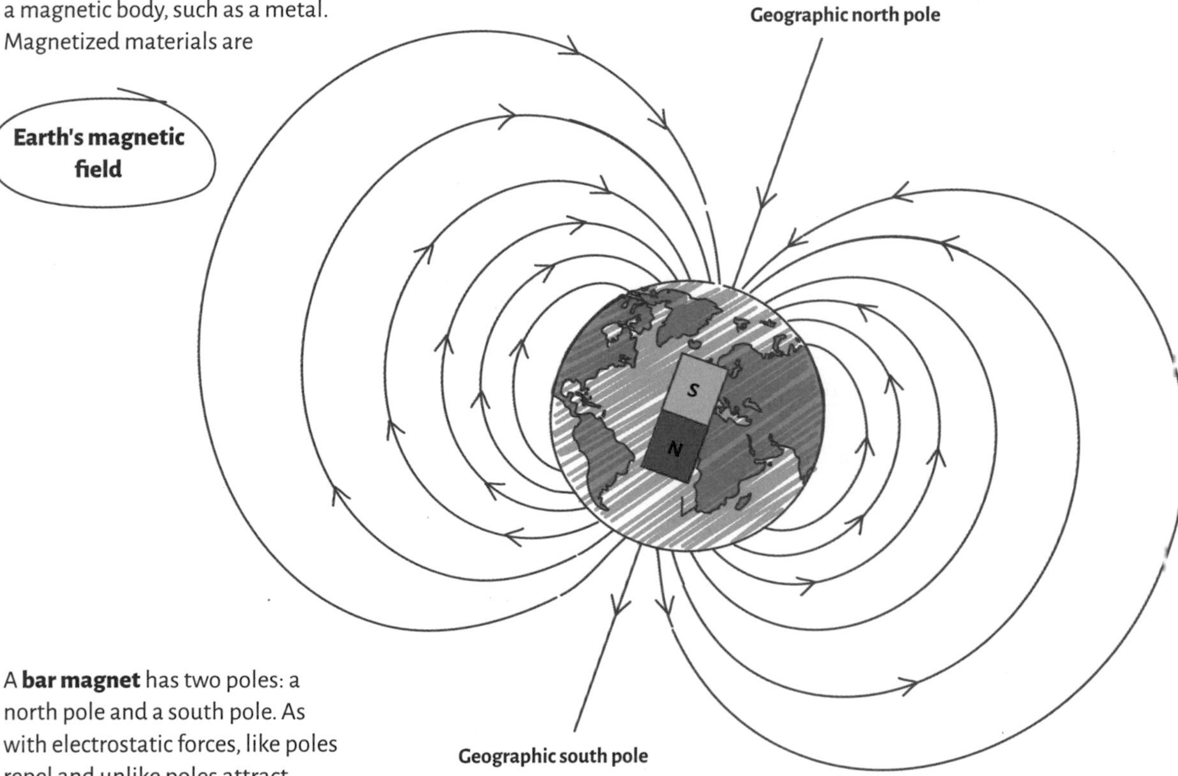

A **bar magnet** has two poles: a north pole and a south pole. As with electrostatic forces, like poles repel and unlike poles attract. Earth acts as a giant bar magnet, and so the magnetic north pole of a compass needle is attracted to the south pole of Earth.

MAKE YOUR OWN MAGNETIC FIELD

You can demonstrate the existence of a magnetic field using a bar magnet, iron filings, and a sheet of paper. Lay the paper over the magnet, and sprinkle iron filings over it. Tap the paper gently, and watch as the filings are attracted to the poles, their position revealing the shape of the magnetic field being generated.

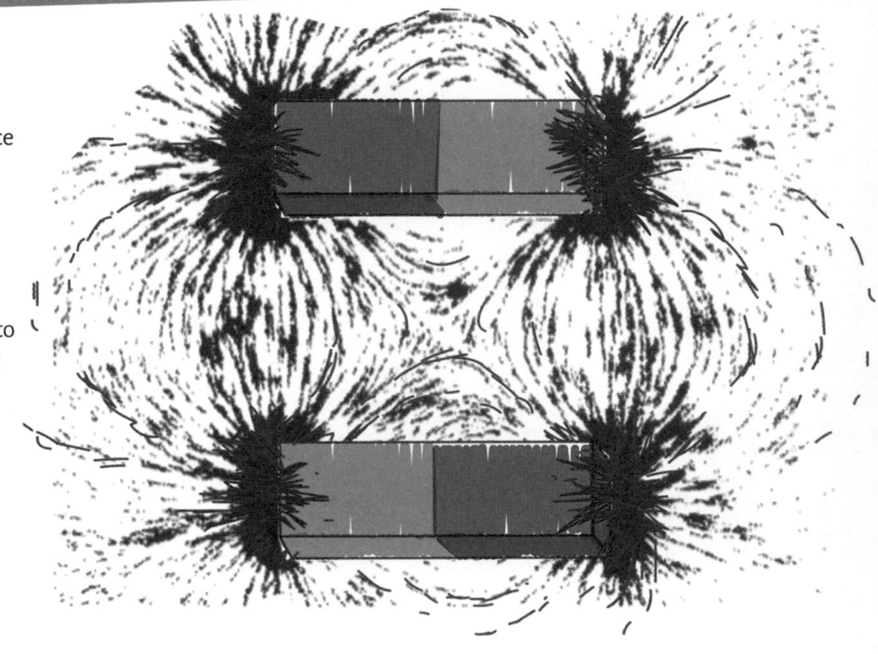

MAGNETIC ATTRACTION

Magnets can be **permanent** (always have a magnetic field) or temporary. **Electromagnets** only possess a magnetic field when an electric current is passed through them.

Magnetic field strength is measured in tesla, T, and varies enormously. A fridge magnet is 0.00005T, whereas the magnets used in Magnetic Resonance Imaging (MRI) are 1.5T.

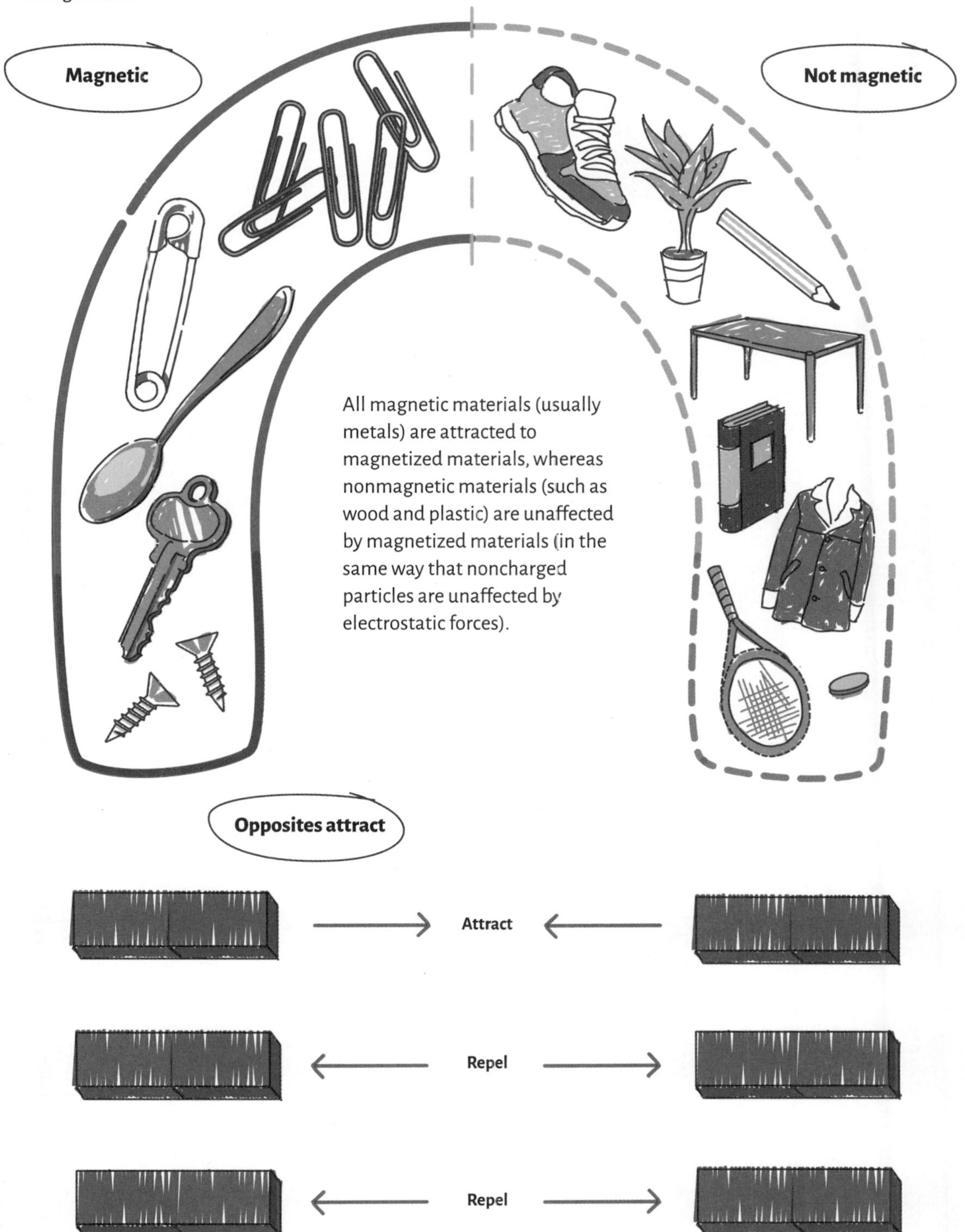

All magnetic materials (usually metals) are attracted to magnetized materials, whereas nonmagnetic materials (such as wood and plastic) are unaffected by magnetized materials (in the same way that noncharged particles are unaffected by electrostatic forces).

NEWTON'S LAWS

Newton formulated his three fundamental laws that govern how forces affect motion. They are as follows:

1. A body will remain at rest (stationary) or continue moving with constant velocity unless acted upon by an external force.

2. The acceleration of a body is directly proportional to the magnitude of the total force applied and is in the same direction as the force. This means that a larger mass requires a larger force to accelerate at the same rate.

3. Every action has an equal and opposite reaction. This means if a force is applied to a body, the body will provide a force equal in magnitude in the opposing direction.

You have seen that if an overall force is applied to a body, it will accelerate. **Acceleration** is how quickly a body's velocity changes (how quickly it speeds up).

In the diagram, the rocket engines provide thrust (T), which is greater than the combined weight and drag forces (W + D). This imbalance of forces provides an upward acceleration. Its speed (blue vector) increases, as shown in the speed-time graph. Each snapshot of the rocket is a fixed time gap apart; the distance (shown by yellow arrows) it has traveled after each time period is shown in the distance-time graph.

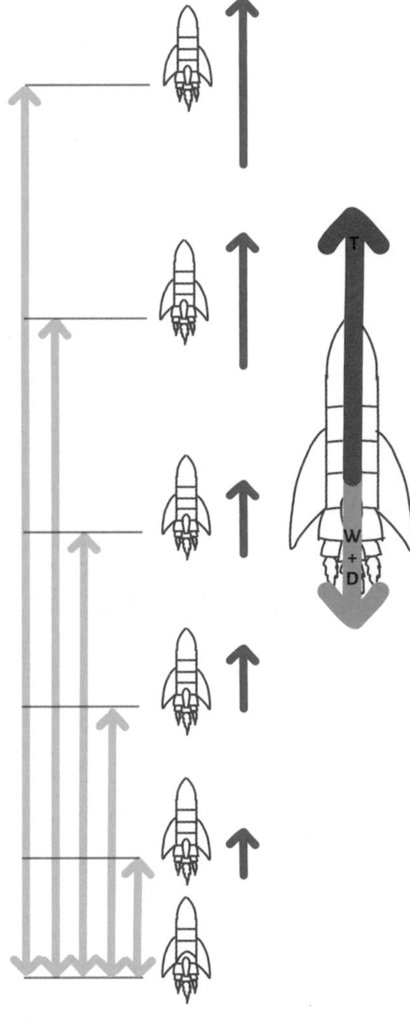

A rocket accelerates upward as a result of its thrust (T) being greater than its weight (W) plus drag (D). The blue arrows indicate greater velocity; the yellow arrows indicate increasing distance.

The rate of increase in velocity will depend directly on two factors: The size of the force and the mass of the body. It makes sense that it requires a larger force to accelerate a larger body.

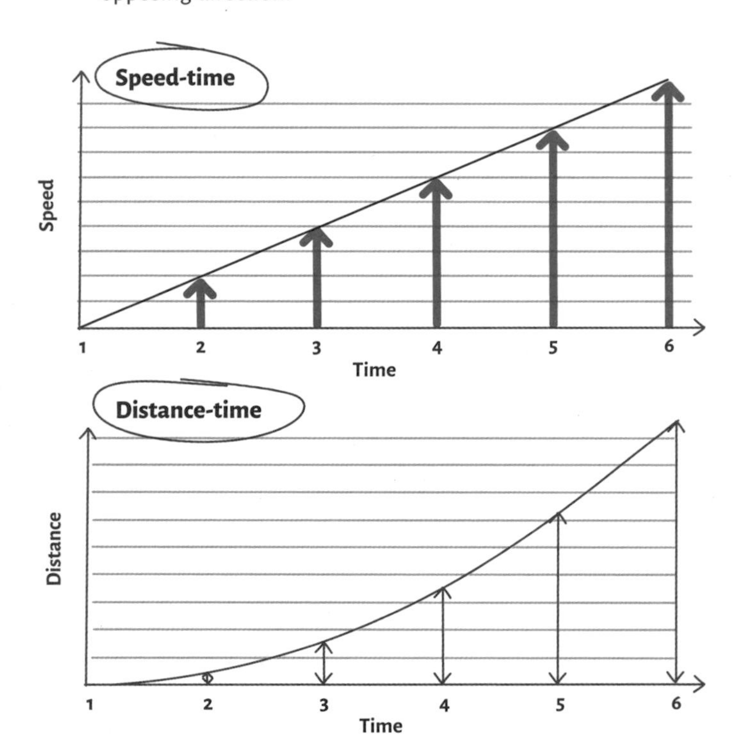

The newton

Rate of change of momentum

Newton formulated his second law that defines what the newton actually is. One newton will change the velocity of one kilogram by one meter per second in every second. This statement is simplified by the formula

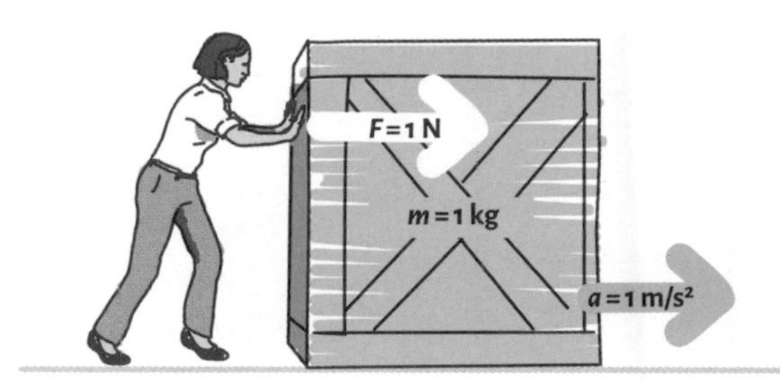

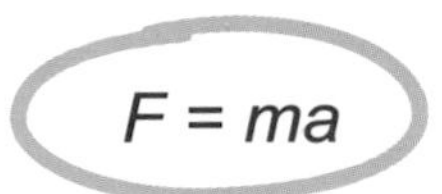

F is the force applied in newtons; m is the mass of the body in kilograms (1 kg = 2.2 lbs); and a is the resultant acceleration (measured in m/s^2). A larger force is required to accelerate a larger mass at the same rate.

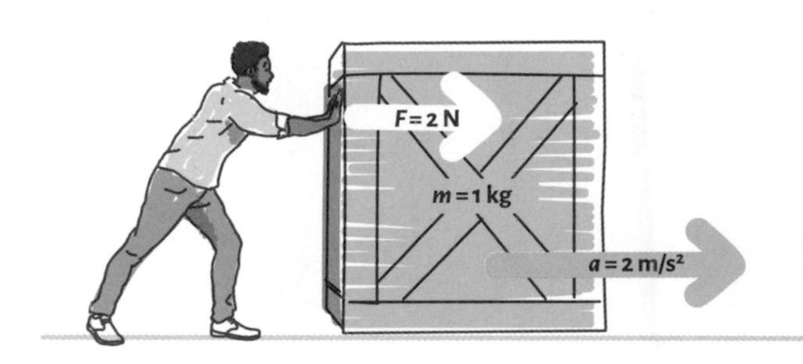

The scenarios above show that the **rate of change of momentum** (how quickly momentum changes) is doubled when the force is 2N compared to a force of 1N in the same time period, providing there is no friction between the crate and the ground in each case.

Here is another definition of the newton. **Momentum** *(p)* is defined as the product of the mass of a body with its velocity and is measured in kg m/s. As a formula, $p = mv$, where p is the momentum, m is the mass of the body (in kg), and v is the velocity (m/s).

A diving kingfisher has a large momentum despite its small mass, as it has a high velocity. A slow-moving bear also has a large momentum as it has a very large mass. A kingfisher has a small momentum compared to a bear, but its mass is 6,000 times smaller, whereas its momentum is only 1,200 times smaller.

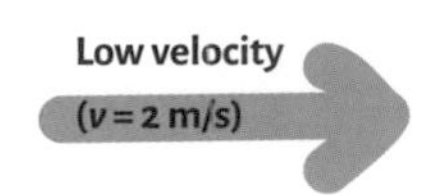

$p = 2 \times 200 = 400$ kg m/s

RECAP

FORCES

CONTACT FORCES

APPLIED FORCE

A pushing force generated by propulsion.

PULLING FORCE

Creates tension in the pulling medium.

FRICTIONAL FORCE

This is caused by the resistance between two surfaces or is experienced by a body moving through a fluid.

SPRING FORCE

Restoring force provided by elastic material stretched by an external force.

HOOKE'S LAW

A force that stretches or compresses a spring over a distance is proportional to that distance.

NORMAL REACTION FORCE

Supplied by a rigid surface on which a weight rests.

VECTOR

Describes magnitude and direction.

LAWS OF MOTION

MOMENTUM

Mass (m), × velocity, (v) measured in kg m/s

THE NEWTON

Measure of force; one newton accelerates 1 kg by 1 m/s^2.

NEWTON'S THREE LAWS OF MOTION

FIRST LAW

A body will remain stationary or move at constant velocity until acted on by an external force.

SECOND LAW

Acceleration is directly proportional to the total magnitude of the force applied.

THIRD LAW

Every action has an equal and opposite reaction.

The force of attraction between any two bodies in the universe; always attractive.

GRAVITATIONAL FORCE

NEWTON'S LAW OF GRAVITY

The magnitude of the force between two masses is proportional to their product and inversely proportional to the distance between them.

$$F = \frac{Gm_1 m_2}{r^2}$$

WEIGHT

The force due to gravity, measured in newtons, N. (1 N = 0.1097 kg).

MASS

The measure of how much of something there is, expressed in kilograms.

$$F = kx$$

NONCONTACT FORCES

NUCLEAR FORCES

Bind atoms together to make different elements; always attractive.

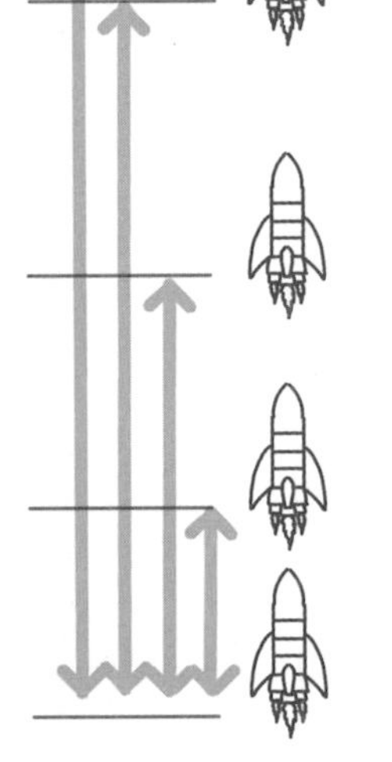

MAGNETIC FORCE

Only affects certain materials; requires a magnetic field; both attracts and repels.

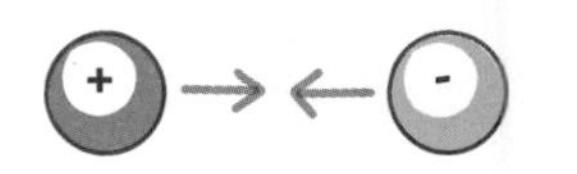

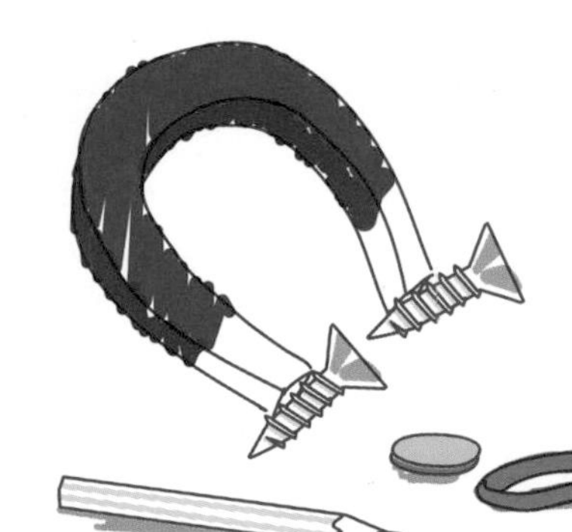

ACCELERATION

A measure of how quickly a body's velocity changes.

ELECTROSTATIC FORCE

Only experienced by charged particles; can be attractive or repellent.

CHAPTER 2

LINEAR MOTION

Physicists can predict the motion of a body if they make some assumptions. A body's motion is described by a variety of parameters that govern where the body will be and when. These parameters are initial velocity, final velocity, displacement, acceleration, and time.

Throughout the motion of the body, it is assumed that its acceleration is constant and occurs in straight lines (it is linear) and that the size of the body—called a particle—is negligible (it has no physical dimensions). This treatment of the body's motion greatly simplifies the math involved and allows us to make predictions about a particle's journey.

PARTICLE POSITION

When considering the movement of a body (referred to here as a particle), you track its position relative to an arbitrary point in space, known as the origin. This is usually set as the point at which the particle begins its journey (at time $t = 0$s). In reality, movement occurs in three dimensions, but here it is simplified.

Displacement and distance

In everyday life, we talk of journeys in terms of distances traveled, but when using equations of linear motion, **displacement** is used, given the symbol *s*.

Displacement, like force, is a vector quantity and refers to the distance moved in a specific direction. Unlike distance, displacement measures the position of a particle at any time, *t* (measured in seconds), with reference to the origin. This is usually achieved by setting up a coordinate system in one, two, or three dimensions.

Displacement can have a negative value if the particle that is being positioned is in the negative region of the coordinate system. However, the distance from the origin can never be negative, as it represents a physical length.

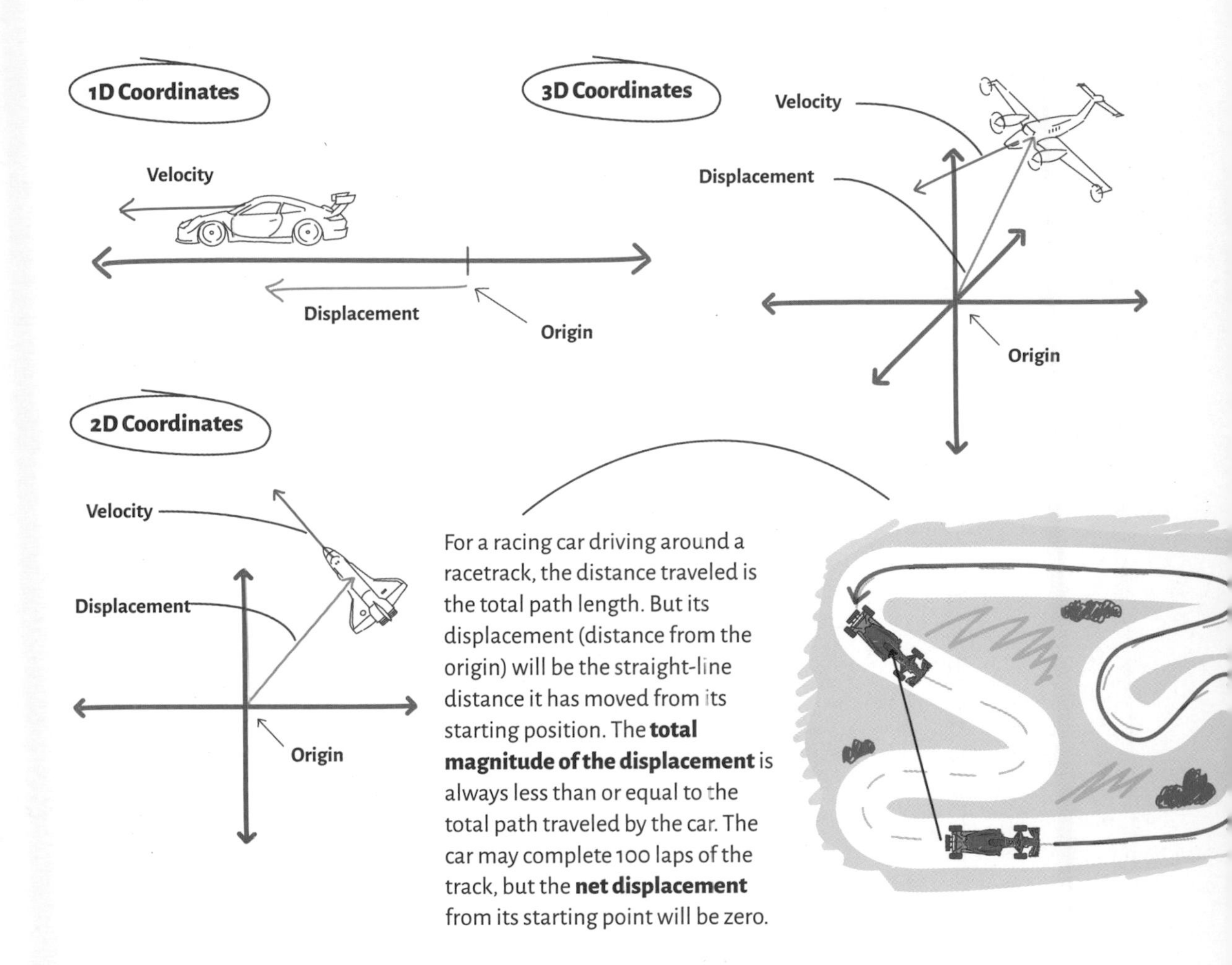

For a racing car driving around a racetrack, the distance traveled is the total path length. But its displacement (distance from the origin) will be the straight-line distance it has moved from its starting position. The **total magnitude of the displacement** is always less than or equal to the total path traveled by the car. The car may complete 100 laps of the track, but the **net displacement** from its starting point will be zero.

PARTICLE MOTION

As well as establishing a particle's position, physicists need to keep track of its velocity (*v*). Velocity, another vector, represents a particle's speed (measured in m/s) and the direction in which it is traveling. Velocity change is measured by acceleration—how rapidly the speed or direction (or both) is changing.

Velocity and speed

A **velocity** can be given either by an angle measured relative to an axis or in **vector form**, with its components resolved along each axis.

1 The motion of a rocket in two dimensions is described by its horizontal velocity in the x direction and its vertical velocity in the y direction. As the rocket moves upward, both components of velocity are positive.

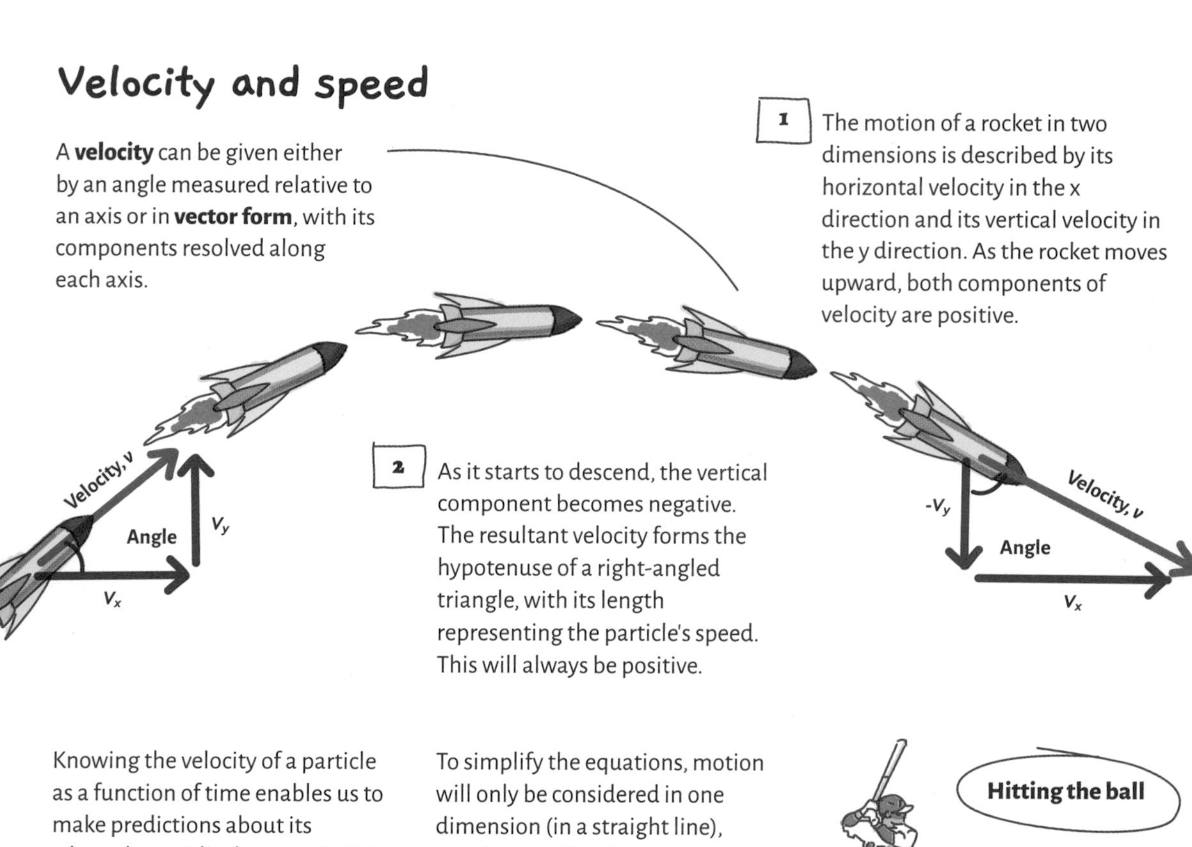

2 As it starts to descend, the vertical component becomes negative. The resultant velocity forms the hypotenuse of a right-angled triangle, with its length representing the particle's speed. This will always be positive.

Knowing the velocity of a particle as a function of time enables us to make predictions about its whereabouts (displacement). But, as you have seen, according to Newton's second law ($F = ma$), if an external force acts upon the particle, its velocity will change.

The most simplistic treatment is to assume that any external force is constant, and therefore, the resultant acceleration is also constant. This greatly simplifies the mathematics, allowing us to easily calculate the change in velocity and track the particle's displacement at any given time.

To simplify the equations, motion will only be considered in one dimension (in a straight line), so velocity will only have two possible directions: forward (a **positive velocity**) and backward (a **negative velocity**).

Hitting the ball

The bat connects with the ball, reversing its direction by providing an opposing force to its motion.

The ball leaves the bat as a projectile with an initial velocity that depends upon the angle, speed, and time of contact with the bat.

Acceleration

As you saw in Chapter 1, forces give rise to acceleration: a change in velocity. A force in a specific direction can either **accelerate** (speed up) or **decelerate** (slow down) a particle in the same direction. **Acceleration** is a vector and measures the rate at which a particle changes velocity—this can be a change in speed, direction of motion, or both. An acceleration acts in precisely the same direction as the external force, and the effect on the particle will depend on the size of the force and the mass of the particle.

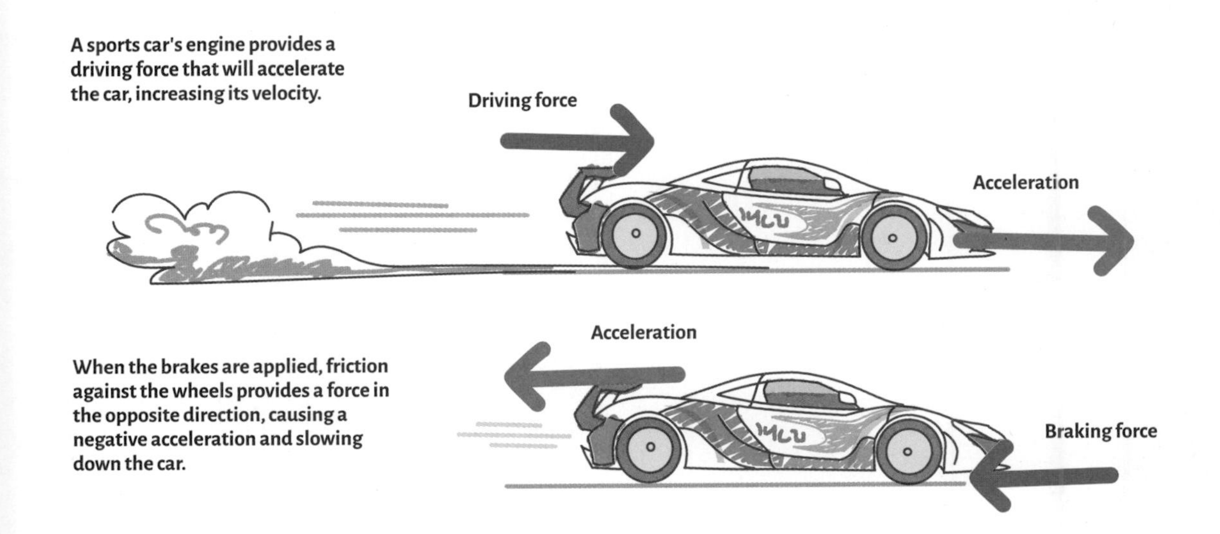

A sports car's engine provides a driving force that will accelerate the car, increasing its velocity.

When the brakes are applied, friction against the wheels provides a force in the opposite direction, causing a negative acceleration and slowing down the car.

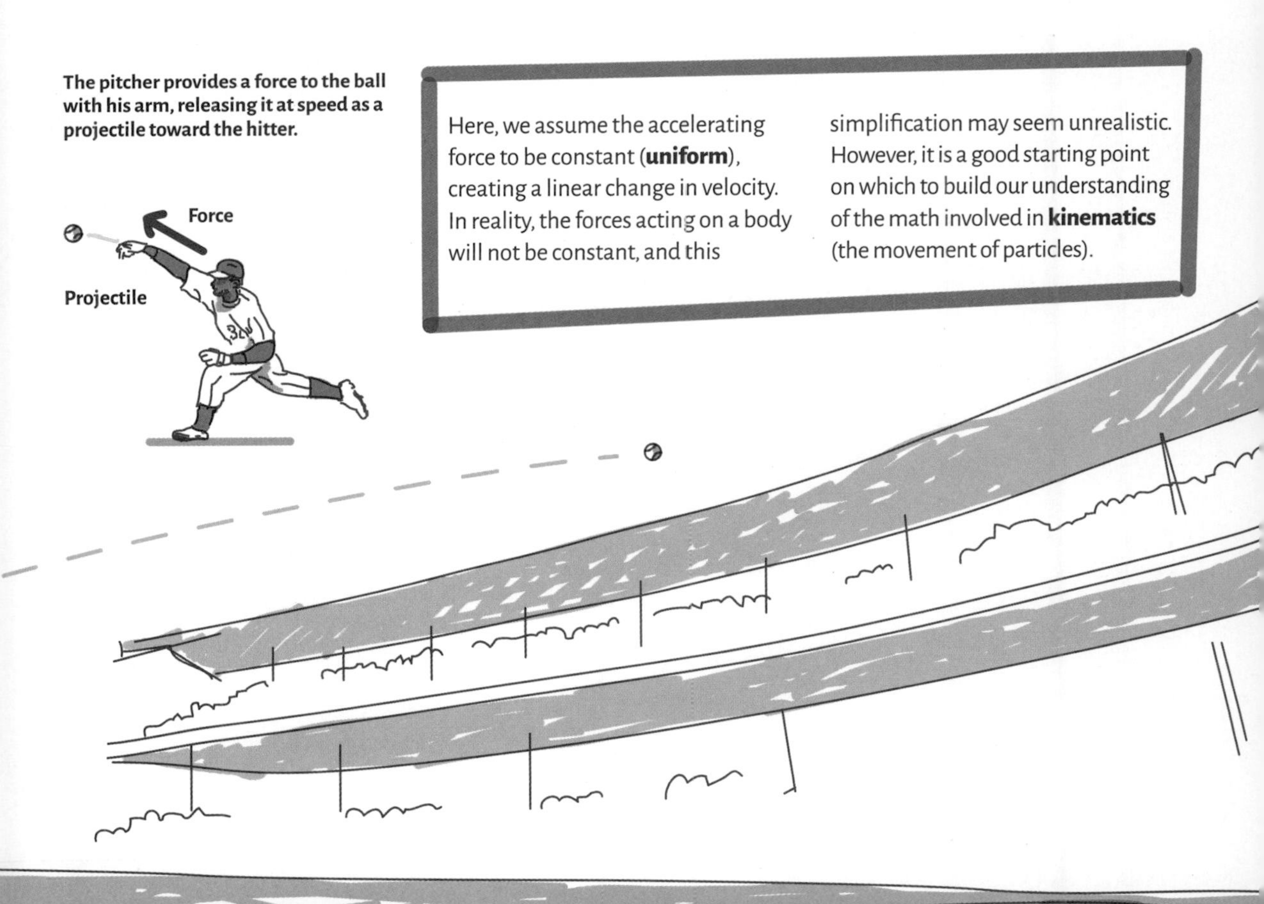

The pitcher provides a force to the ball with his arm, releasing it at speed as a projectile toward the hitter.

Here, we assume the accelerating force to be constant (**uniform**), creating a linear change in velocity. In reality, the forces acting on a body will not be constant, and this simplification may seem unrealistic. However, it is a good starting point on which to build our understanding of the math involved in **kinematics** (the movement of particles).

MOTION GRAPHS

The movement of a particle can be illustrated graphically by tracking its velocity (*v*) or displacement (*s*) as time (*t*) progresses. You can measure various properties, such as acceleration (*a*) and total distance traveled, using motion graphs. Motion graphs also allow equations of motion to be derived, enabling calculations to be made.

Velocity-time graphs

Velocity-time graphs are a visual means by which to monitor the change in velocity of a particle (y-axis, labeled *v*) as time progresses (along the x-axis, labeled *t*). In one dimension, the particle's velocity can only be positive or negative and will therefore be recorded either as a positive value above the x-axis or a negative value below the x-axis.

Increasing velocity

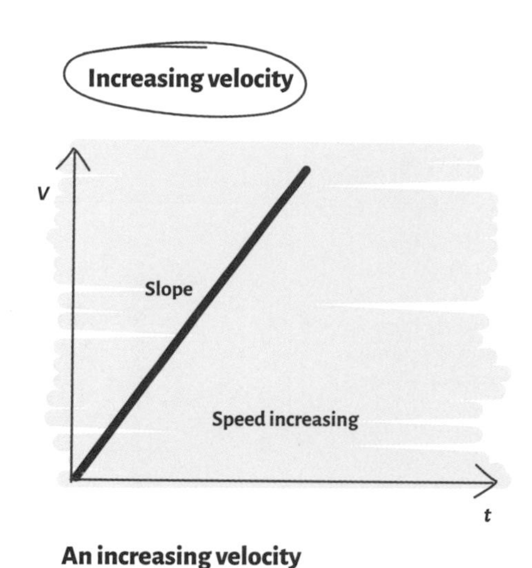

An increasing velocity (accelerating particle) will show a positive slope (sloping upward) on the graph.

Decreasing velocity

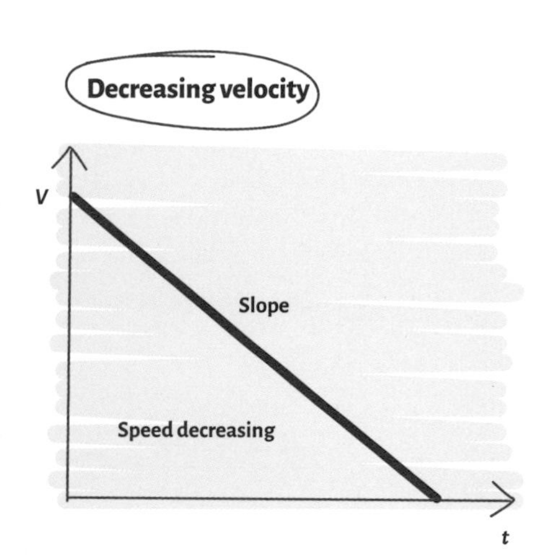

A decreasing velocity (a decelerating particle or a negative acceleration) will show a negative slope (sloping downward).

Uniform velocity

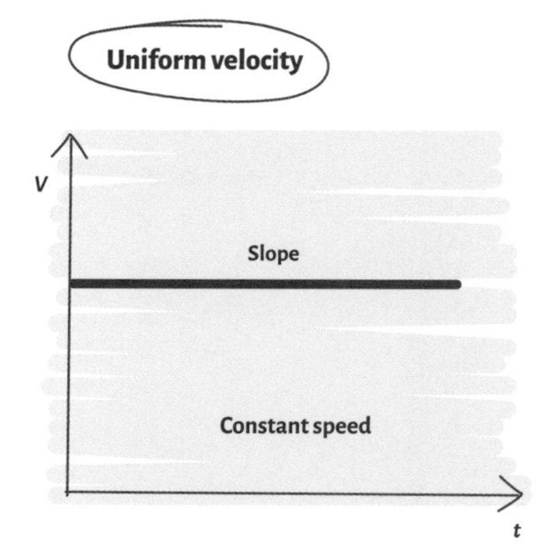

Constant or uniform velocity is represented by a horizontal line with a slope of zero, indicating no acceleration.

A **multistage velocity-time graph** is a representation of the relationship between distance traveled, speed, and time (distance = speed × time). It allows us to calculate the distance traveled and the displacement of the particle, measured from the origin.

Under **uniform acceleration**, the graph consists of a series of straight lines, with each line representing a different phase of the particle's journey with varying accelerations.

Acceleration is a measure of how rapidly the velocity changes. This is represented by the steepness of the slope. A steeper slope indicates a larger change in velocity over a fixed time period (change in velocity divided by time taken for that change).

Any stage below the time axis (x-axis) represents the particle moving in the opposite, or negative, direction, with a larger negative value representing a higher speed.

Multistage velocity-time graph

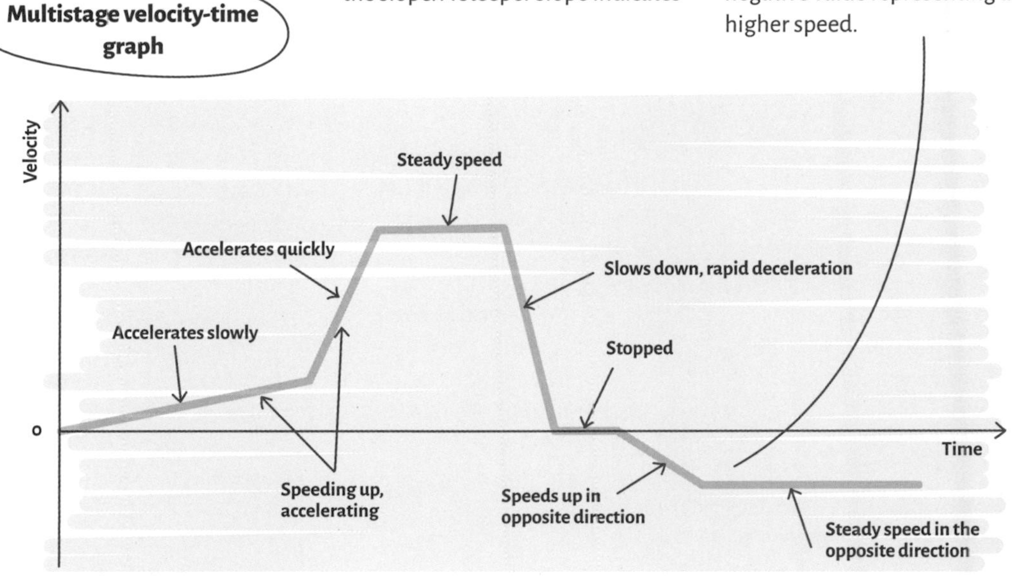

CALCULATING DISTANCE AND DISPLACEMENT

Summing the areas of the graph and using the convention that an area below the time axis is negative will give the displacement of the particle as measured from the origin. The total distance is represented by the total positive area enclosed by the graph and the time axis, summing the positive values of each region.

As the graph consists only of straight lines, it is usually straightforward to calculate the total length of the journey by breaking up the graph into separate shapes and calculating the area of each.

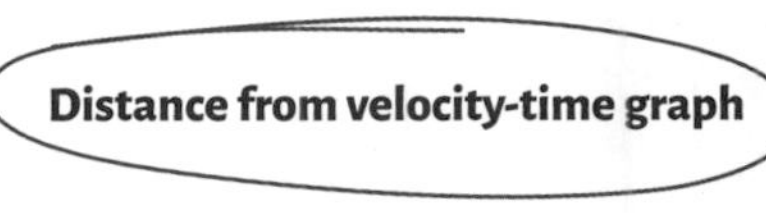

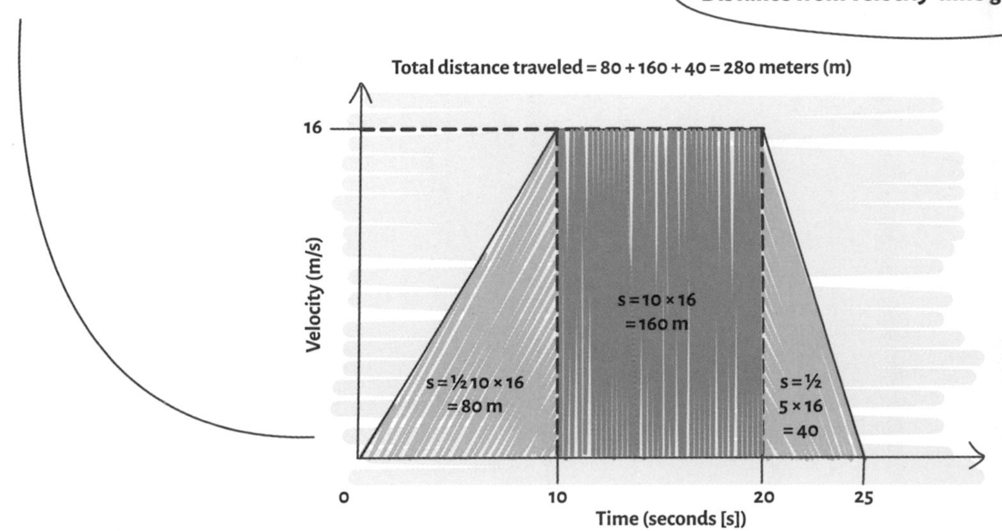

Displacement-time graphs

Displacement-time graphs show the location of a particle relative to the origin at any given time, t. As speed is given by the distance traveled divided by the time taken, the gradient or slope of a displacement-time graph gives the particle's velocity.

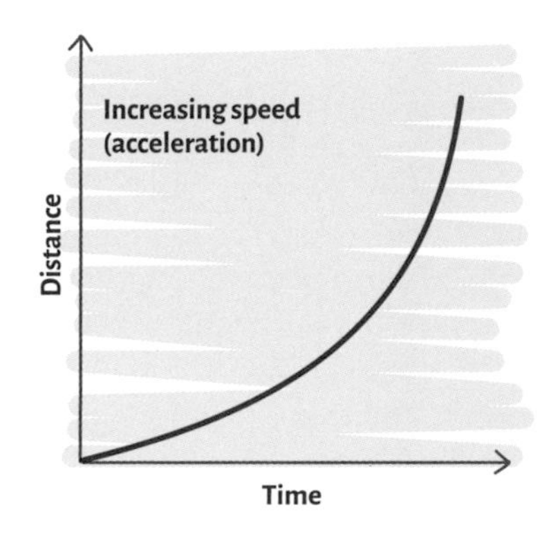

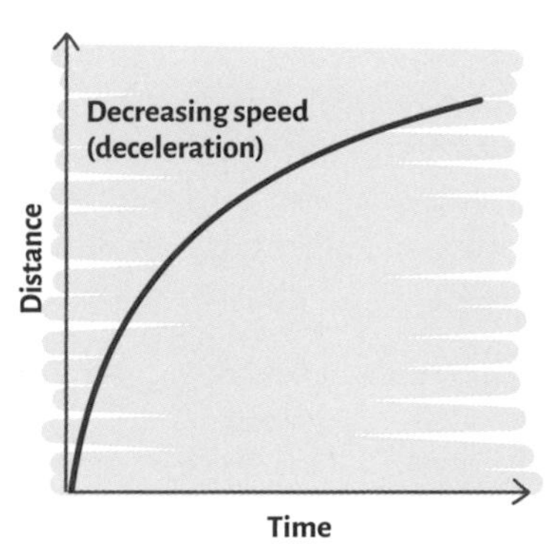

If the particle is accelerating, its velocity is changing, so the slope of the graph will also change—shown as a curve. An accelerating body shows an **increasing slope** or an upward steepening curve, whereas a decelerating body shows a **decreasing slope** with a flattening curve.

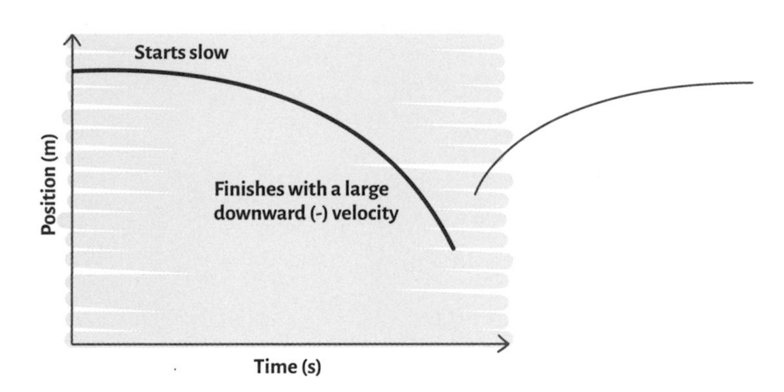

An accelerating particle covers more distance per fixed time period. An object dropped from a height above the ground will start with **zero velocity** (zero slope) and begin to accelerate toward the ground (increasing negative slope).

A particle moving at **constant speed** is represented by a straight line with a nonzero slope. The steeper the line, the faster it is moving. A positive slope represents a particle moving forward; a negative slope signifies a particle moving back toward (and past) its starting position. A horizontal line shows that the particle is stationary, as the displacement from the origin remains constant.

If the line on a displacement-time graph falls below the time axis, the particle has moved away from its initial position but in the opposite direction: a negative displacement.

Displacement-time graph

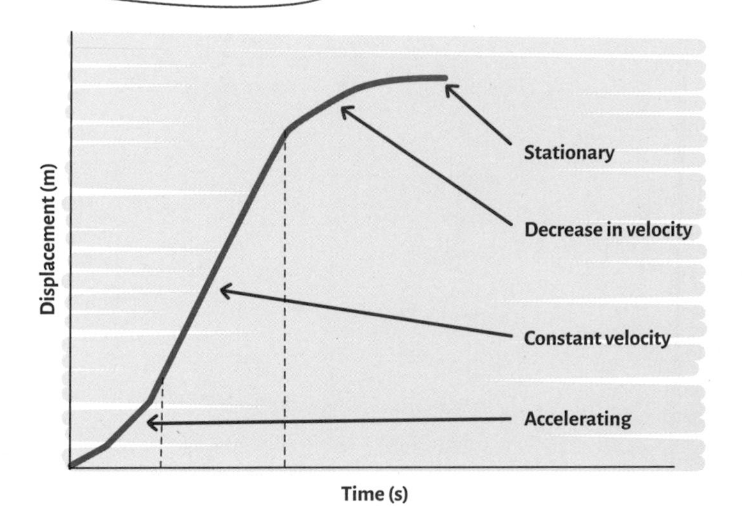

Bouncing ball

A bouncing ball dropped from a height of *h* meters will accelerate uniformly due to gravity at a rate of 9.8 m/s². Velocity increases vertically downward, indicated by an increasing (negative) slope. In this example, the origin represents the ground, not the starting position of the ball—this is where the ball will eventually stop as some of its energy is lost through each bounce. It will hit the ground when the displacement is zero and rebound with a positive velocity (moving vertically upward) with a magnitude less than that of impact as it loses energy.

The slope of the **displacement-time graph** fluctuates between positive and negative as the velocity direction changes from moving up (positive) to moving down (negative). There is an instantaneous change in direction as the ball hits the ground and bounces back. This is not entirely accurate, as, in reality, the ball will compress as it slows and expand again as it reverses its direction.

The displacement-time graph looks like the path followed by a ball bouncing along the time axis.

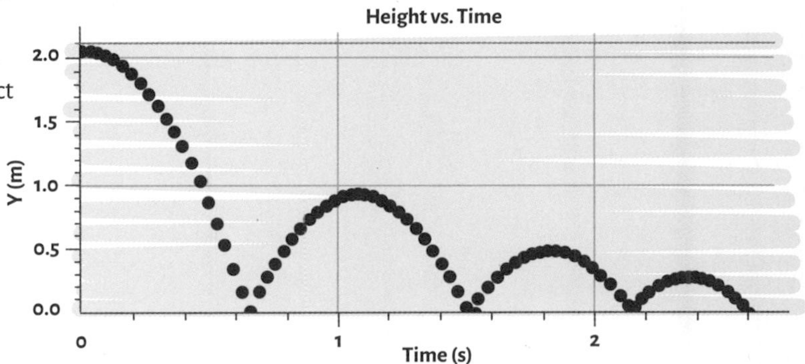

Ball's velocity-time graph

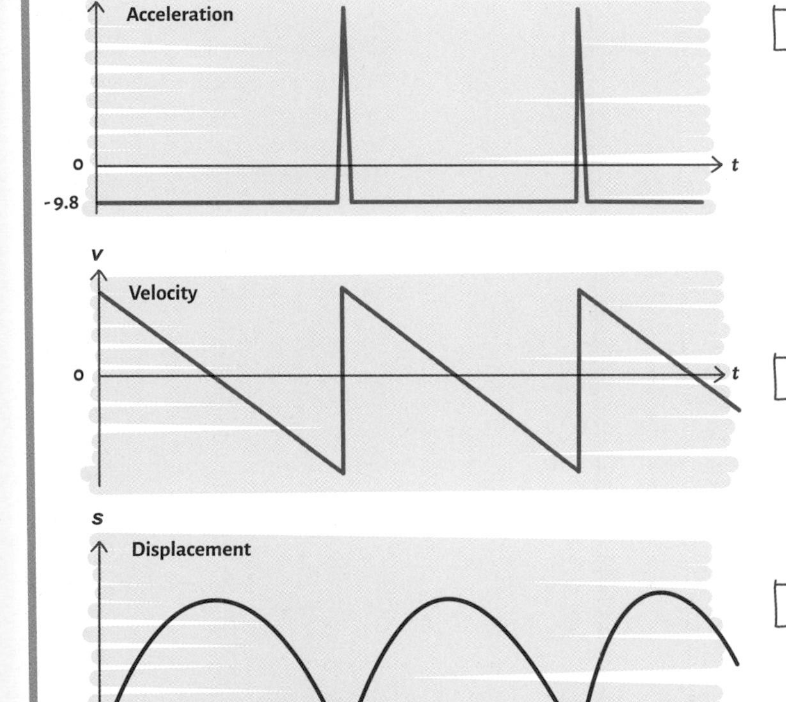

1. The **velocity-time graph** of a bouncing ball shows how the velocity of the ball completely reverses on impact and becomes positive as it bounces back up. Its velocity then decreases, becomes zero at its highest point, then starts to accelerate downward, as shown by an increasing negative velocity.

2. The maximum rebound speed decreases as energy is lost as sound and heat. The slope of the velocity-time graph is always negative and constant unless the ball is rebounding.

3. This slope shows the acceleration due to gravity, which is constant apart from the small time periods during impact.

CONSTANT ACCELERATION

When acceleration is constant throughout the motion of a particle, simple equations can be used to predict this motion. A uniform increase in velocity represents constant acceleration, as does motion under the influence of gravity.

Equations of motion

Equations of motion are derived from a simple velocity-time graph, where a particle is accelerating uniformly from an initial velocity, *u*, to a final velocity, *v*, in *t* seconds.

In many countries, these equations are known as SUVAT equations; the letters stand for the following variables: *s* – distance, *u* – initial velocity, *v* – velocity at time *t*, *a* – acceleration and *t* – time. This is quicker to say and is a very handy revision mnemonic.

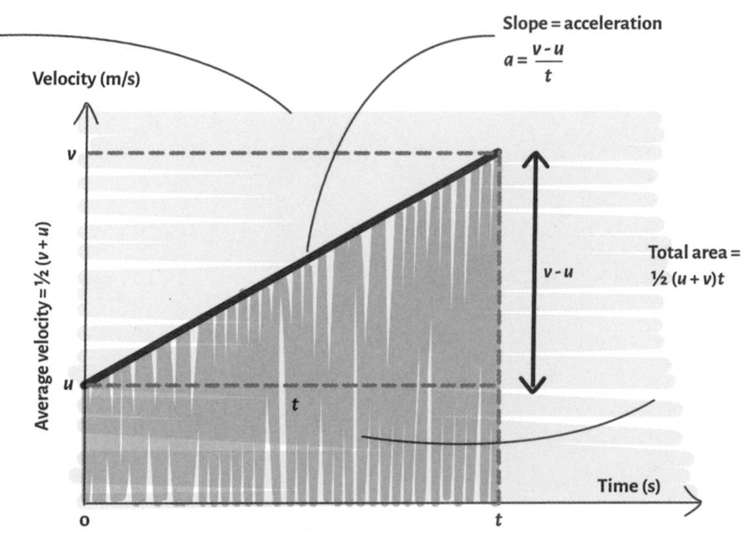

The **acceleration**, *a*, can be calculated from the slope:

$$a = \frac{v - u}{t}$$

The **displacement**, *s*, can be calculated from the area of the shape under the line (a trapezoid). To calculate the area of a trapezium, model it as a rectangle, by finding the average of the two different side lengths and multiplying this by its length. In terms of the graph, this translates to finding the particle's average velocity and multiplying it by the time, *t*, for which it is traveling.

This is the resulting equation:

$$s = \frac{1}{2}(u + v)\,t$$

These two can be mathematically combined to create the following two formulas:

$$s = ut + \frac{1}{2}at^2$$

$$v^2 = u^2 + 2as$$

The four formulas are collectively known as the **equations of motion**. These allow physicists to predict the motion of a particle given some information about its initial state, such as its initial velocity and acceleration and the time for which it is traveling. These formulas are only applicable if the acceleration of the particle is uniform and three pieces of information are initially known.

Under the influence of gravity, acceleration, *a*, is exchanged for acceleration due to gravity, *g* (9.8 m/s²).

Projectile motion

A **projectile** is any particle that is accelerated by gravity alone. This could be a body that has been **dropped** from a height, **thrown** upward, or **launched** at an angle other than 90° to the horizontal. Once in motion, the body's path is controlled by gravity alone (air resistance is ignored at this level). The path of the projectile is referred to as a **trajectory** and will vary according to the angle and speed at which the body is launched.

If a body is dropped from a height, it will accelerate downward. Due to gravity, its vertical velocity, *vy*, increases at a rate of 9.8 m/s².

If it is launched horizontally from the same height, the time taken to hit the ground is identical, but the body will also have a horizontal velocity component, *vx*, which remains constant throughout its flight. The resulting trajectory as *vy* increases is a smooth curve called a **parabola**.

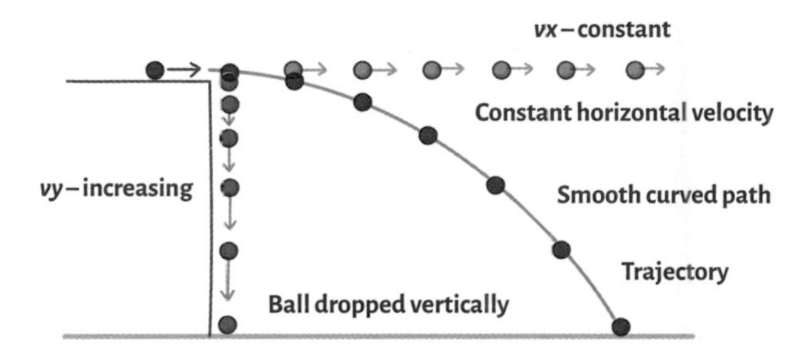

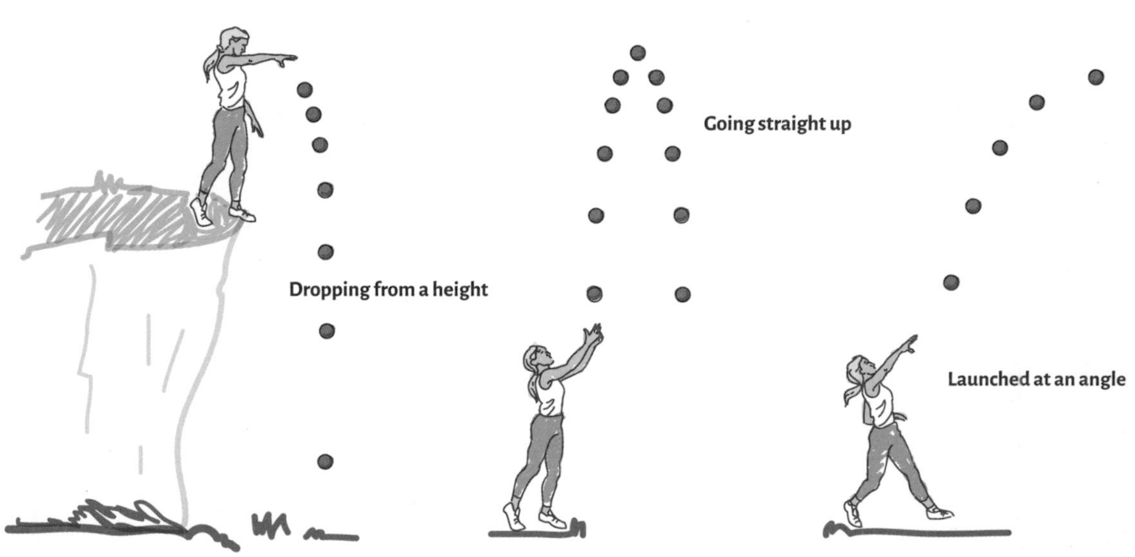

The horizontal distance the projectile travels is known as the **range**, *x*, and is directly connected to the horizontal speed of the body and its time in the air. The **ascent** (upward motion) and **descent** (downward motion) times for a projectile are identical and depend on the maximum height reached—the total is called the **flight time**.

If a body is launched with the same speed but differing angles, the range will vary. The range of the projectile is found by multiplying the flight time by the horizontal speed of the body and is a maximum for a fixed launch speed at an angle of 45°.

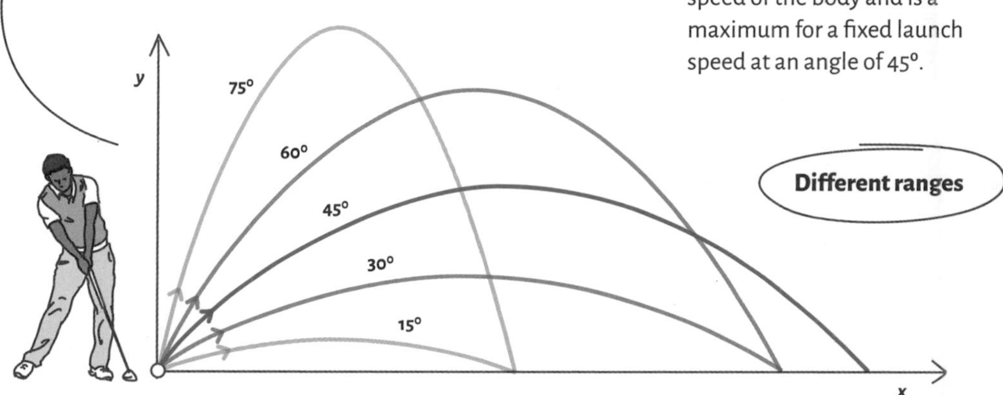

RECAP

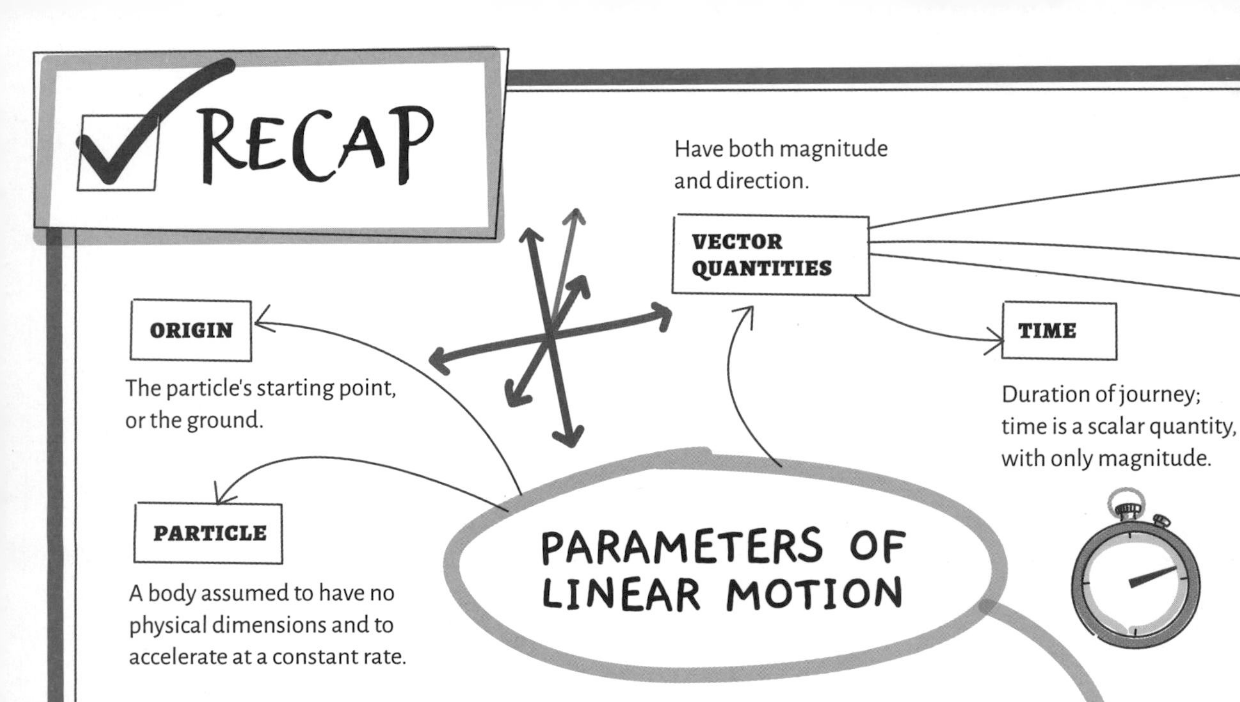

LINEAR MOTION

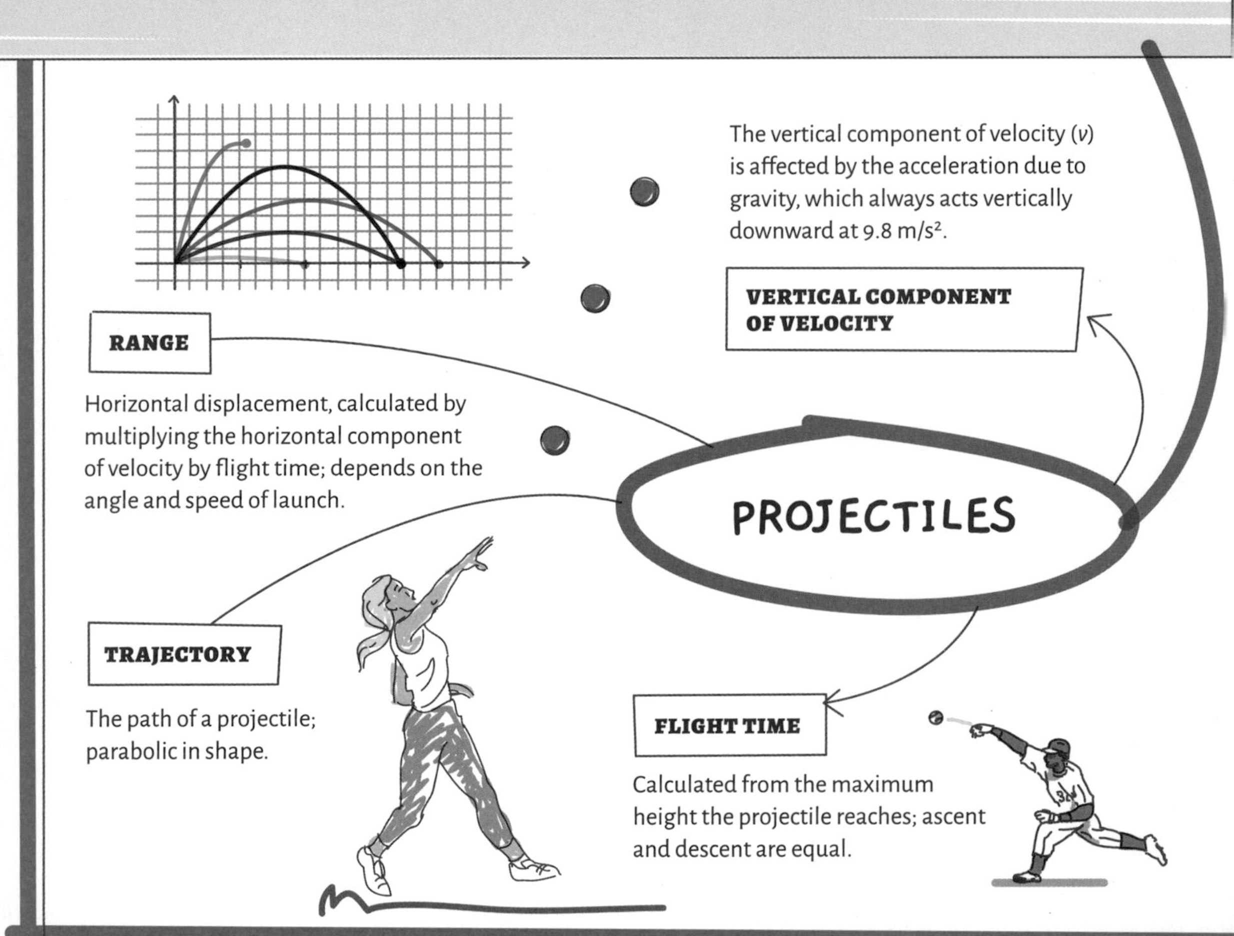

VELOCITIES (U AND V)

Speeds attained by the particle during its journey.

ACCELERATION (A)

The change in velocities throughout the particle's journey.

DISPLACEMENT (S)

The particle's position relative to the origin.

MOTION-TIME GRAPHS

VELOCITY-TIME GRAPH

Slope indicates the acceleration of a particle. A negative velocity in one dimension indicates that the particle is moving backward.

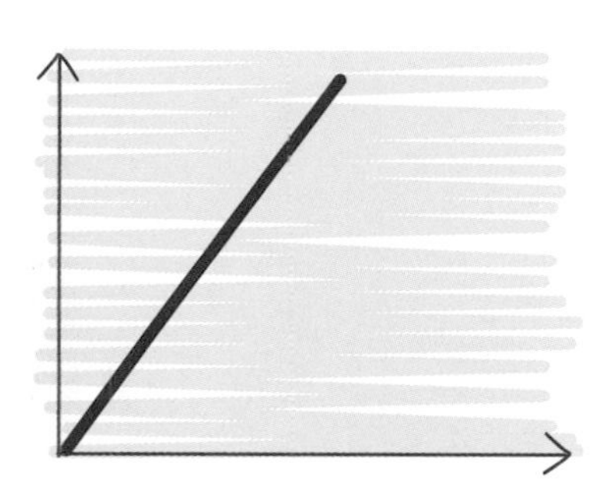

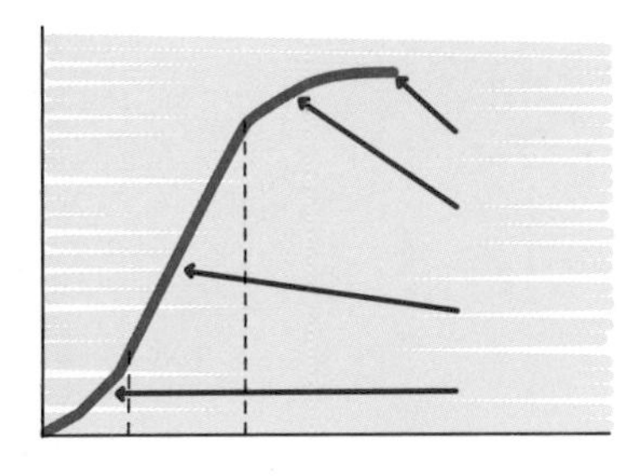

EQUATIONS OF MOTION

The four main equations of linear motion, derived from motion graphs.

$$a = \frac{v - u}{t}$$

$$s = \frac{1}{2}(u + v)t$$

$$s = ut + \frac{1}{2}at^2$$

$$v^2 = u^2 + 2as$$

Where u is initial velocity, v is final velocity, a is acceleration, and t is time.

DISPLACEMENT-TIME GRAPHS

Slope indicates the velocity of a particle. A negative displacement in one dimension shows the particle is left of its origin.

CHAPTER 3

ROTATIONAL MOTION

A body will move in a straight line unless acted upon by an external force. As you have seen, if the force is along the line of motion, the body will either accelerate or decelerate according to Newton's second law: $F = ma$. For a body to move around in a circle, there must be a force acting toward the center of the circular path, which is always at 90° to the velocity of the body. This force can be provided by a variety of mechanisms and can be either a contact or a noncontact force.

EXAMPLES OF ROTATIONAL MOTION

There are many examples of bodies that move in a circular path, from those in the quantum mechanical world, such as electrons moving around a nucleus, to the macroscopic, such as planets orbiting a star.

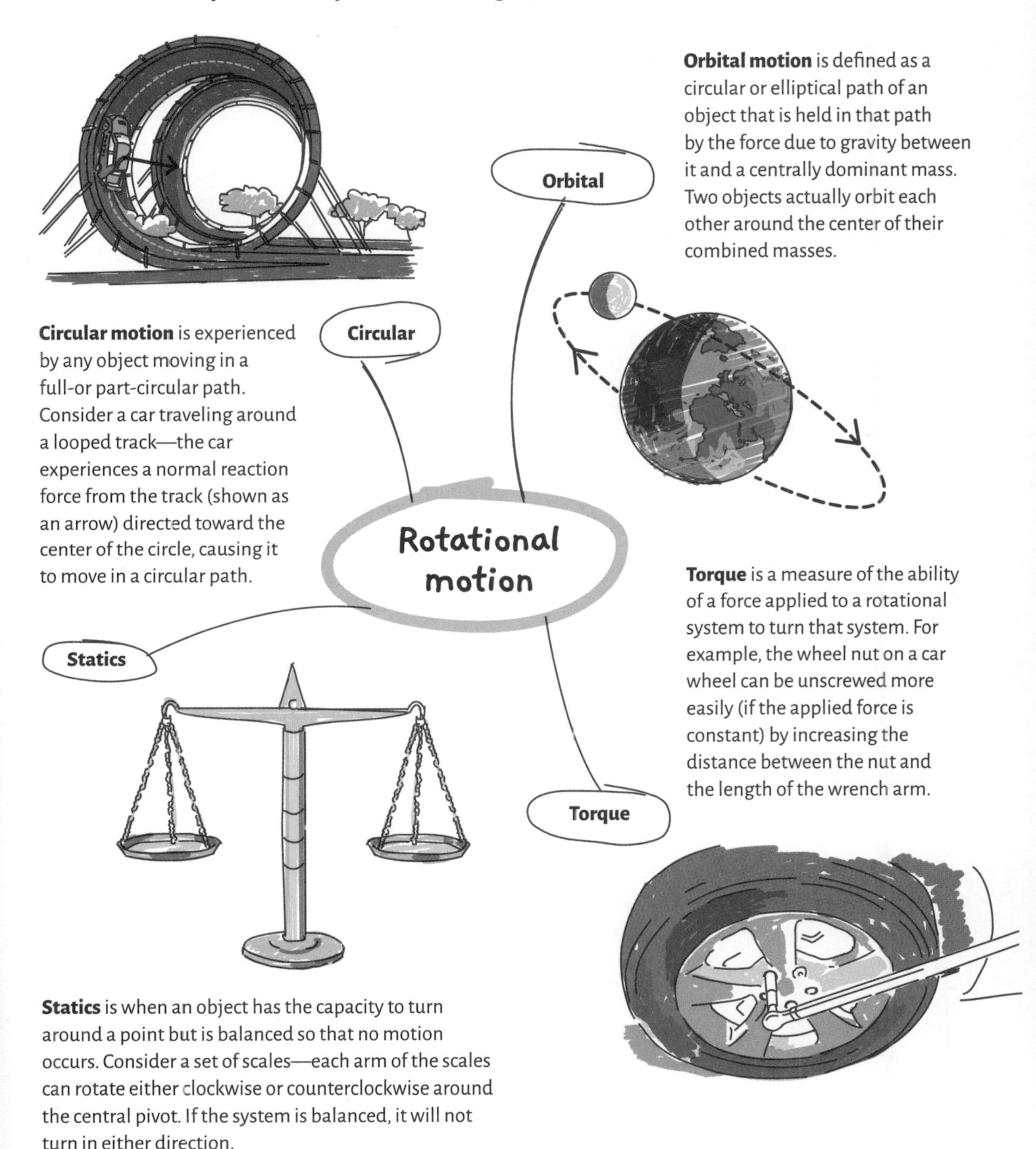

Orbital motion is defined as a circular or elliptical path of an object that is held in that path by the force due to gravity between it and a centrally dominant mass. Two objects actually orbit each other around the center of their combined masses.

Circular motion is experienced by any object moving in a full-or part-circular path. Consider a car traveling around a looped track—the car experiences a normal reaction force from the track (shown as an arrow) directed toward the center of the circle, causing it to move in a circular path.

Torque is a measure of the ability of a force applied to a rotational system to turn that system. For example, the wheel nut on a car wheel can be unscrewed more easily (if the applied force is constant) by increasing the distance between the nut and the length of the wrench arm.

Statics is when an object has the capacity to turn around a point but is balanced so that no motion occurs. Consider a set of scales—each arm of the scales can rotate either clockwise or counterclockwise around the central pivot. If the system is balanced, it will not turn in either direction.

CIRCULAR MOTION

All motion in a circular path has continually changing direction. Velocity is a vector, defined by both magnitude and direction. Acceleration is defined by the rate of change of velocity (*v*). Since one component (the direction) of the velocity is always changing when a body is moving in a circular path, it is always accelerating, even though its speed may remain constant.

This is called **centripetal** (center-seeking) acceleration. At this level, you will assume that the body's speed remains unchanged.

Since the body is accelerating, there must be a force (from $F = ma$). The origin of this force will vary depending upon the system and is referred to as the centripetal force. The force responsible for the acceleration is always directed toward the center of the circle.

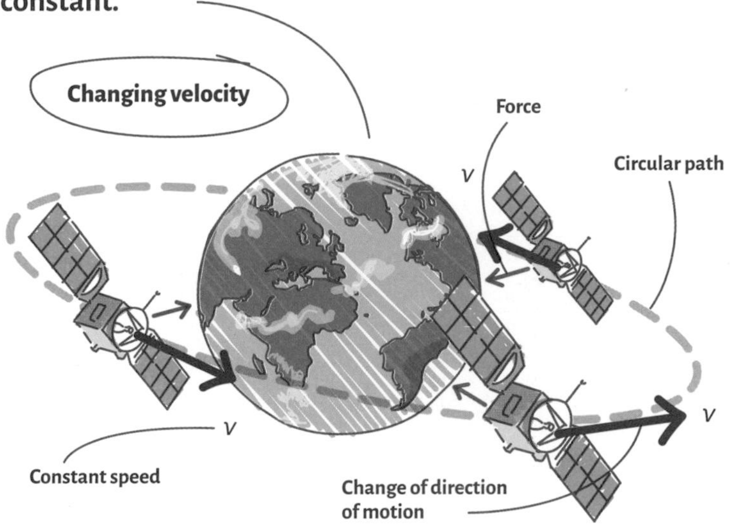

DAYTONA RACING

The Daytona track allows cars to travel much faster as it takes place on a banked circular-shaped bend. In a similar way to how a plane turns in the air, a Daytona race car is angled slightly on the bends, providing a **nonvertical normal reaction force** (contact force) between the car and the road. A component of that force acts toward the center of the circular path, providing much of the force needed to keep the car from sliding. The **friction** between the tires and the road provide the rest of this force.

The steeper the track, the larger the force acting toward the center of the circle, although the car would need to travel much faster so that it didn't slide down the slope.

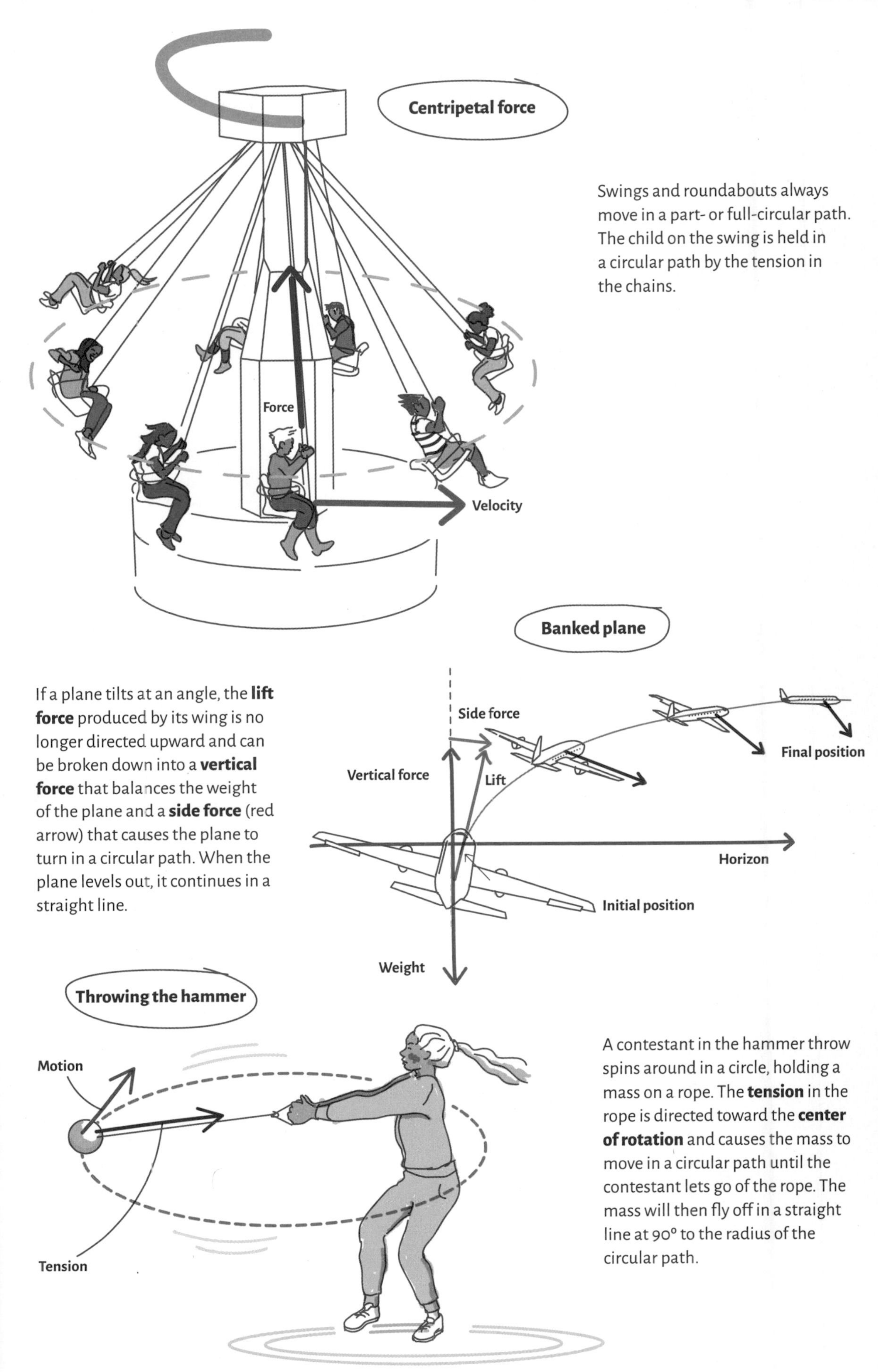

Swings and roundabouts always move in a part- or full-circular path. The child on the swing is held in a circular path by the tension in the chains.

If a plane tilts at an angle, the **lift force** produced by its wing is no longer directed upward and can be broken down into a **vertical force** that balances the weight of the plane and a **side force** (red arrow) that causes the plane to turn in a circular path. When the plane levels out, it continues in a straight line.

A contestant in the hammer throw spins around in a circle, holding a mass on a rope. The **tension** in the rope is directed toward the **center of rotation** and causes the mass to move in a circular path until the contestant lets go of the rope. The mass will then fly off in a straight line at 90° to the radius of the circular path.

ORBITAL MOTION

When a planet's orbit around a star is perfectly circular, its speed remains constant. However, its velocity is always changing due to the variation of the planet's direction vector. The planet is therefore always accelerating toward the star at the center of its orbit. The gravitational attraction between the star and the planet provides the centripetal force required to keep the planet in orbit.

Orbital speeds

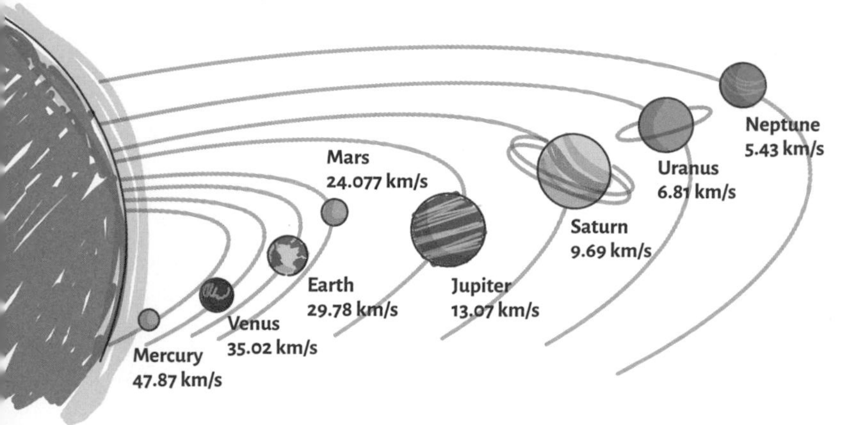

The time taken for the planet to complete one revolution—the **orbital period**—depends on the orbit radius and the mass of the central star. For Earth, this period is approximately 365 days, which defines our year.

Earth's **orbital path** is very close to perfectly circular, so the distance from Earth to the sun remains unchanged. Because of this, the average global temperature has remained relatively stable over Earth's lifetime.

EARTH'S ORBIT

Seasonal temperature variation on Earth is governed by the position of its orbit within the yearly cycle, because the planet is spinning on an axis tilted at approximately 23.5° to the vertical. This means that more solar radiation strikes the northern hemisphere from April through September, and this is reversed in the months from October to March.

Earth rotates on this axis once every twenty-four hours, causing day and night as the sun illuminates half of its surface at one time.

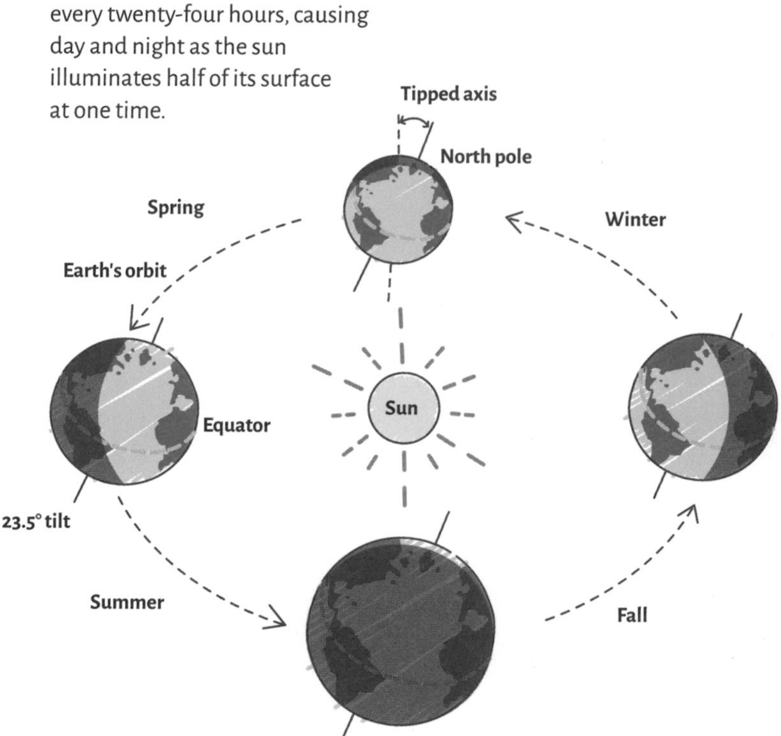

Orbital period

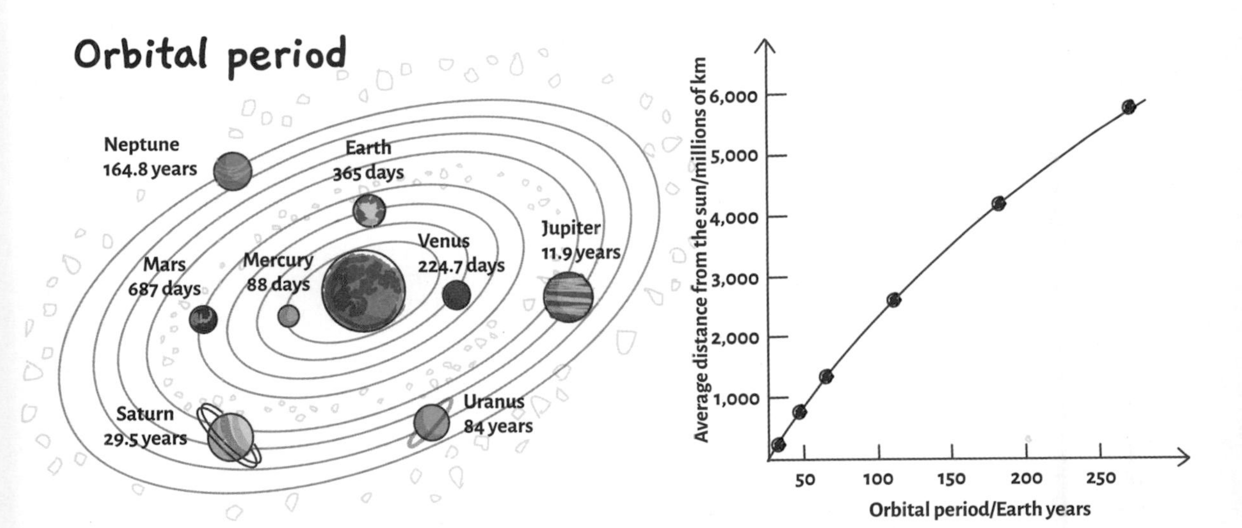

The German astronomer Johannes Kepler (1571–1630) formulated the exact relationship between **orbital radius** and time period that accurately predicts the motion of the planets as they are observed. He was a contemporary of Newton's and used Newton's law of gravitation, combined with his own observations, to formulate the following relationship between orbit period, T, sun's mass, M, and orbit radius, r. The Universal Gravitational Constant, G, equals 6.67×10^{-11}:

$$T^2 = \frac{4\pi^2}{GM} r^3$$

This formula allows physicists to determine the orbital periods of all the solar system's planets with a high degree of accuracy.

Retrograde motion

Years before Kepler formulated his laws of planetary motion, it was the Polish mathematician and astronomer Nicolaus Copernicus (1473–1543) who began the astronomical revolution by suggesting that the solar system was sun-centered (**heliocentric**) rather than Earth-centered (**geocentric**). He published his theory in 1543, based primarily on the difficulty of explaining the behavior of the observed motion of Mars.

When observed from Earth in the night sky, Mars exhibits what is known as **retrograde motion**: It appears to move forward then backward in a small loop. This could not be easily explained by scientists before orbit paths and the speed of the orbiting body were well understood, thanks to Kepler's observations.

Looking at the concentric orbit paths of Earth and Mars, and their observed relative speeds, you can easily see why this once mystifying phenomenon takes place.

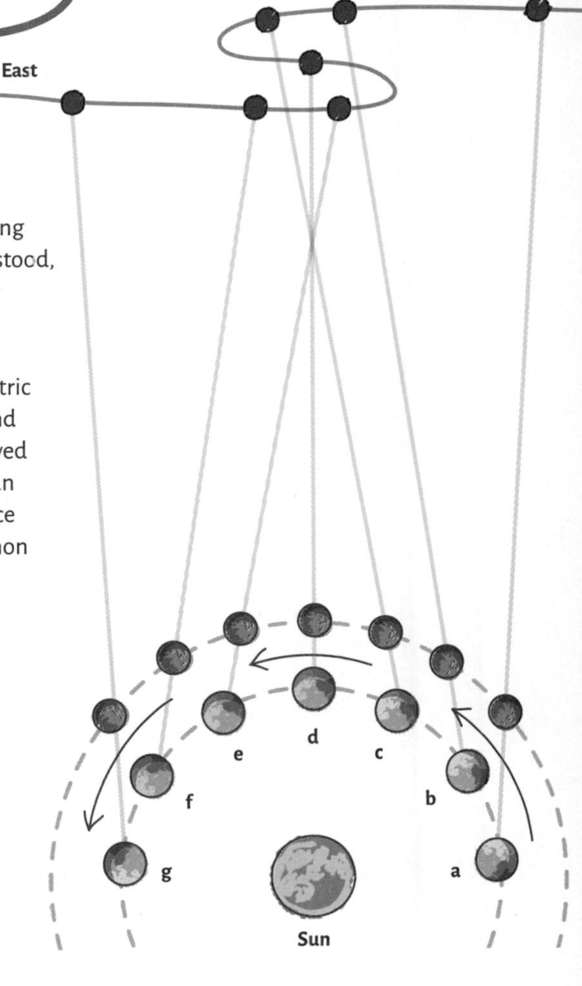

Torque

If a force is applied via a rigid lever around a point that can rotate, the resultant quantity is referred to as **torque.** The **work done**, W, in moving a body is defined as the **force applied**, F, multiplied by the **distance**, s, moved ($W = Fs$).

If the force applied is perpendicular to the radius of a circular path, the greater the distance from the center, the larger the path length. This means that for the same work done in rotating the system, the force required to rotate the body is reduced as s is increased.

This is **the principle of levers**, and it allows large loads to be moved with relatively small forces, albeit with the compromise of distance. That is why a wrench with a long handle is easier to rotate.

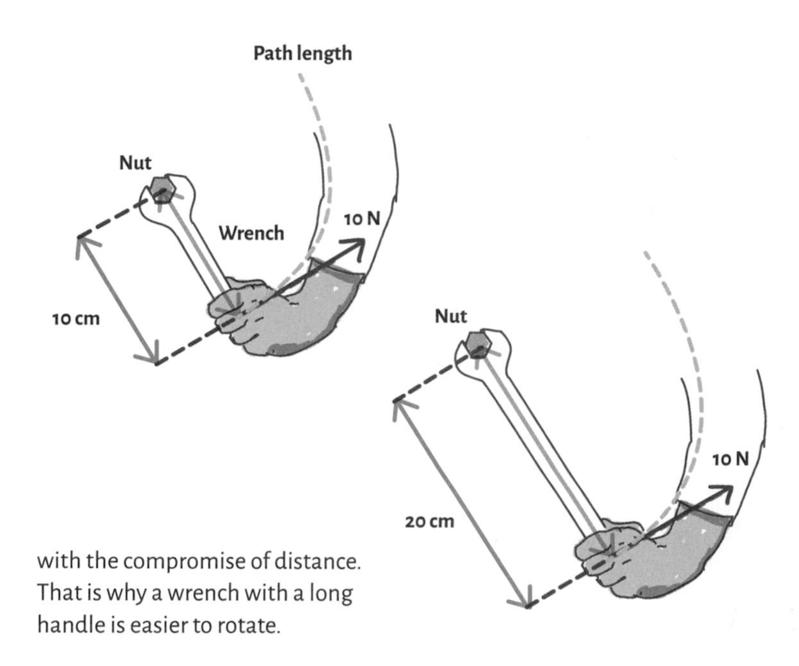

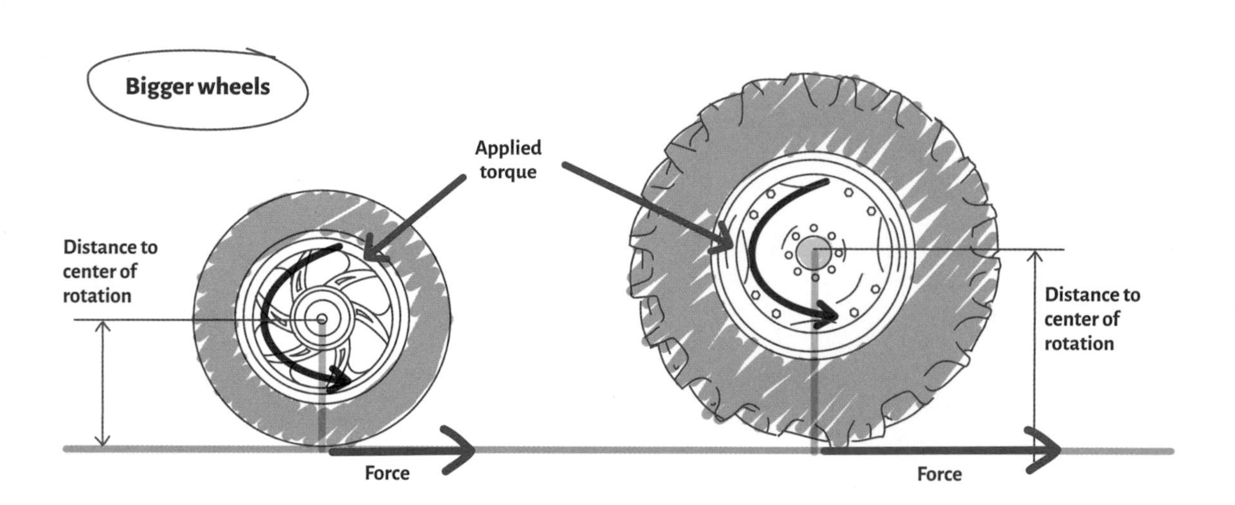

A car provides a rotational force delivered to its wheels via the drive shaft from the engine. The forward propulsion of the car is supplied via the friction of the wheels against the road and acts at 90° to the axis of motion. If the wheel is much larger, such as a tractor wheel, the force from the engine may be the same (assuming the engines are identical), but the torque will be much higher due to the increased wheel radius. The forward motion is much slower, however, as the circumference of the wheel scales with the radius.

In physics, energy is always conserved and requires a force to transfer it. In a rotary system, the energy transferred is proportional to the force applied multiplied by the perpendicular distance (at 90°) from the rotational axis.

ROTATIONAL KINEMATICS AND DYNAMICS

A rotational system will turn around its center point if it is free to do so. However, there may be balance around the point of rotation, preventing movement. This is called statics.

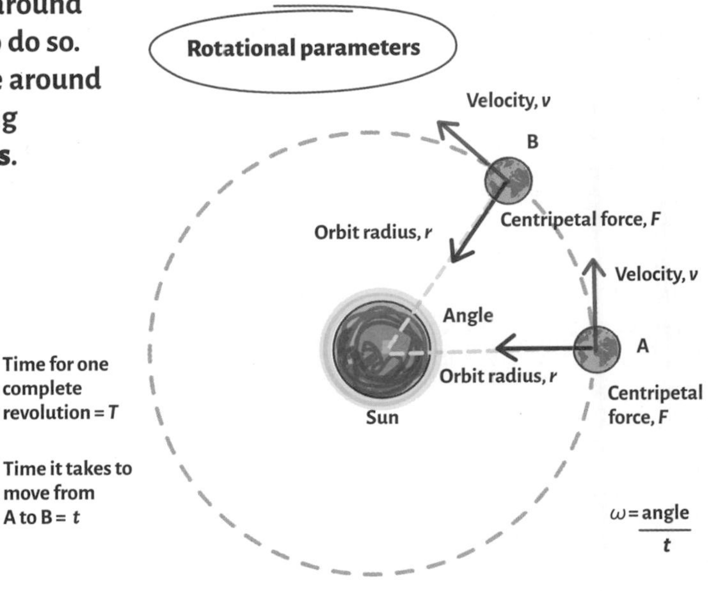

Moving bodies

Rotational motion can be described by a set of parameters.

Physicists measure **instantaneous velocity**, v, the **radius of the circular path**, r, the **centripetal acceleration**, a, and the **angular speed**, ω (the angle swept out per second). The time taken for one complete revolution is known as the **time period**, referred to as T. The centripetal acceleration of a body moving in a circular path is given by the formula:

$$a = \frac{v^2}{r} \text{ or } a = r\omega^2$$

From Newton's second law ($F = ma$), the associated centripetal force is:

$$F = \frac{mv^2}{r} \text{ or } F = mr\omega^2$$

where m is the mass of the body measured in kilograms.

This relationship strictly governs the circular path of the body. For a larger mass, the force will scale linearly for a fixed radius, whereas an increase in the body's speed requires a much larger increase in force. For a body of fixed mass moving in a larger circle, its speed will decrease, as will the centripetal force required.

Imagine a car moving in a circle at a constant speed. Friction between the road and the tires provides the centripetal force. If the circles get smaller, the force needed to keep the car in circular motion increases. When the force required exceeds the maximum friction, the car will skid out control.

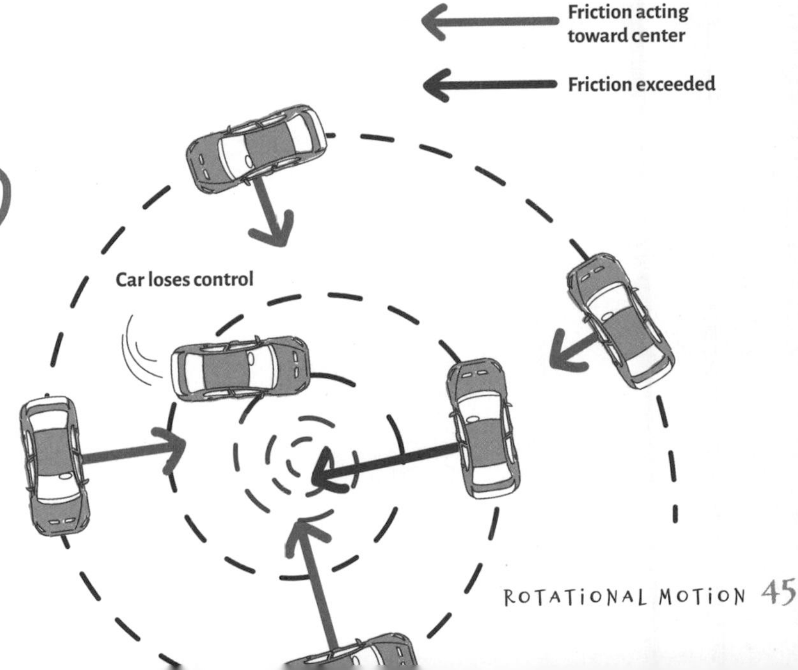

Statics

As you have seen, in a rotational system, a body is able to turn around its center point if acted on by a force. However, if a second opposing force is present and balances the first exactly, movement is prevented. The body is said to be in equilibrium.

Imagine a seesaw in a playground. If a child sits on one end, the seesaw will rotate around the midpoint (called the fulcrum or pivot) until it is supported by the ground. If a heavier child sits on the other end, there is an imbalance, and the seesaw will rotate in the opposite direction.

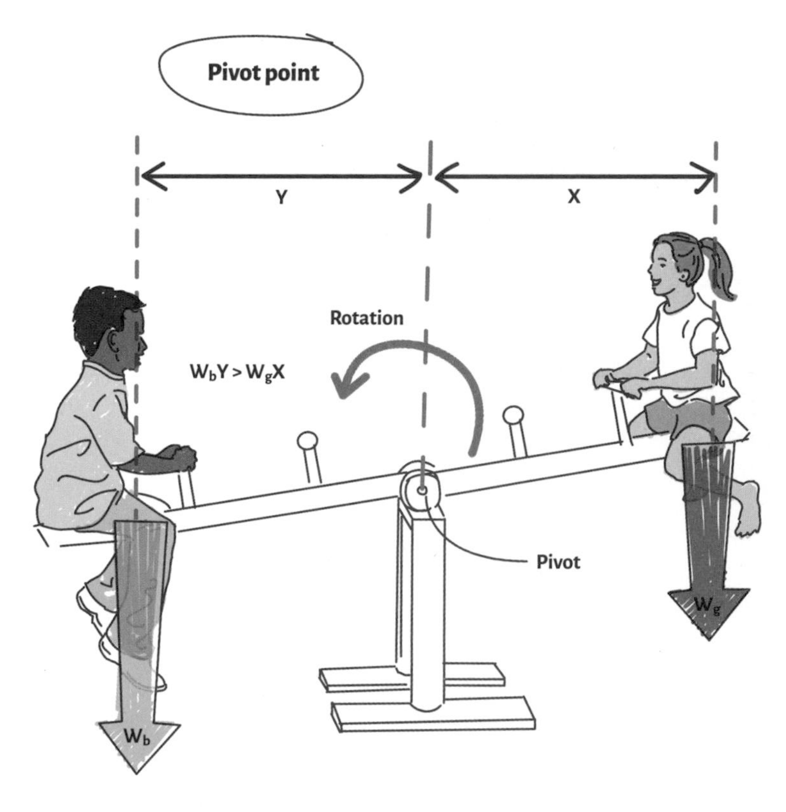

It is a combination of the weight of each child and their relative distances from the **pivot point** that governs this rotation. This is called the torque and is defined by the product of the force, *F*, and its perpendicular distance, *x*, from the rotation point ($T = Fx$).

A **turning moment** can be either clockwise or counterclockwise. It is measured in Nm (newton meters) and has the same base units as energy.

The principle of **balanced torques** states that if the sum of the clockwise torques is equal to the sum of the counterclockwise torques, there will be no imbalance and therefore no rotation. Using this principle, if the heavier child on the seesaw moves toward the center point, his **perpendicular distance** from the pivot decreases, so his torque around the pivot of the seesaw will decrease. At some point, both clockwise and counterclockwise torques are equal, and the seesaw will be balanced.

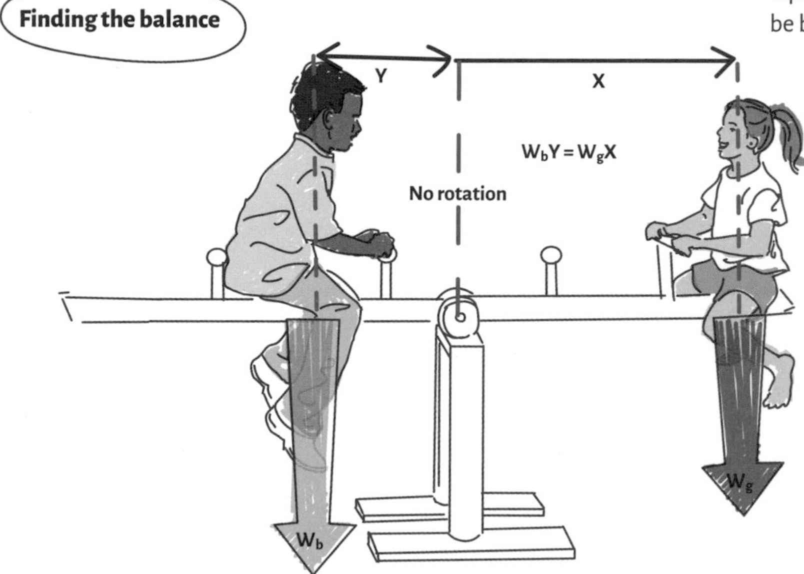

Balanced torques

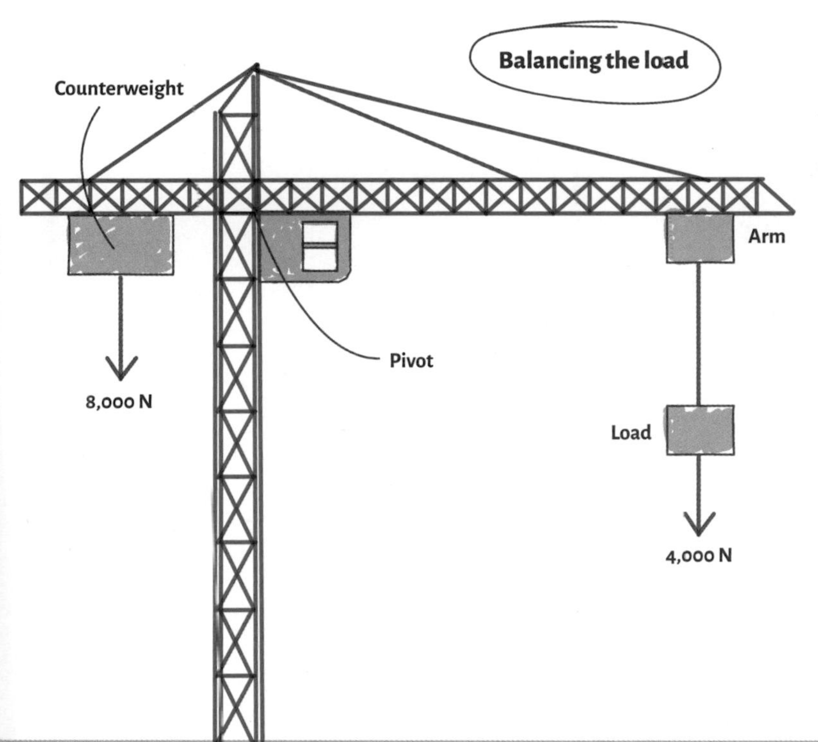

A crane uses the **principle of balanced torques** to balance itself when lifting a load. It does so by varying the distance of a large counterweight from the pivot as it lifts a load, so that the clockwise torque produced by the load is exactly matched by the counterclockwise torque produced by the counterweight.

A truck crossing a bridge is also an example of a **rotational system in balance**, although no rotation takes place.

Each support of the bridge is a pivot point that supports the weight of the bridge and the crossing truck.

Consider the point P as a pivot. The weight of the bridge and the truck both produce counterclockwise (CCW) torques. These are both balanced by the clockwise (CW) torque produced by the reaction force, F_A, of the first support multiplied by the bridge length.

As the truck crosses the bridge, the reaction forces F_A and F_B vary to maintain equilibrium, so the bridge remains stable.

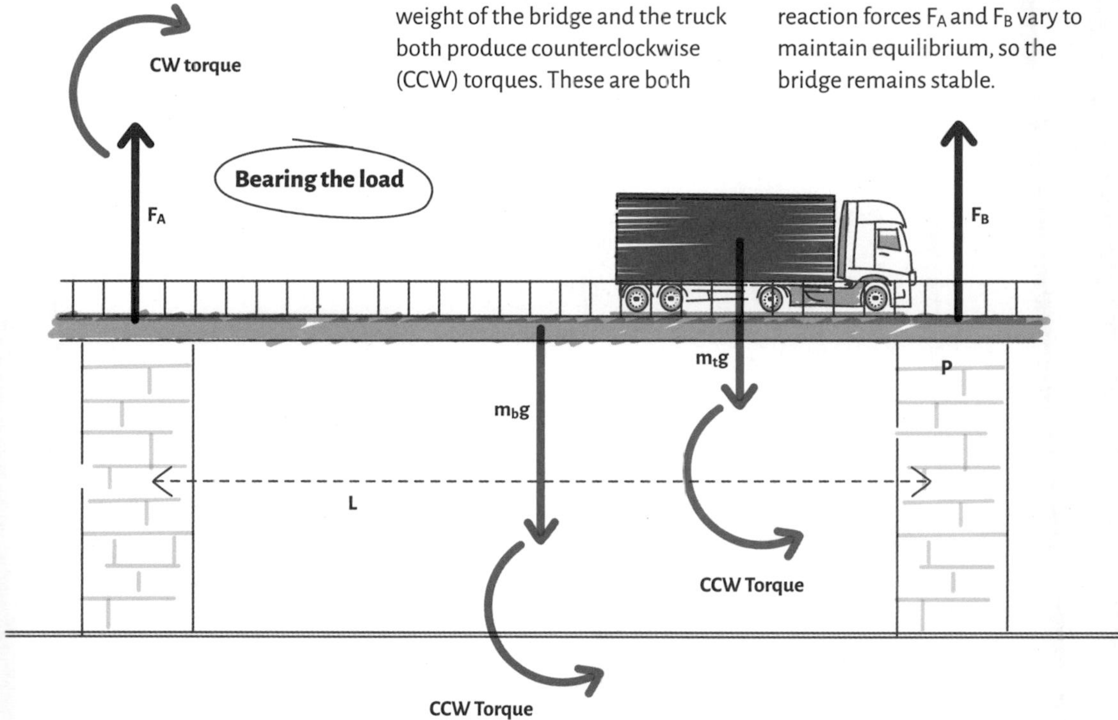

TYPES OF FORCE

CENTRIPETAL FORCE

A child on a swing is held in a circular path by the tension in the chains.

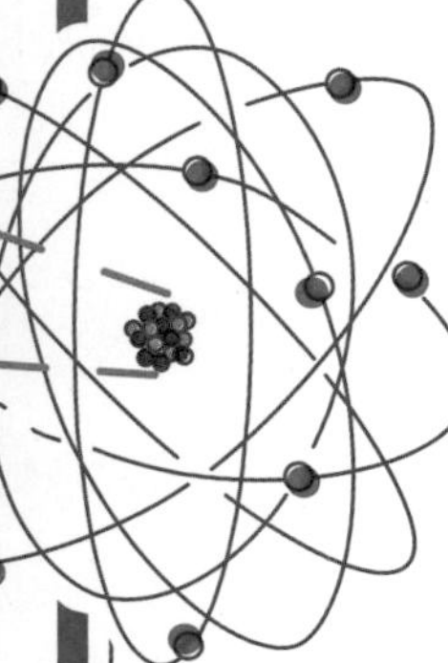
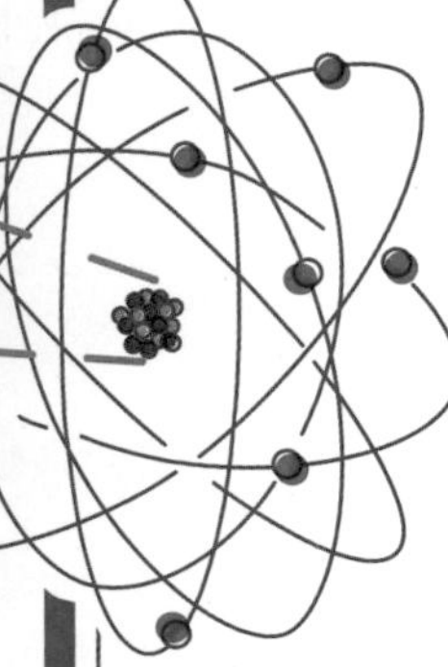

ELECTROSTATIC FORCE

Electrons are kept in orbit by their electrostatic attraction to a proton.

CIRCULAR MOTION

FRICTION

Friction between its tires and the road allow a car to turn.

GRAVITY

Planets orbit stars due to the gravitational attraction between their masses.

ROTATIONAL MOTION

TURNING MOMENT

The product of the force applied to a rigid body that is free to rotate and the perpendicular distance at which it acts.

CENTRIPETAL ACCELERATION

A body will accelerate toward the center of a radius, *r*, if there is an external force acting toward the center of motion.

$$a = \frac{v^2}{r} \text{ or } a = r\omega^2$$

STATICS

ROTATIONAL DYNAMICS

STATIC EQUILIBRIUM

Occurs when the sum of the clockwise torques is equal to the sum of the counterclockwise torques.

CENTRIPETAL FORCE

The force acting toward the center of a circular path is provided by mechanisms such as tension in a rope.

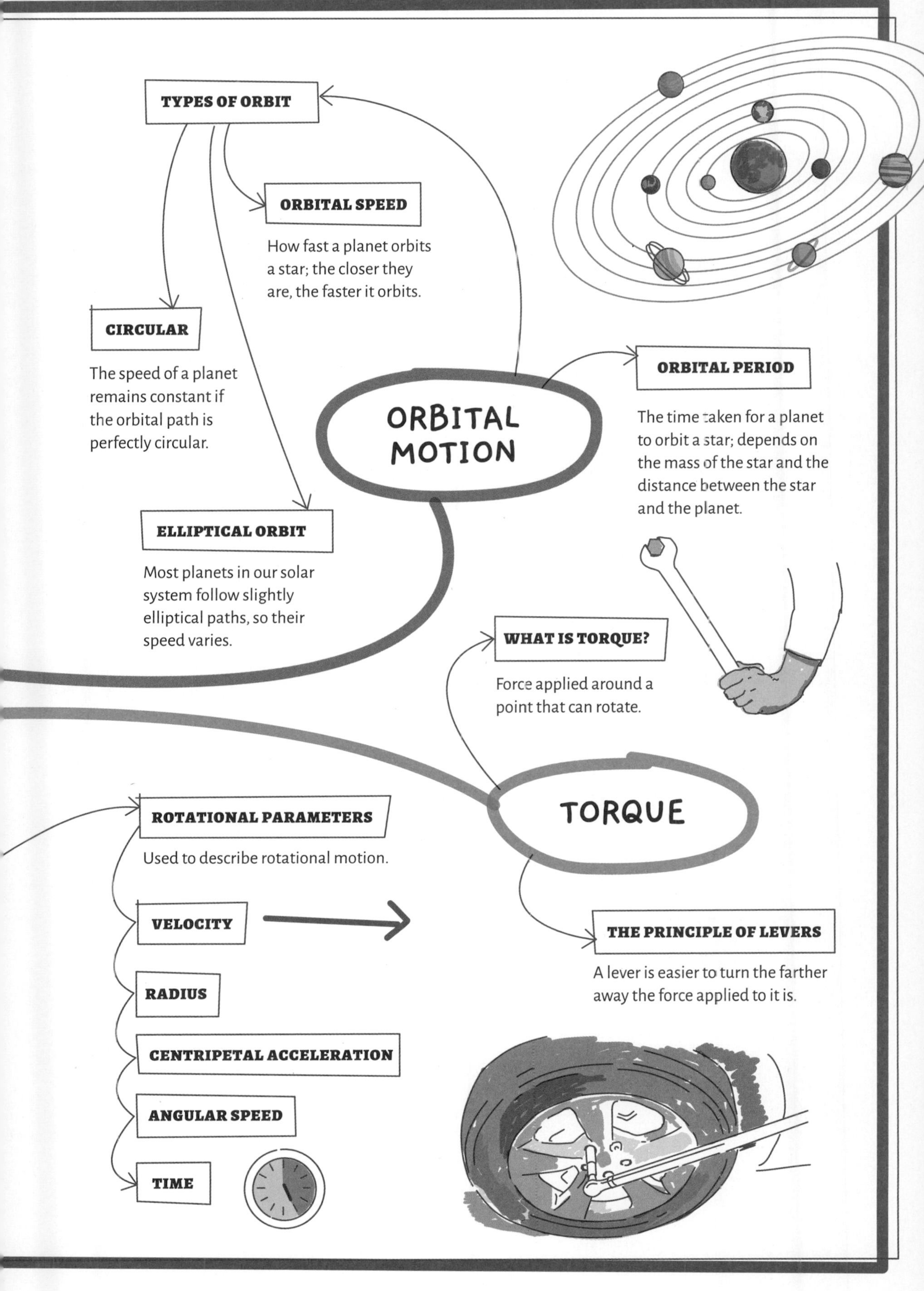

TYPES OF ORBIT
ORBITAL SPEED
How fast a planet orbits a star; the closer they are, the faster it orbits.
CIRCULAR
The speed of a planet remains constant if the orbital path is perfectly circular.
ORBITAL MOTION
ORBITAL PERIOD
The time taken for a planet to orbit a star; depends on the mass of the star and the distance between the star and the planet.
ELLIPTICAL ORBIT
Most planets in our solar system follow slightly elliptical paths, so their speed varies.
WHAT IS TORQUE?
Force applied around a point that can rotate.
TORQUE
ROTATIONAL PARAMETERS
Used to describe rotational motion.
VELOCITY
RADIUS
CENTRIPETAL ACCELERATION
ANGULAR SPEED
TIME
THE PRINCIPLE OF LEVERS
A lever is easier to turn the farther away the force applied to it is.

CHAPTER 4

CONSERVATION LAWS

Everything in the universe is governed by the laws of physics. Our understanding of these laws allows physicists to make predictions based upon observing different quantities, such as energy, momentum, and electric charge. A fundamental law that is unbreakable in physics states that, within a closed system, certain quantities remain unchanged or are conserved. For example, the universe contains a fixed quantity of energy. This may be converted to different types of energy but never created or destroyed.

TYPES OF CONSERVATION LAWS

There is a set of absolute rules in physics that dictates the outcome of many different interactions. Certain physical quantities are always conserved. This section addresses the most important of them: energy, linear momentum, angular momentum, and electric charge (see also pages 63, 70, and 155).

Angular momentum is basically defined as the product of an object's rotational speed—this is measured by change in angle per second—and its moment of inertia, *I*. The body's moment of inertia is a measure of its resistance to a change in rotation speed (either to speed it up or slow it down).

Linear momentum is always conserved when objects collide. The sum of each mass in a system multiplied by its velocity is constant in a closed system and is the same before and after a collision. A white cue ball striking a set of pool balls will transfer its momentum to them.

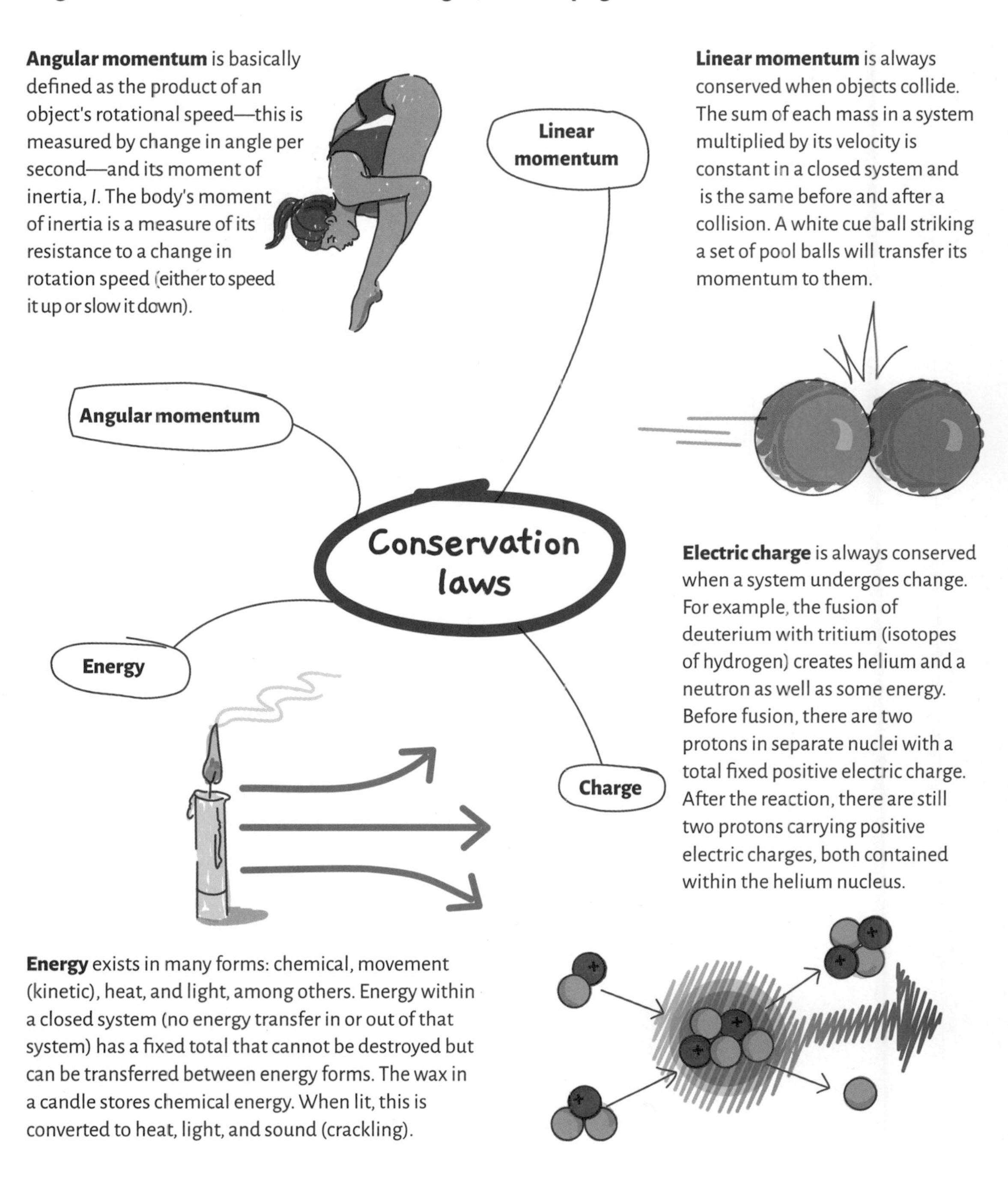

Electric charge is always conserved when a system undergoes change. For example, the fusion of deuterium with tritium (isotopes of hydrogen) creates helium and a neutron as well as some energy. Before fusion, there are two protons in separate nuclei with a total fixed positive electric charge. After the reaction, there are still two protons carrying positive electric charges, both contained within the helium nucleus.

Energy exists in many forms: chemical, movement (kinetic), heat, and light, among others. Energy within a closed system (no energy transfer in or out of that system) has a fixed total that cannot be destroyed but can be transferred between energy forms. The wax in a candle stores chemical energy. When lit, this is converted to heat, light, and sound (crackling).

CLOSED SYSTEMS

Many quantities in physical systems can be measured, allowing us to predict the outcome of the event or events being observed. These may include the position, momentum, and total energy of all the particles within the system. A system is considered to be closed—there can be no transfer of energy or mass—and it may include multiple bodies, such as gas particles, or a single body, such as a basketball.

The total energy, electric charge, and linear and angular momentum within a closed system is fixed, although transfer of any of these quantities can (and does) occur between particles within the system.

An example of a closed system could be a perfectly insulated container full of air where no heat can be transferred from or to the moving particles within the container. Energy can be transferred between particles, but the overall net energy of the system remains constant.

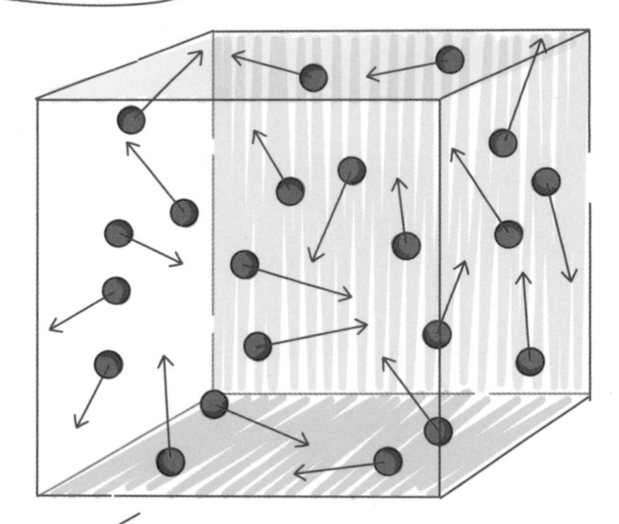

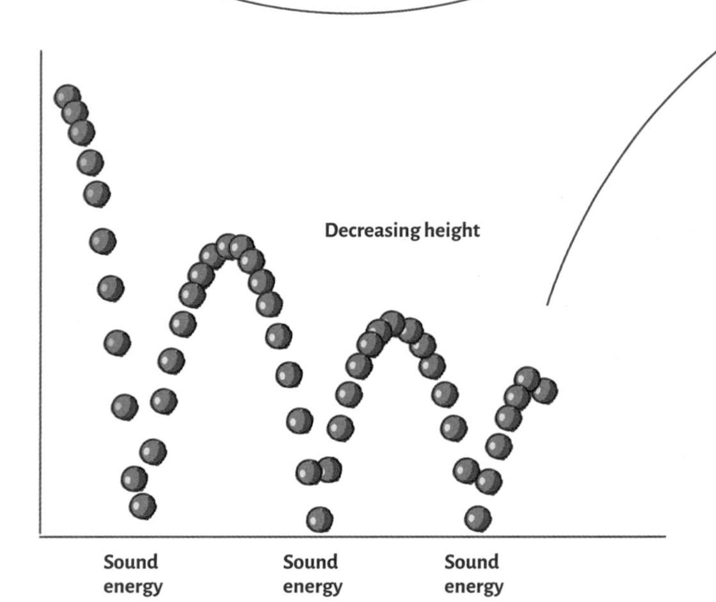

A ball bouncing on a surface can be modeled as a closed system, including the surrounding air and the floor. As the ball bounces, energy is lost to its surroundings via soundwaves and heat. The ball's momentum therefore decreases and eventually becomes zero.

In reality, these theoretical situations are not enclosed or isolated. Exchange of energy does occur because the entire universe is connected. But the assumption of isolation allows models to be constructed in order to make predictions about the system as a whole.

Conservation of energy

Energy exists in many forms, such as kinetic (movement), potential (stored), heat, light, and chemical. Energy is readily converted between different forms and exchanged between particles. It is a fundamental property of energy that it can never be created or destroyed.

A moving body has **kinetic energy** (KE) and momentum. Unless it is in a vacuum, the body will be traveling through a fluid (such as water) containing particles that cause a frictional force acting against the direction of motion.

A diver transfers energy to the water by contact with the particles. The KE of the diver's body transfers to the KE of the fluid's particles, increasing its temperature. With no forward force, the body will slow as it loses energy.

KE of diver turns to KE and heat in the water

Drag

Forward motion

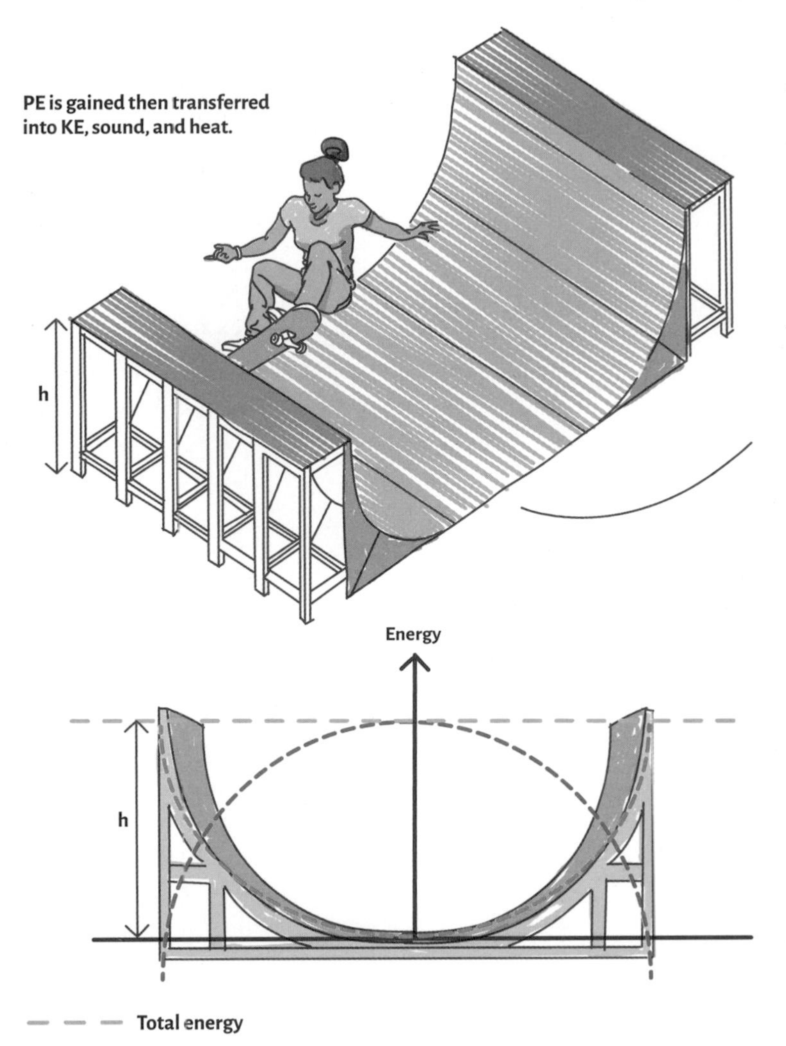

A skater of mass m who moves up a ramp of height h meters above the ground gains **potential energy** ($PE = mgh$). This potential energy is transferred into kinetic energy ($KE = \frac{1}{2} mv^2$) when she accelerates back down the ramp due to gravity g. Some of this energy is transferred to the air particles as sound. All of the potential energy is transferred to the motion of the skater and heating of the air through friction.

All systems convert energy into multiple forms (usually heat and sound) as time progresses, distributing the energy more widely. Energy conversion processes usually cannot be reversed, and the entire system becomes more disorganized. This is called **entropy**. However chaotic a system becomes as its energy is converted to different forms, the total of all energies remains constant.

COLLISIONS

The laws of conservation allow us to predict the results of physical processes involving energy exchanges, heat transfer, or the radioactive decay of an unstable nucleus.

When bodies interact within a system, there is an exchange of both energy and momentum. Such interactions are complex, but physicists simplify the dynamics by considering the interacting bodies to be particles that are nonrotating and which collide head-on. In reality, collisions will invariably be **oblique** (not head-on), but averaged out over many collisions, it is a good approximation.

Interacting bodies

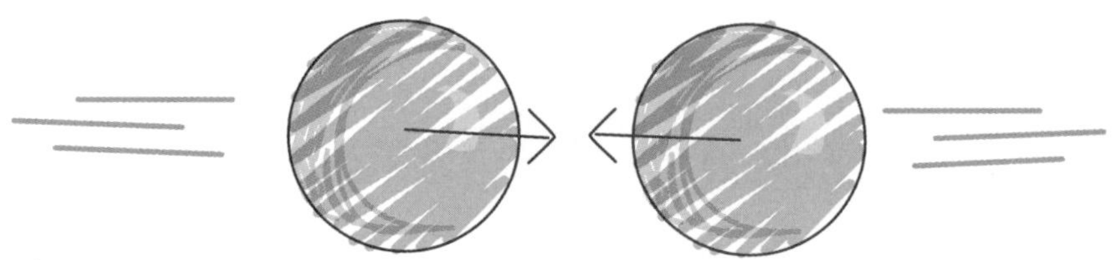

The particles are also considered to be **incompressible**—meaning they are rigid and do not deform. Again, this is a greatly simplified version of reality but a good starting point from which to develop more rigorous models.

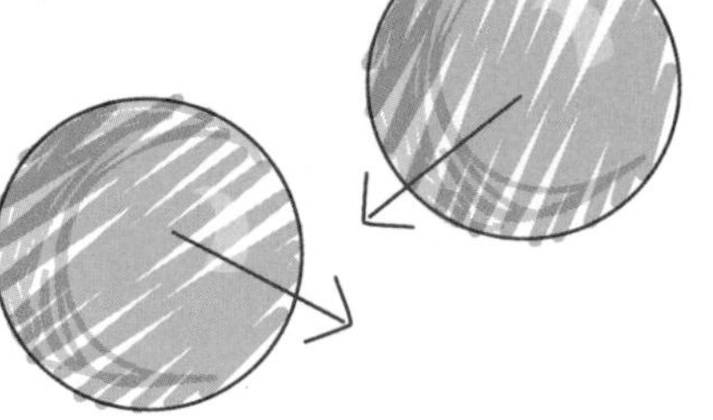

Compressing the ball

STORED POTENTIAL

As momentum is transferred from a swinging golf club to the ball resting on the tee, some of its energy is stored as **elastic potential** in the ball, which compresses to receive it. When the ball leaves the tee, it expands and regains its shape, releasing the stored potential as kinetic energy that pushes the ball forward. The ball is now carried by a momentum equal to the momentum in the golf cub swing, with a minor amount being lost through heat.

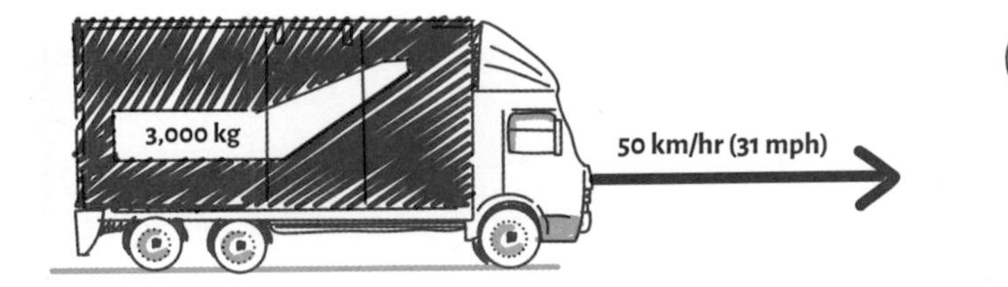

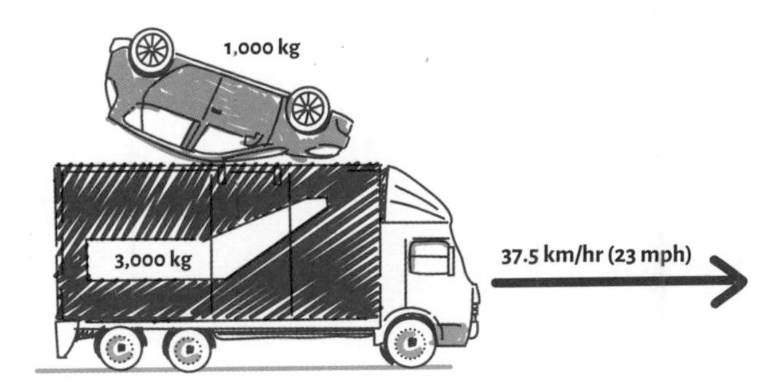

There are two types of collision to consider: **perfectly elastic** and **inelastic**.

For a perfectly elastic collision, it is assumed that the particles exchange energy perfectly and there is no loss of total kinetic energy (KE) of all the particles within the system. For an event that exhibits perfectly elastic collisions, heat and sound transfer do not occur as a result of the collision, so the particles continue at the same speed but in a different direction.

For an inelastic collision, this is not true and there is a degeneration of particle KE as sound and heat energy radiate from the impact. This means that the total KE of all the particles before a collision is greater than that after the collision. This creates an average decrease in the velocities of the particles as energy is converted to other forms. Sometimes in an inelastic collision, such as a collision between two vehicles, both particles join together and they move as one, with a reduced final velocity.

Perfect particles

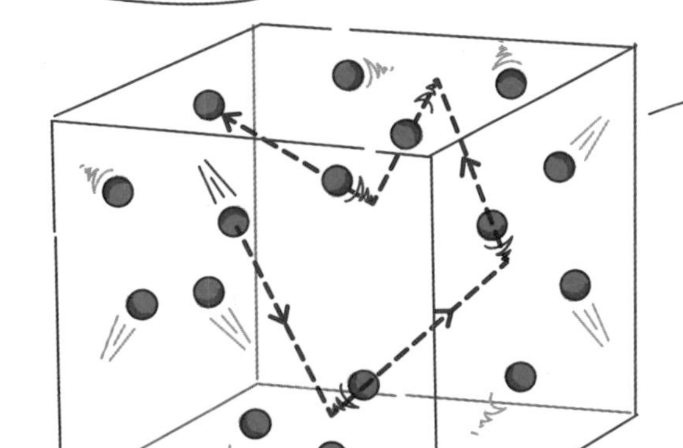

If the collision is perfectly elastic, both kinetic energy and momentum are conserved. An example of a perfectly elastic collision is the interaction between **ideal gas** particles enclosed in a container and colliding with each other, with their velocities governed by the heat of the gas.

If the collision is inelastic, only momentum remains unchanged.

In both types of collision, linear momentum is always conserved.

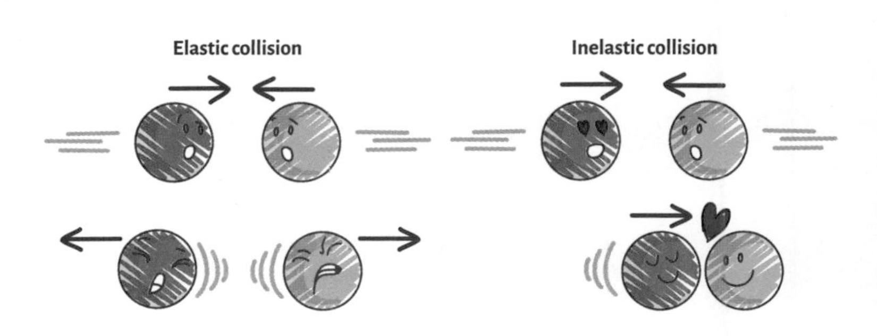

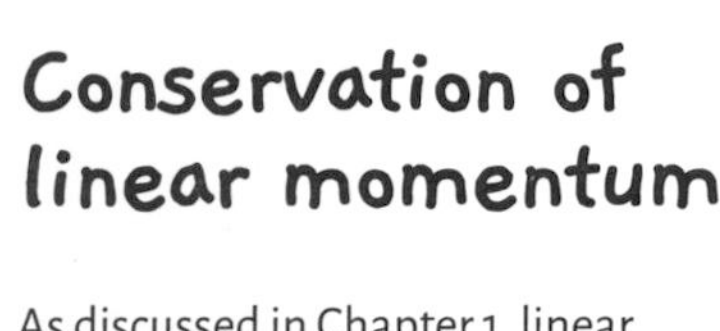

Conservation of linear momentum

As discussed in Chapter 1, linear momentum, p, is the product of a body's mass and its velocity ($p = mv$). Linear momentum is conserved when two (or more) bodies interact, such as during a collision (which may be either perfectly elastic or inelastic). This principle is known as the **conservation of linear momentum**.

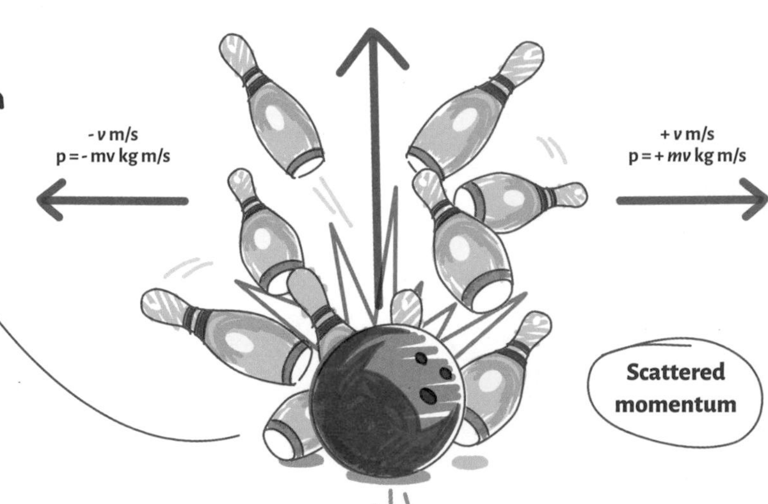

Momentum is a vector quantity, and its direction in one dimension is determined by being either **positive** (moving right) or **negative** (moving left).

Consider a pool ball that is struck in a positive direction at another stationary ball of equal mass. The nature of the motion of each particle after the collision will depend on the mass of each particle and the efficiency of energy transfer.

On the rebound

First ball

Before collision

Second ball

Collision

The combined momentum of both balls after the collision will be equal to the momentum of the first ball before the collision.

Energy is transferred to the second ball, which will move in a positive direction.

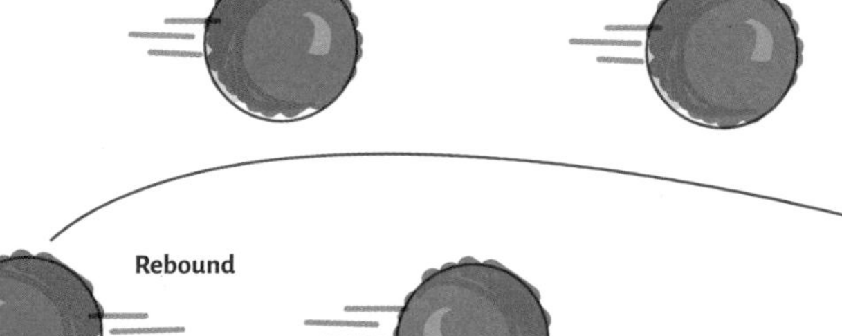

When they collide, the first ball may continue moving forward, but at a reduced velocity, or may rebound.

If the first ball rebounds on collision with the second ball, it will have a negative velocity, and hence a negative momentum, and will move in the opposite direction.

RECOIL

A bullet loaded into a rifle is stationary before it is fired, and therefore has zero momentum, as it has zero velocity. When the trigger is pulled, the bullet is fired and will almost instantaneously gain momentum as its velocity rapidly increases because of the energy stored in the gunpowder. The rifle recoils in the opposite direction, but at a much lesser speed because of its mass, and so has negative momentum. The **sum of momenta** of both bullet and rifle remains zero, as they are equal in magnitude but opposite in direction.

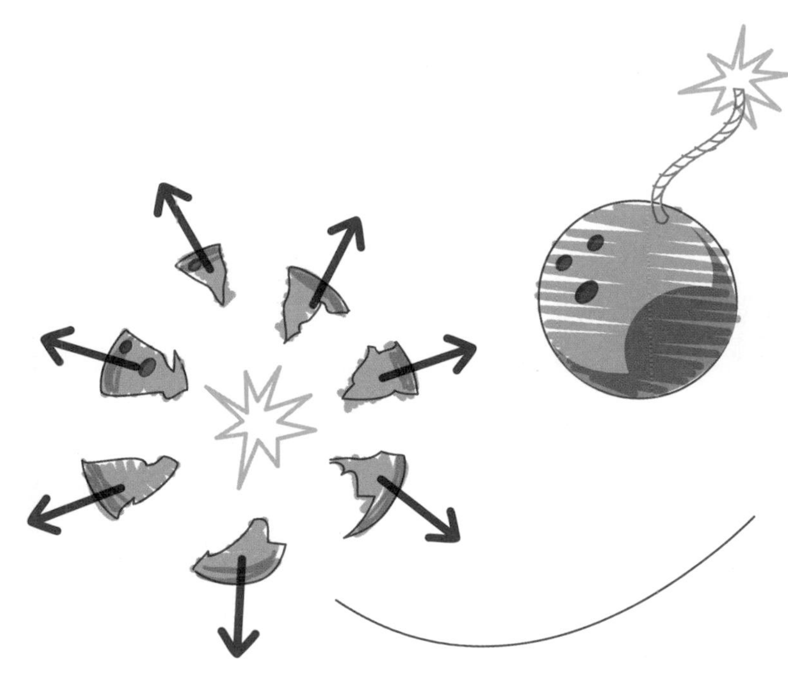

EXPLOSION

A firework or any exploding object also conserves momentum. The initial intact body will have no velocity relative to its upward movement. When it explodes, however, shards of material are equally accelerated in all directions, resulting in a sum of zero momentum because the velocity directions of each piece are all in opposition.

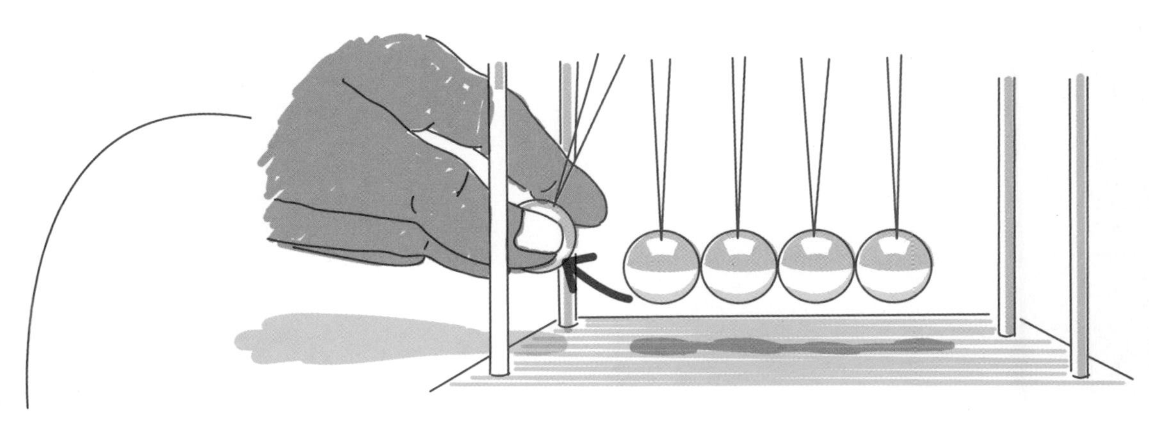

NEWTON'S CRADLE

Newton's cradle demonstrates this conservation law perfectly. As the moving ball strikes the rigid structure of the stationary balls, it transfers its momentum to the ball of equal mass on the other side. Energy is lost via sound, so the system ultimately slows down.

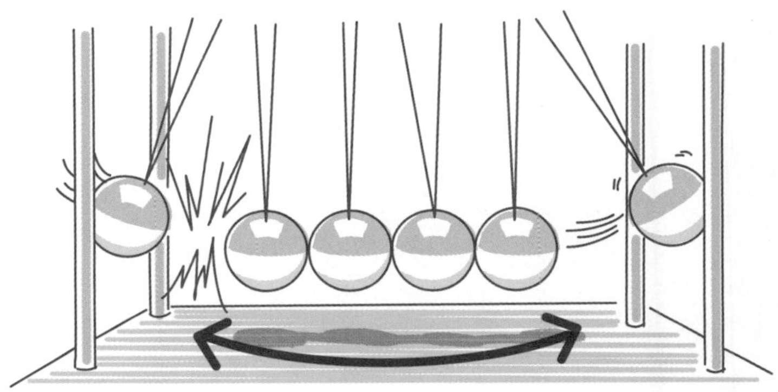

Conservation of angular momentum

Angular momentum, L, is a property of a rotating or spinning body. Its magnitude is affected by the total mass of the spinning body and how spread out the mass is relative to the axis of rotation. A body with a mass concentrated toward its center of rotation has a smaller angular momentum than one with the same time period of rotation but a mass concentrated at a larger distance.

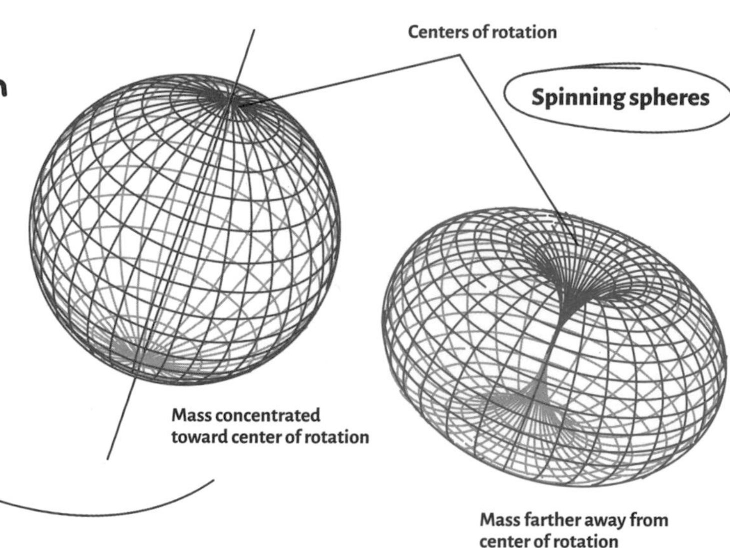

Angular momentum is always conserved within an isolated system as the distribution of mass changes.

Angular momentum is conserved when a skater changes her shape while she spins. As she moves her arms inward, she will spin faster as the average distance of her total mass to her axis of rotation decreases.

If a star that is rotating begins to collapse at the end of its life cycle, it will start to rotate faster. Some stars, if they are massive enough, collapse to form a neutron star or a black hole.

They emit radio waves that can be observed and used to determine rotation speed.

As a star's mass becomes more centralized (assuming it loses no mass into space), the rotation speed also increases, in order to conserve angular momentum. Some dead stars rotate in excess of 40,000 times per minute as a result of their radius decreasing.

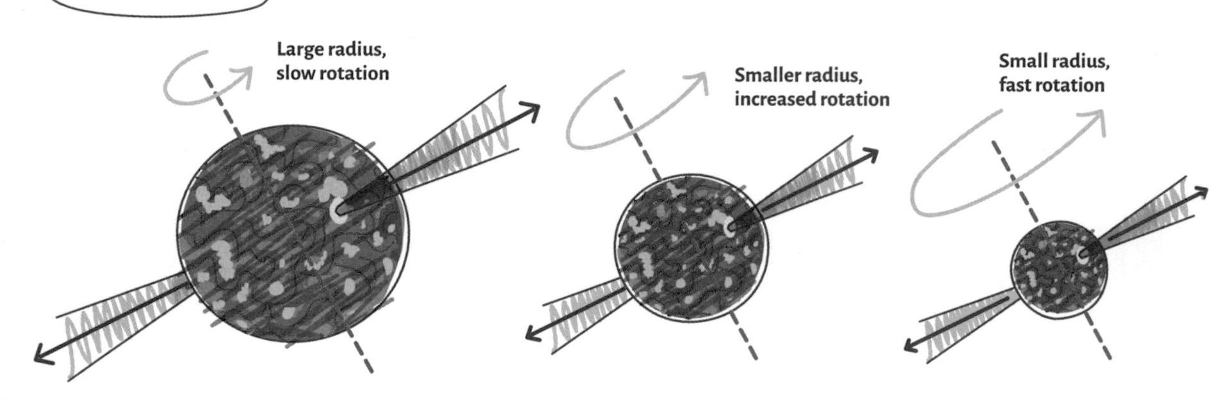

Angular momentum is the product of an object's rotational speed and its **moment of inertia**, *I*. A body's moment of inertia is a measure of its resistance to a change of any kind in rotation.

When a body is rotating or spinning, it possesses angular momentum that is determined by the rotation speed and mass distribution of the body. This property is conserved even if the shape of the rotating body changes. A spinning diver rotates more quickly as she concentrates her mass by making herself as small as possible.

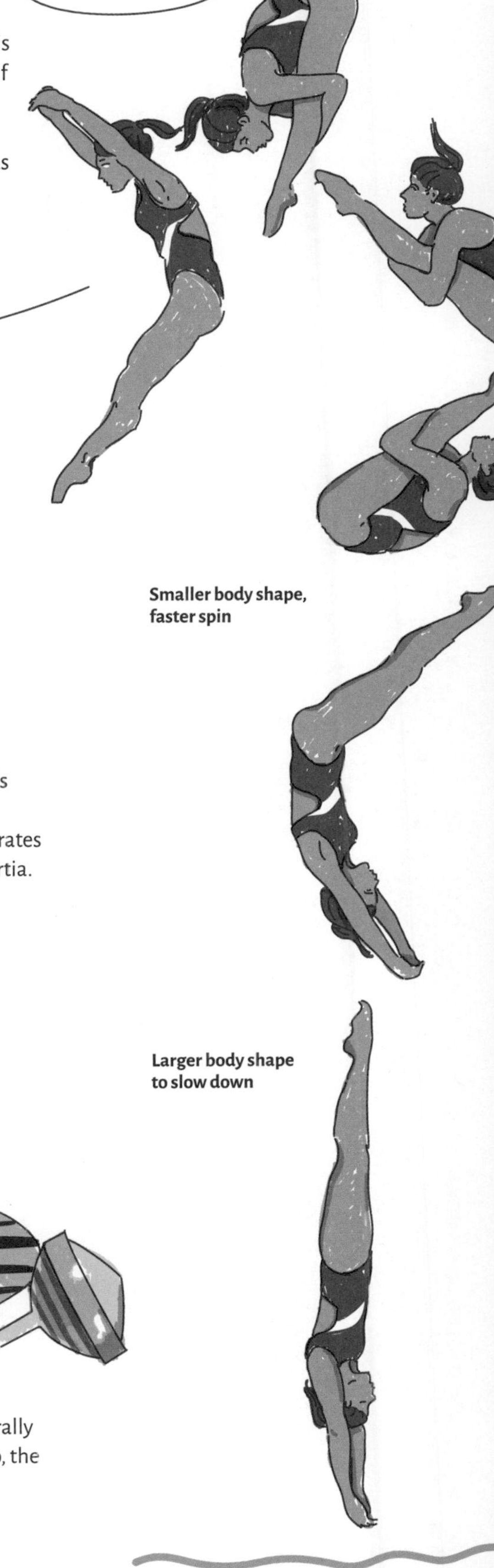

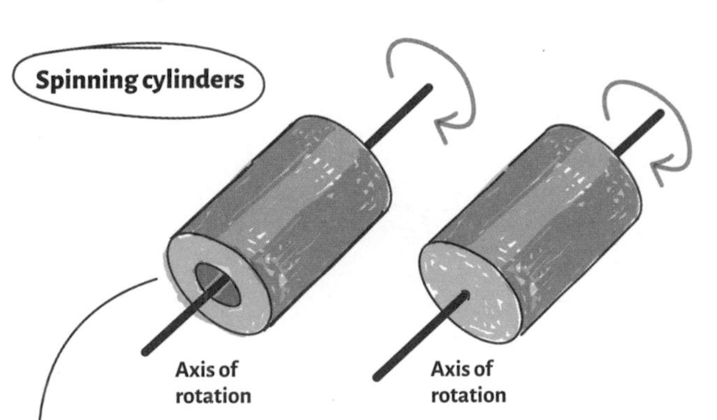

Take two cylinders of equal size and density. A hollow cylinder would be easier to slow down than a solid one as it has less mass but occupies the same amount of space (if the hollow center is included). This is a greatly simplified view, but it illustrates the idea of moments of inertia.

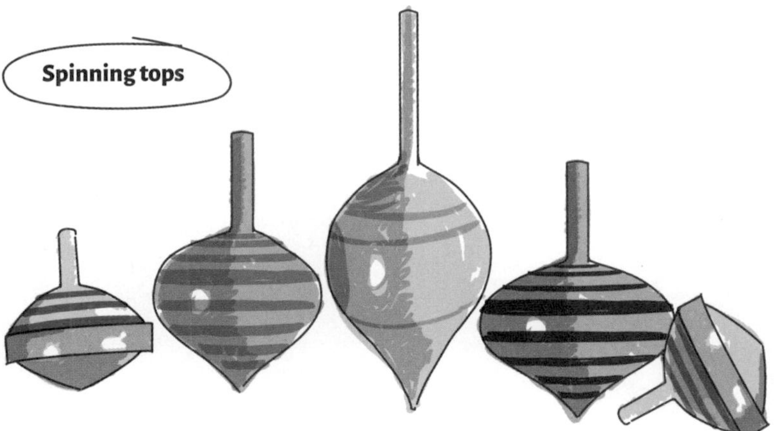

Spinning tops are designed to spin for as long as possible, and the design of their wide tapered shape maximizes angular momentum in order to achieve this. Generally speaking, the wider the top, the longer it will spin.

CONSERVATION LAWS

UNIVERSALITY OF CONSERVATION LAWS

Energy, momentum, and electric charge within a closed system are always conserved.

TYPES OF ENERGY

MOMENTUM

A vector quantity, momentum, *p*, is the product of mass, *m*, and velocity, *v*; there are two types, linear and angular.

$$p = mv$$

ENERGY

Never destroyed but transfers between forms: kinetic, chemical, gravitational, potential, heat, sound, light.

ELECTRIC CHARGE

Is conserved even when the system undergoes change.

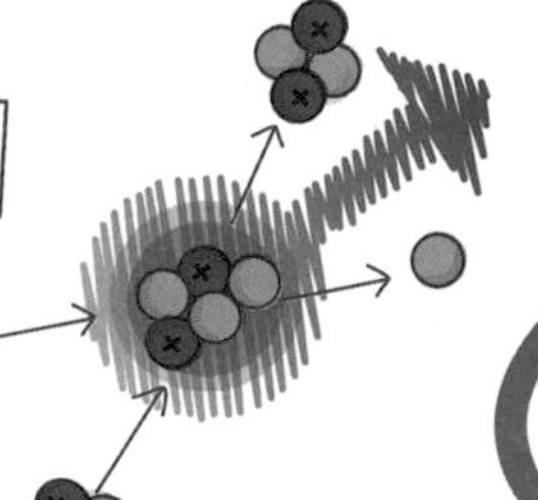

MOMENTUM

ANGULAR MOMENTUM

The product of an object's rotational speed and its moment of inertia.

WHAT IS MOMENTUM?

Momentum is a vector, so two identical bodies moving in opposite directions at the same speed will have a total momentum of zero.

THE MOMENT OF INERTIA

A body's moment of inertia is a measure of its resistance to a change in rotation.

DISTRIBUTION OF MASS

The moment of inertia changes as the distribution of mass changes.

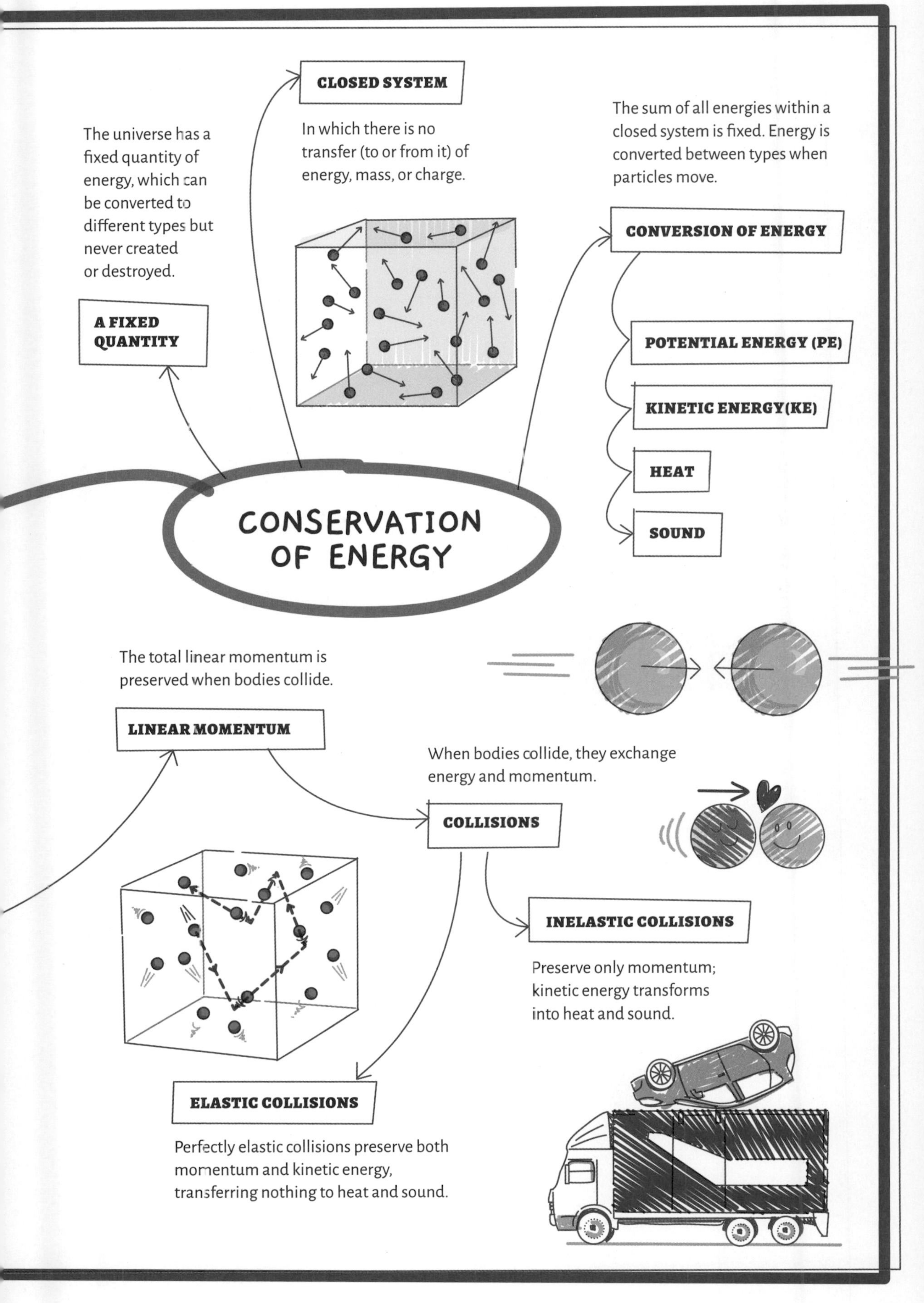
CLOSED SYSTEM
In which there is no transfer (to or from it) of energy, mass, or charge.
The universe has a fixed quantity of energy, which can be converted to different types but never created or destroyed.
A FIXED QUANTITY
The sum of all energies within a closed system is fixed. Energy is converted between types when particles move.
CONVERSION OF ENERGY
POTENTIAL ENERGY (PE)
KINETIC ENERGY(KE)
HEAT
SOUND
CONSERVATION OF ENERGY
The total linear momentum is preserved when bodies collide.
LINEAR MOMENTUM
When bodies collide, they exchange energy and momentum.
COLLISIONS
INELASTIC COLLISIONS
Preserve only momentum; kinetic energy transforms into heat and sound.
ELASTIC COLLISIONS
Perfectly elastic collisions preserve both momentum and kinetic energy, transferring nothing to heat and sound.

CHAPTER 5

ELECTRICITY

Electricity features in our lives every day: lighting our homes and streets, powering the Internet, and charging our cellphones. At the flick of a switch, a mysterious flow of energy transforms the way we live. It is a relatively new invention, which has harnessed the power of lightning and delivered it into our homes. It's hard to imagine how our lives would be without it.

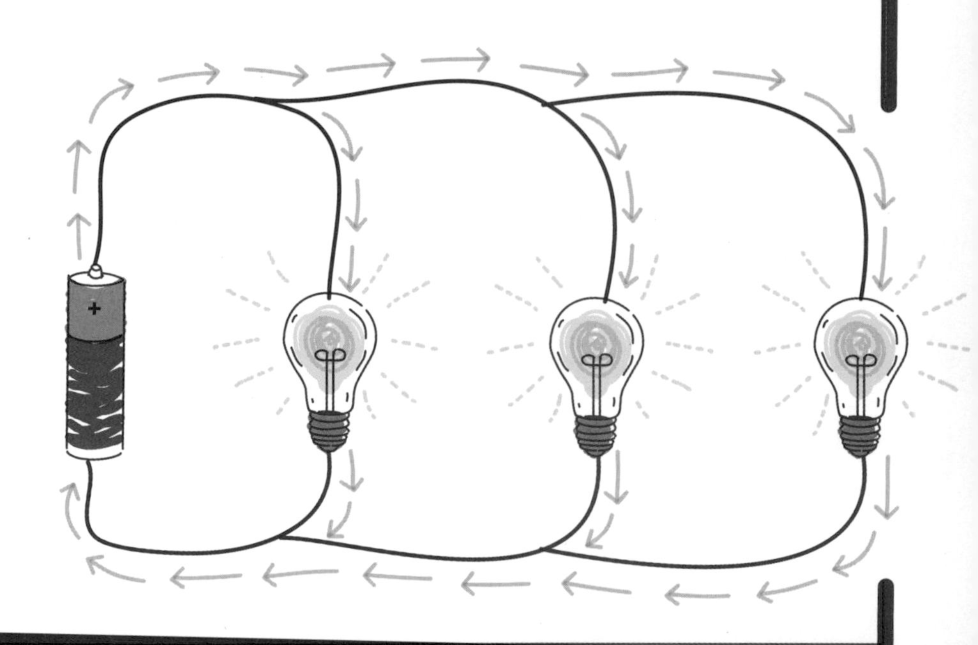

ELECTRIC CHARGE AND CHARGE TRANSFER

Electricity is caused by charged particles: electrons and ions. Electrons carry a negative charge, and ions a positive or a negative charge. Charge is measured in coulombs (C) and has the symbol Q. A coulomb is equal to the quantity of electric charge flow in 1 second defined by a current of 1 ampere (A). It is named for French engineer and physicist Charles-Augustin Coulomb (1736–1806).

The **coulomb** is a very large unit of charge by comparison to the particles that carry it. Electrons carry what is known as the **elementary charge**, e, which is approximately 1.6×10^{-19}C. By comparison, lightning bolts carry a typical charge of between 15C and 300C.

1 coulomb equals 60 million, million, million electrons.

An electron or an ion carries exactly one elementary charge and any charged particle carries a whole number of electrons or ions, which gives it its charge.

Electric charge can either be **static** (static electricity) or moving. A charge can be moved from one place to another via a **potential difference** between two points. In electricity, potential difference is usually called **voltage**.

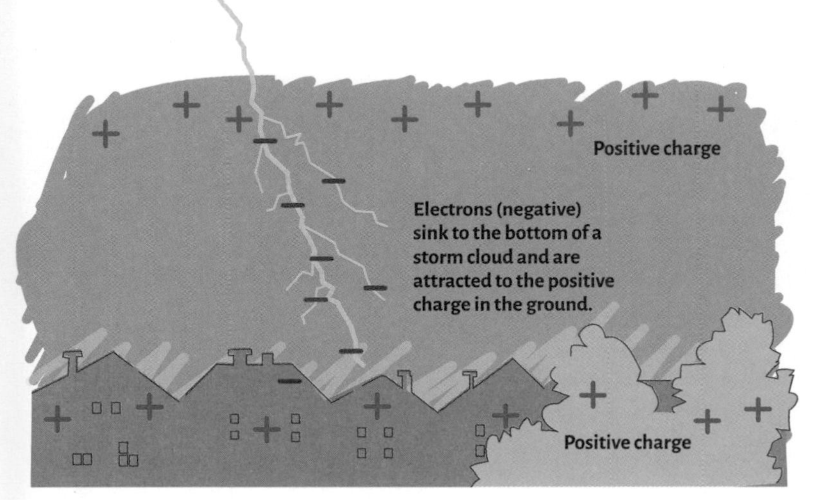

Think about how a river flows. There is a slope down which the water runs, from a high potential energy to a lower one. In a similar way, charged particles flow along a potential difference—the direction of flow depends on whether the charge is positive or negative.

This charge flow is referred to as **charge transfer**, and the rate at which transfer occurs is called the **electric current**. The direction of current flow is defined as the direction in which a positive charge is moving, referred to as **conventional current**.

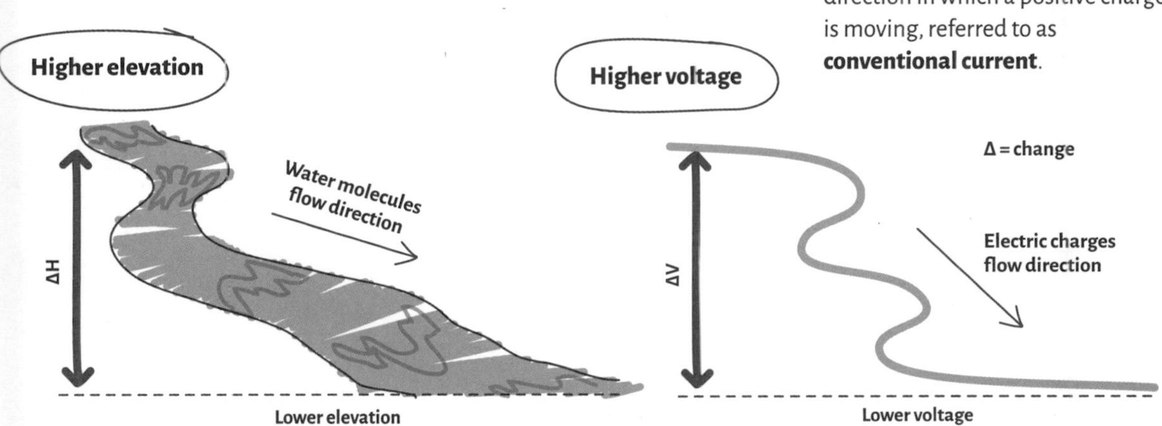

CURRENT, VOLTAGE, AND RESISTANCE

The flow of charged particles (current) is controlled primarily by two factors: the size of the potential difference (voltage), and the resistivity of the medium used to carry the charge transfer (resistance).

Current and voltage

For movement of any charged particles to produce a current, there must be both a supply of charged particles and a potential difference, or voltage, to drive that movement. The electricity you use is transferred by electrons and is conducted from one place to another by wires, usually made of copper. Copper is full of electrons that are free to move along the length of the wire. These are called **free electrons** but will remain static unless a potential difference is provided.

Free electrons from outer shells of metal atoms

Metal ions

The potential difference applied is measured in volts, V. A **volt** is defined as the energy transferred per unit charge (joules per coulomb, J/A). The standard international unit of energy, the **joule** is defined by the energy expelled when moving an object with a force of 1 newton through a distance of 1 meter.

In electrical terms, it is also the heat dissipated when a current of 1 amp flows through a resistance of 1 ohm for 1 second. This in turn defines the **ohm** (unit of electrical resistance; see Ohm's law, page 67).

An **amp**, short for ampere and named for André-Marie Ampère (1775–1836), is a unit of current. One amp is equivalent to 1 coulomb passing a given point in the wire every second.

1V will produce 1A if there is 1 ohm of resistance.

Electric voltage

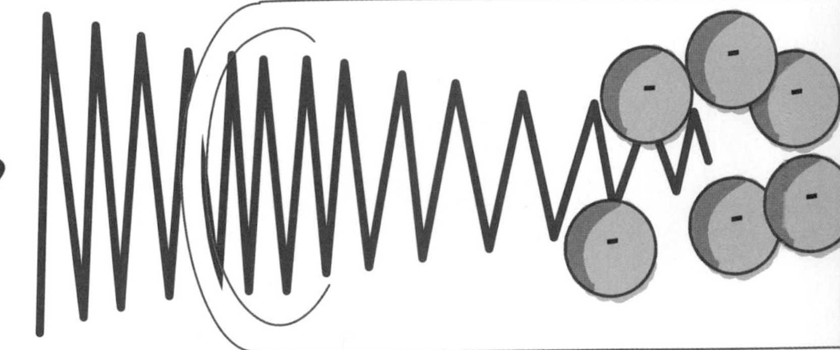

Electrical resistance

Electrical **resistance** within a wire is controlled by a number of factors. Primarily, these are the material and the length and thickness (the cross-sectional area) of the wire. Temperature also has a large effect on resistance but will be ignored for this chapter.

The material used will affect charge flow based on its structure and the availability of free electrons.

Again, imagine a river flowing down a gradient. The flow of water will be affected by the steepness of the slope (external voltage), the depth and width of the river (cross-sectional area), as well as the nature of the riverbed and any obstacles such as rocks (the material of the wire).

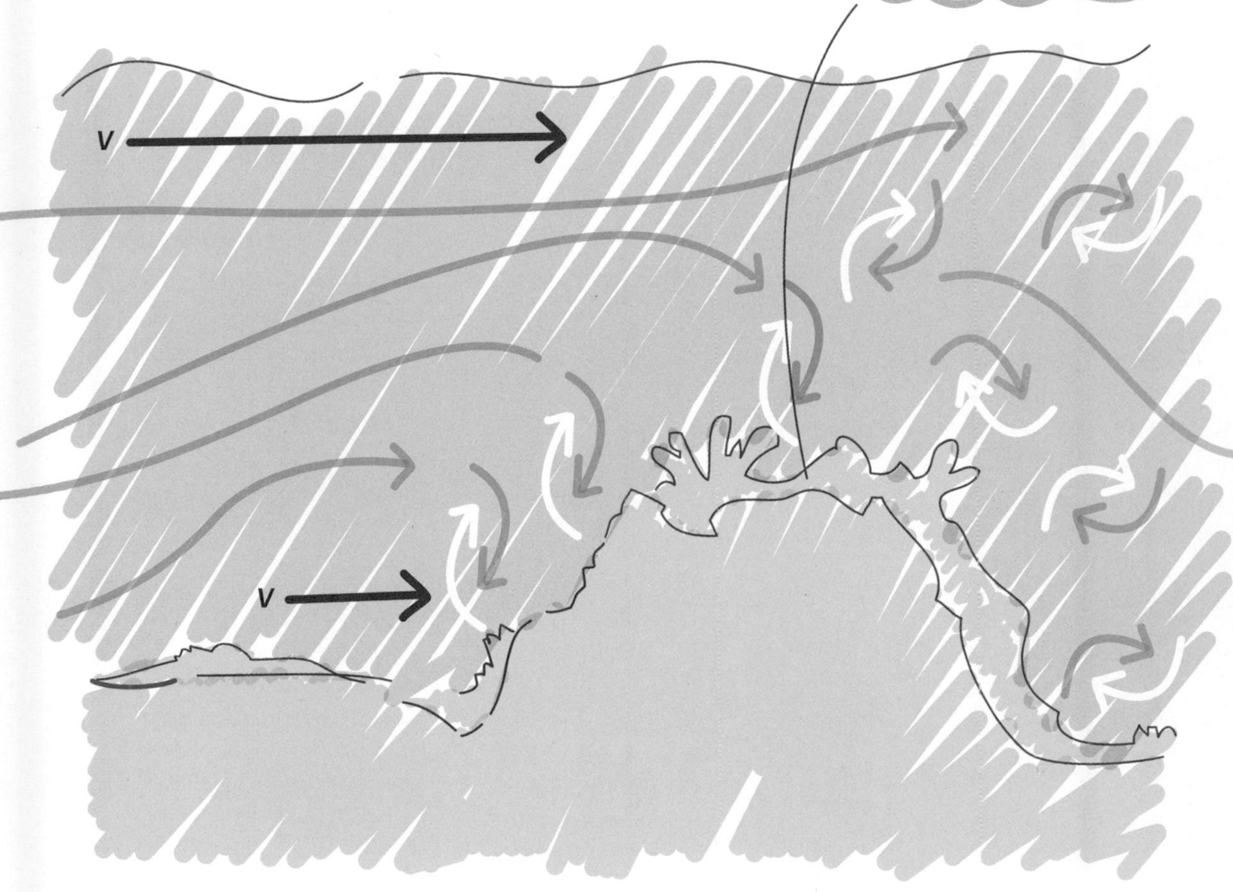

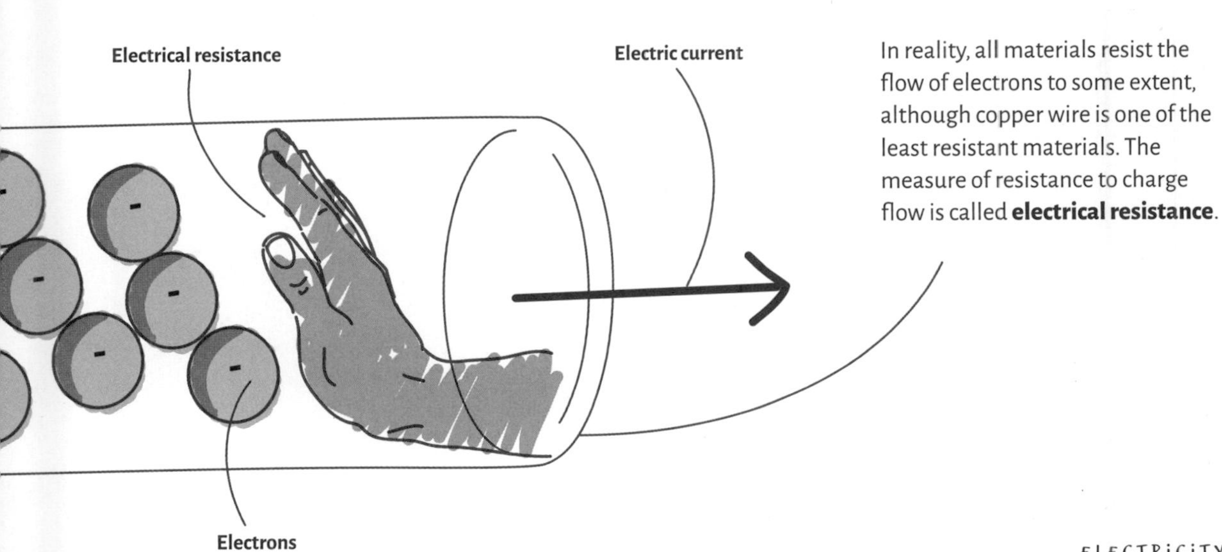

In reality, all materials resist the flow of electrons to some extent, although copper wire is one of the least resistant materials. The measure of resistance to charge flow is called **electrical resistance**.

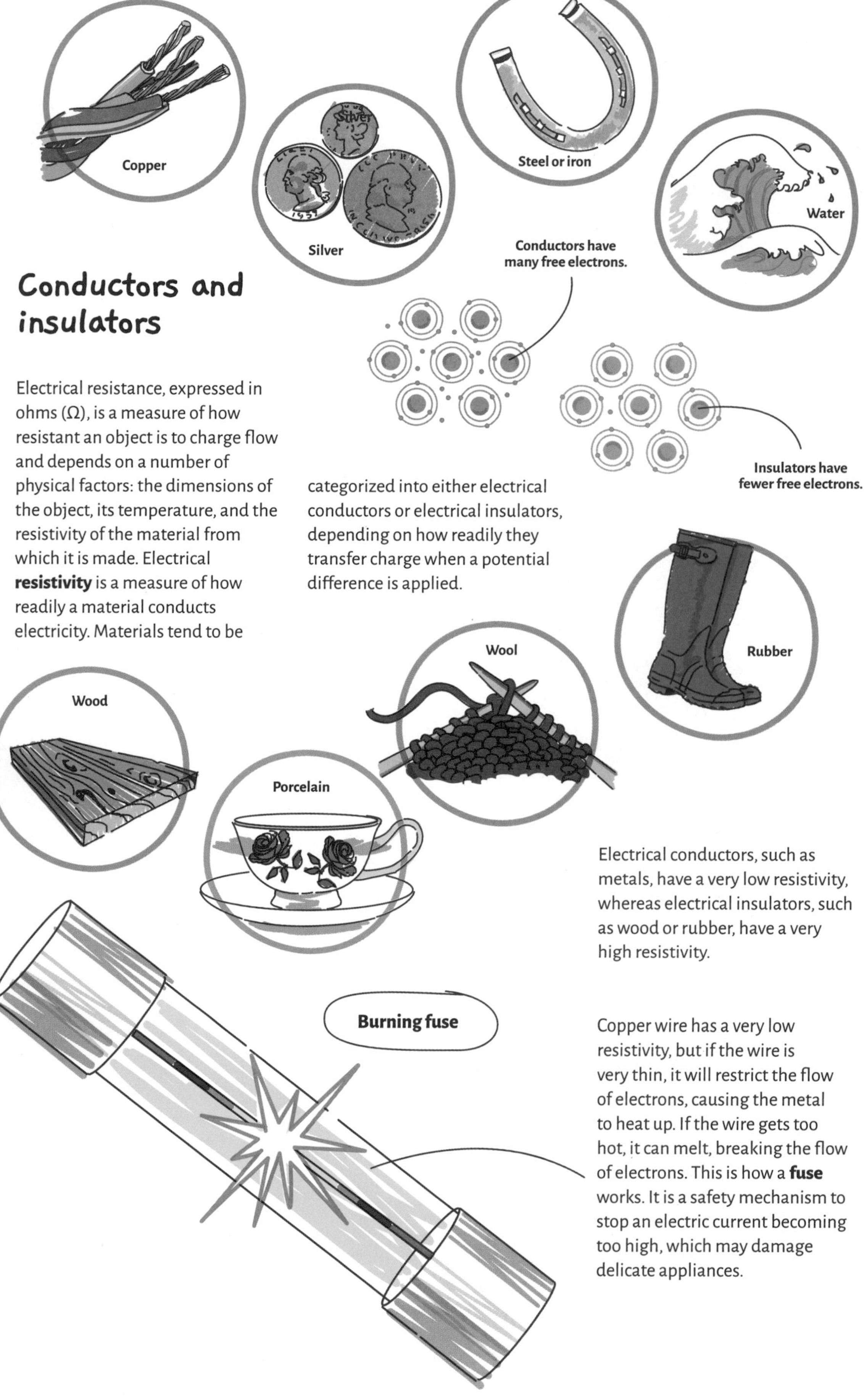

Conductors and insulators

Electrical resistance, expressed in ohms (Ω), is a measure of how resistant an object is to charge flow and depends on a number of physical factors: the dimensions of the object, its temperature, and the resistivity of the material from which it is made. Electrical **resistivity** is a measure of how readily a material conducts electricity. Materials tend to be categorized into either electrical conductors or electrical insulators, depending on how readily they transfer charge when a potential difference is applied.

Electrical conductors, such as metals, have a very low resistivity, whereas electrical insulators, such as wood or rubber, have a very high resistivity.

Copper wire has a very low resistivity, but if the wire is very thin, it will restrict the flow of electrons, causing the metal to heat up. If the wire gets too hot, it can melt, breaking the flow of electrons. This is how a **fuse** works. It is a safety mechanism to stop an electric current becoming too high, which may damage delicate appliances.

Ohm's law

For a material with low electrical resistivity, such as copper, it only requires a small voltage to move electrons. For a material with a high electrical resistance, such as air, wood, or rubber, a much higher voltage is required. This is the basis of **Ohm's law**, formulated by German physicist Georg Ohm (1789–1854) in 1827, to explain the relationship between voltage, current, and resistance.

The current, I, between two points is directly proportional to the voltage, V, across these points and the connecting medium: summarized as $I = kV$, where k is a constant for the connecting medium.

The constant, k, is a constant for a specific material and is defined as how easily that material conducts electricity (its **conductance**)—this is universally defined by $1/R$, or the **reciprocal of the resistance**.

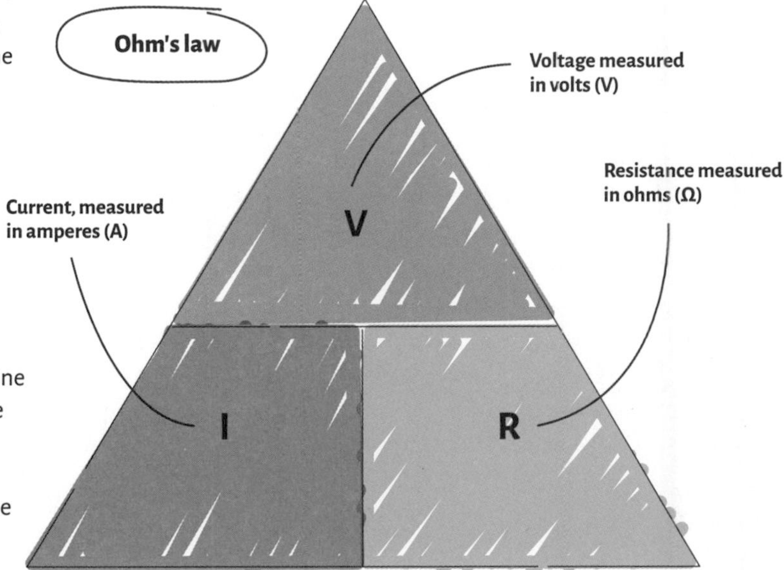

Ohm's law enables us to determine the resistance of a material if the voltage is controlled and the current is measured. The formal triangle helps to remember these relationships.

The three forms of Ohm's law

$$V = IR \qquad I = \frac{V}{R} \qquad R = \frac{V}{I}$$

OHM'S LAW

Voltage across a resistor is directly proportional to the current flowing through the resistor.

This formula defines the ohm: A material with a resistance of 1Ω requires a potential difference of 1V to cause a current of 1 amp to flow. An ohm is a very small quantity for most materials, so often KΩ (1,000 ohms) and MΩ (1,000,000) are used.

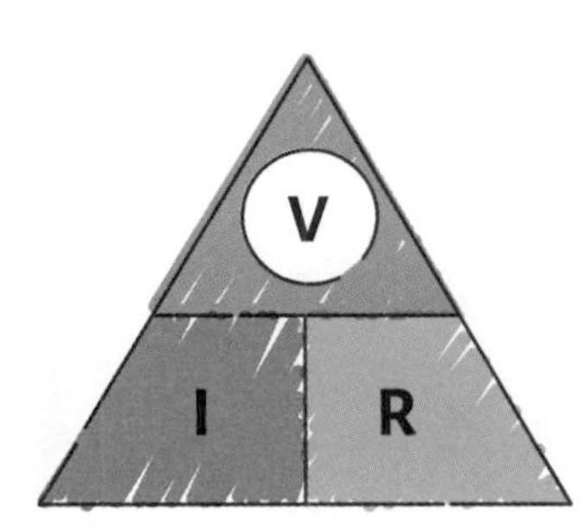

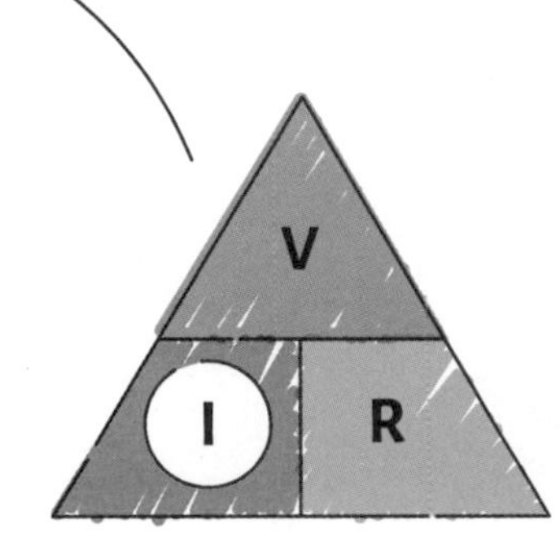

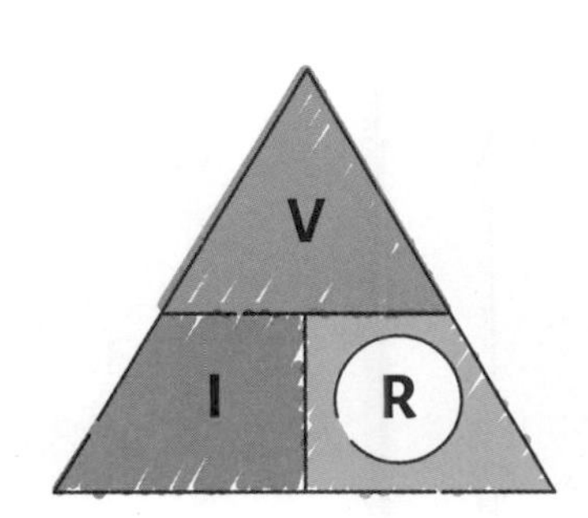

ELECTRICAL CIRCUITS

Electrical circuits allow current to flow. Circuits are comprised of wire connecting various components to a power source, which may be a single cell or a number of cells (known as a battery). A circuit will be a complete loop in order to carry the charge from one terminal of the battery to the other.

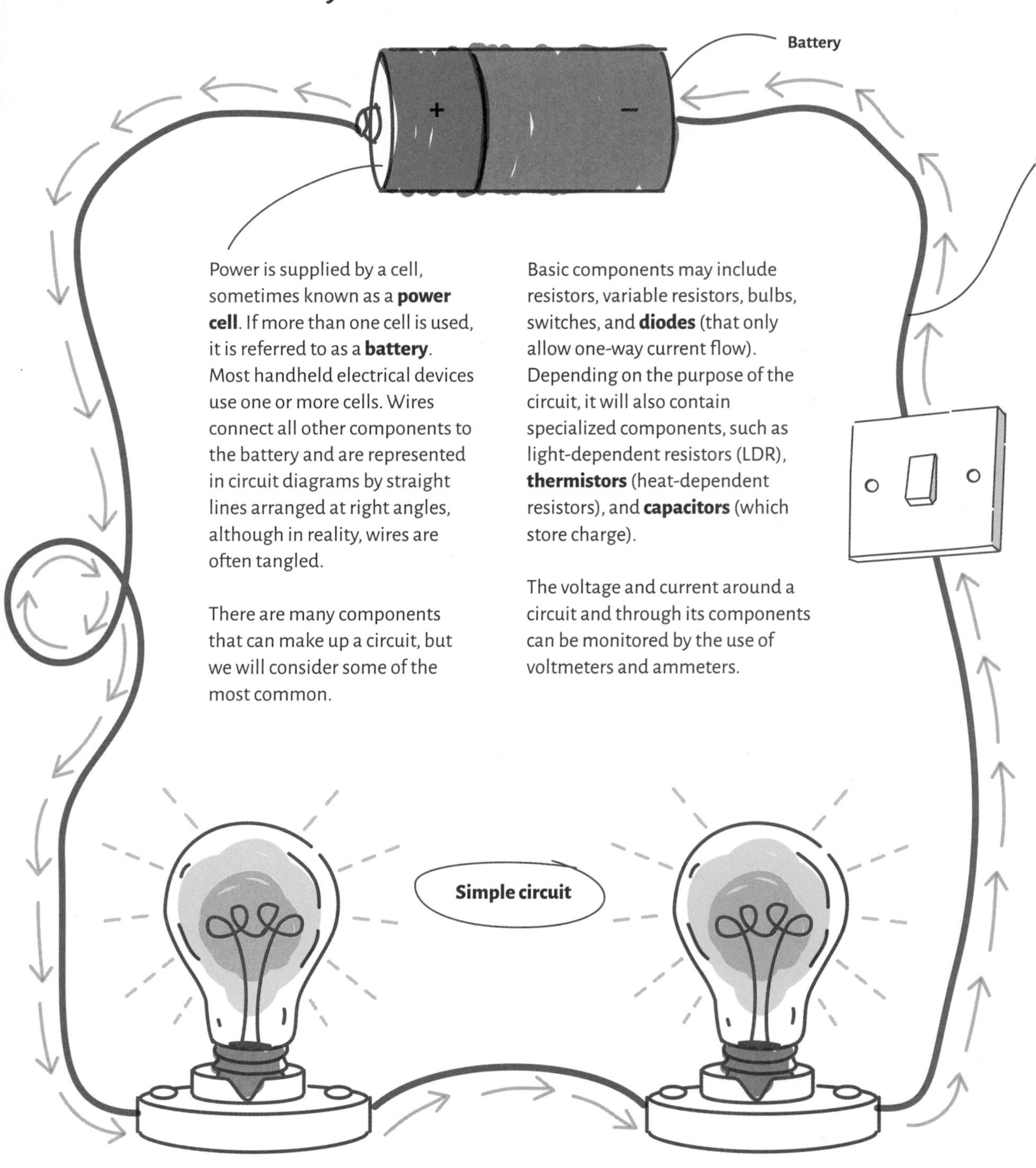

Power is supplied by a cell, sometimes known as a **power cell**. If more than one cell is used, it is referred to as a **battery**. Most handheld electrical devices use one or more cells. Wires connect all other components to the battery and are represented in circuit diagrams by straight lines arranged at right angles, although in reality, wires are often tangled.

There are many components that can make up a circuit, but we will consider some of the most common.

Basic components may include resistors, variable resistors, bulbs, switches, and **diodes** (that only allow one-way current flow). Depending on the purpose of the circuit, it will also contain specialized components, such as light-dependent resistors (LDR), **thermistors** (heat-dependent resistors), and **capacitors** (which store charge).

The voltage and current around a circuit and through its components can be monitored by the use of voltmeters and ammeters.

Circuit symbols

The symbols used when illustrating circuit diagrams are internationally recognized and allow clear guidance to construct even complex circuits.

Diode: A conductor; allows current to flow in one direction and blocks its flow in the other.

Light-emitting diode: LED; emits light when a current passes through it.

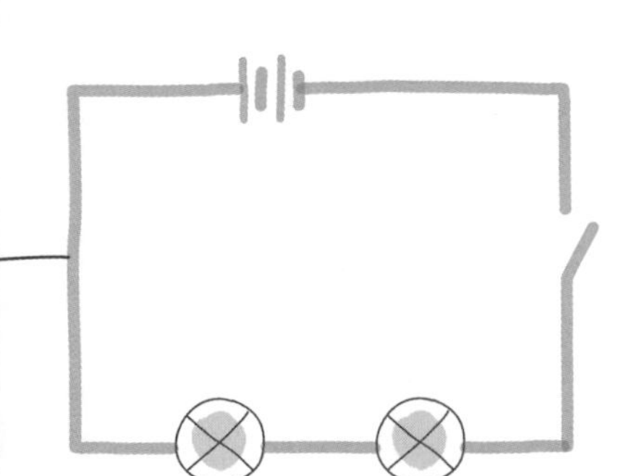

This circuit diagram shows in symbols the elements of the circuit drawn opposite.

Cell: A single unit that converts chemical energy into electrical energy.

Voltmeter: Monitors potential difference between two points in a circuit , measured in volts.

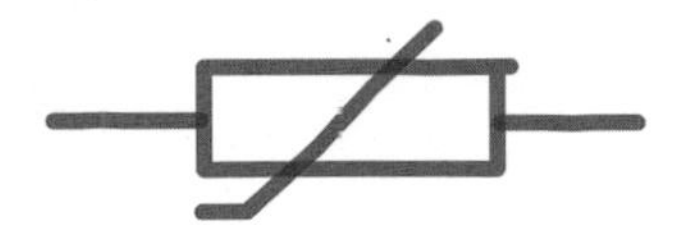

Thermistor: Safety device; resistor governed by temperature; resistance increases or decreases as temperature rises or falls.

Battery: The source of energy in a circuit, made of a collection of cells.

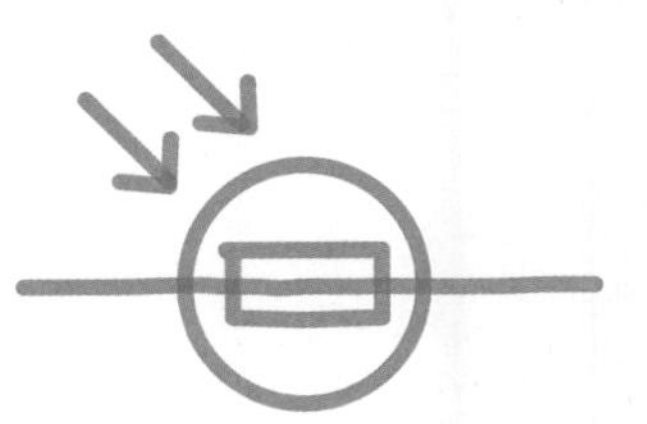

Light-dependent resistor: LDR; detects light levels; resistance decreases as light increases.

Variable resistor: Resistor that can vary the level of resistance it applies, so can increase or decrease current flow.

Fixed resistor: Resistor (used to reduce current flow without switching it off) that provides a constant resistance regardless of environmental changes.

Ammeter: To monitor electric current, measured in amps.

Capacitor: Stores electrical charge and releases it as the circuit requires.

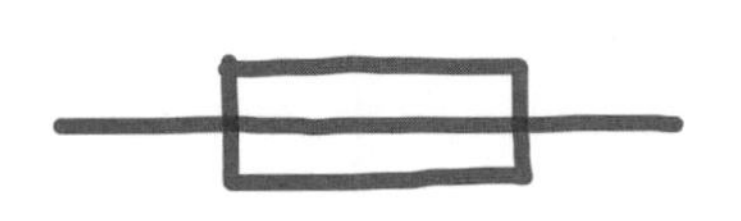

Fuse: Safety device; contains a wire that melts easily, so breaks the circuit if there is a fault.

Lamp: Used to signal whether or not the circuit is working.

Kirchhoff's laws

In 1845, the German physicist Gustav Kirchhoff (1824–1887) first described his laws concerning the conservation of current and potential difference around a **closed-circuit loop**.

Kirchhoff stated that the total charge traveling around the loop is conserved: the **law of conservation of charge**.

As charge is moved around the circuit by the potential difference across the battery terminals (known as the **electromotive force**, or **emf**), its total remains constant around the circuit. If the circuit splits at a junction, the total charge entering the junction is equal to the total charge leaving the junction (conservation of charge). The current (charge per second) will split in a ratio that is governed by the resistances of each path. A higher resistive path will carry less current (from $V = IR$).

KIRCHHOFF'S FIRST LAW
Current flowing into a node (or a junction) must be equal to current flowing out of it.

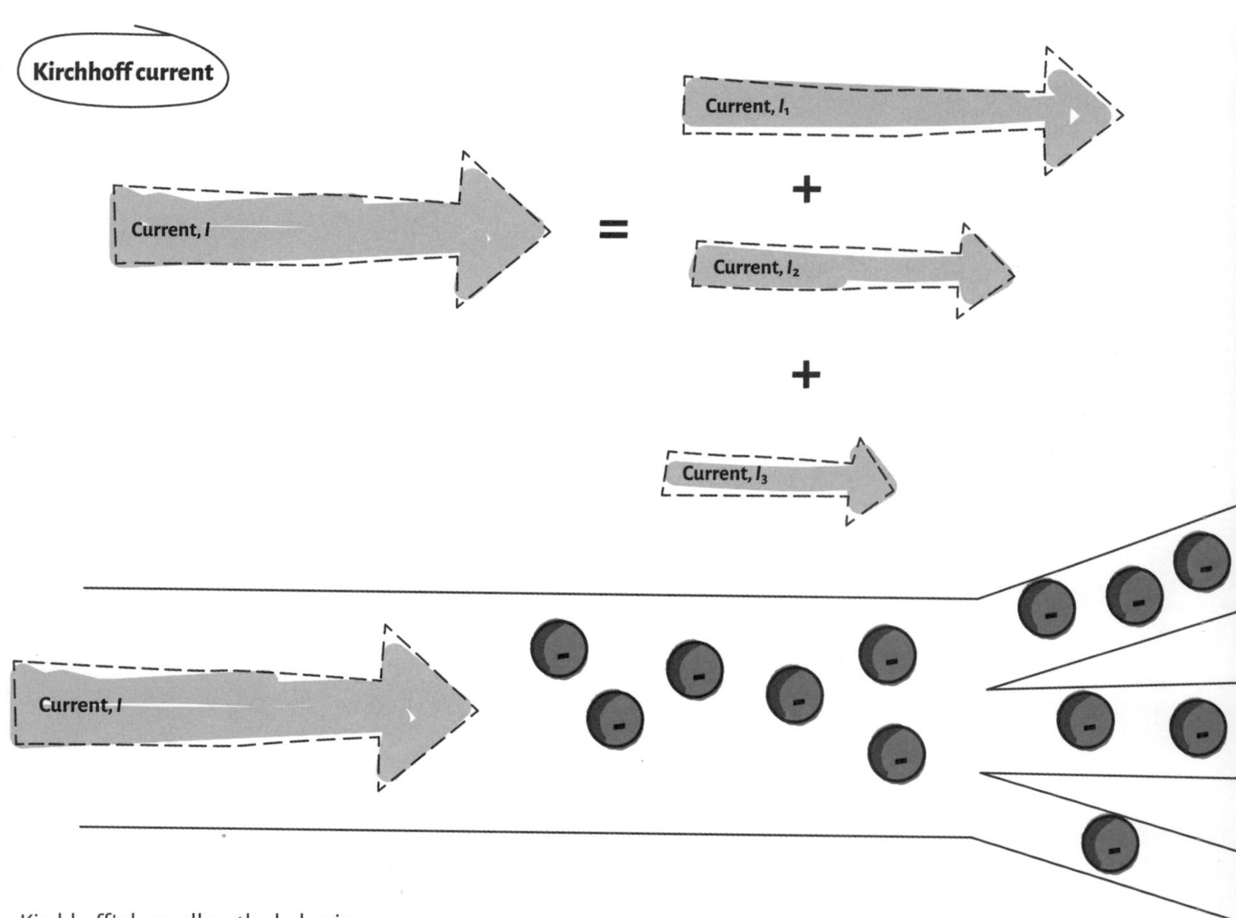

Kirchhoff's laws allow the behavior of circuits to be predicted with some degree of accuracy if the **internal resistance** of the power supply is known.

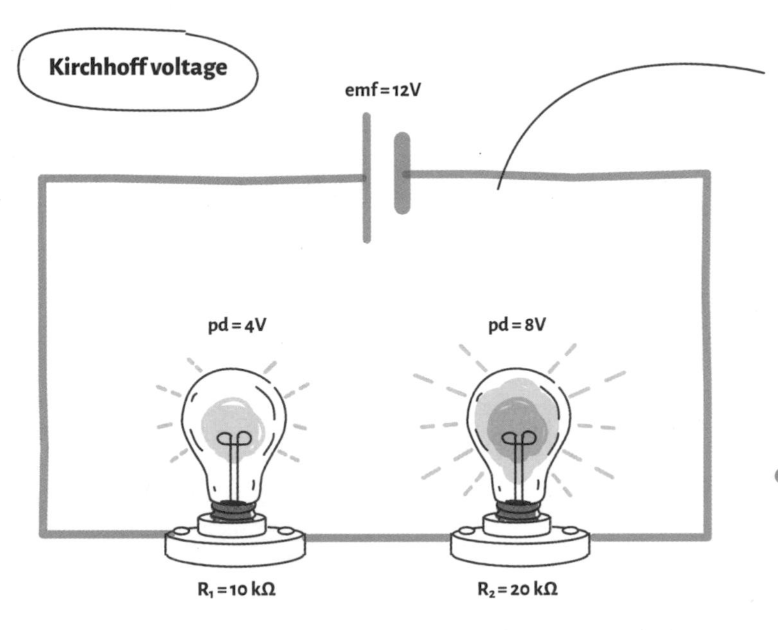

The emf is split between each component and junction according to the ratio of their resistances. A higher resistance across a component or junction requires a higher proportion of the emf to drive the current through. Ignoring internal resistance, the sum of all potential differences (pd) around the circuit will be equal to the emf.

KIRCHHOFF'S SECOND LAW

The sum of all voltages around any closed loop in a circuit must equal zero.

The potential difference across the battery driving the current will be equal to the sum of the potential differences measured across each individual component, assuming there is no resistance in the connecting wires.

Circuit heating occurs in real life because there is resistance in the adjoining wires and through the power supply, although, as you have seen, the resistance in copper wires is very small.

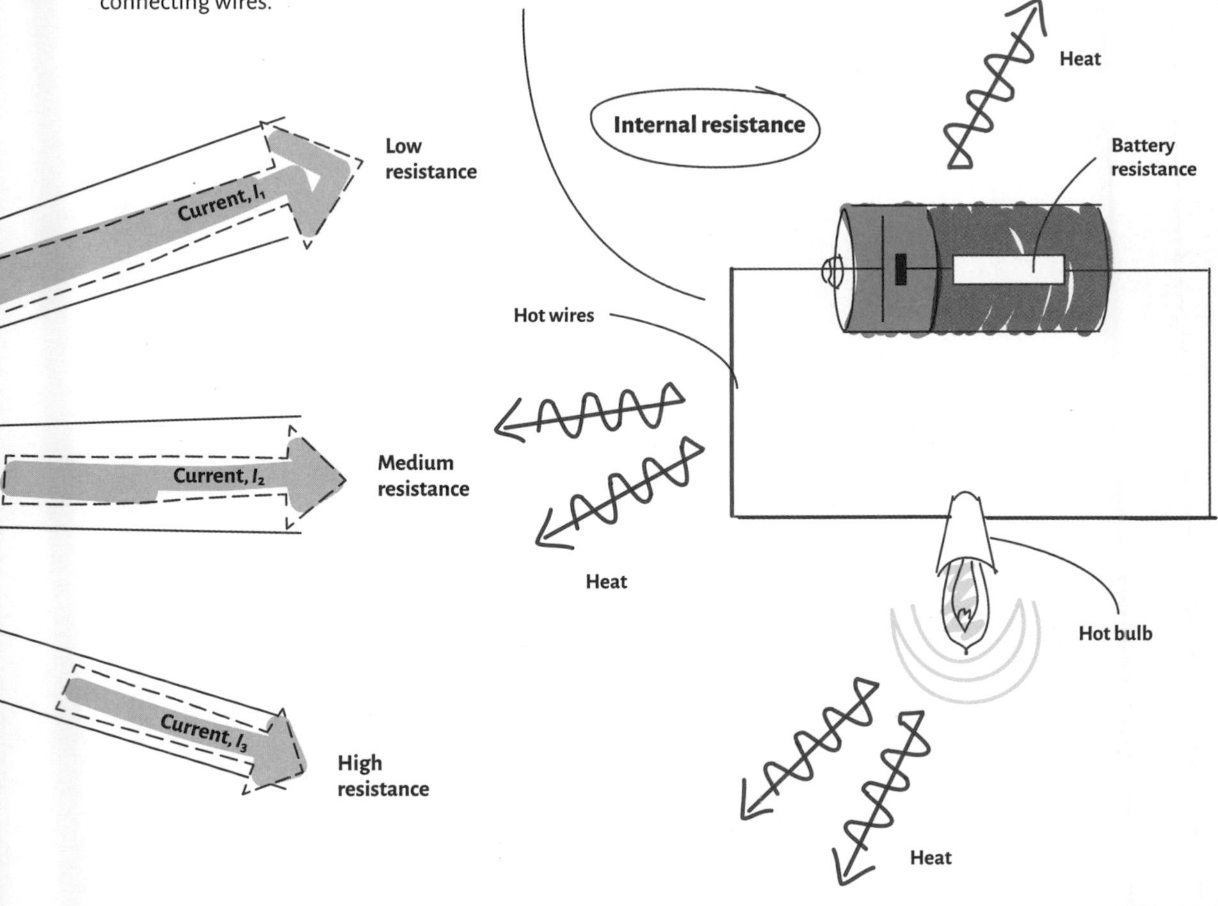

Series and parallel circuits

Depending on how components are connected in their circuits, they will be classified as either **series** or **parallel** circuits. If components are connected together in a line around one loop, this is a series circuit. If the circuit splits and delivers current to its components along separate routes, the components are said to be in parallel.

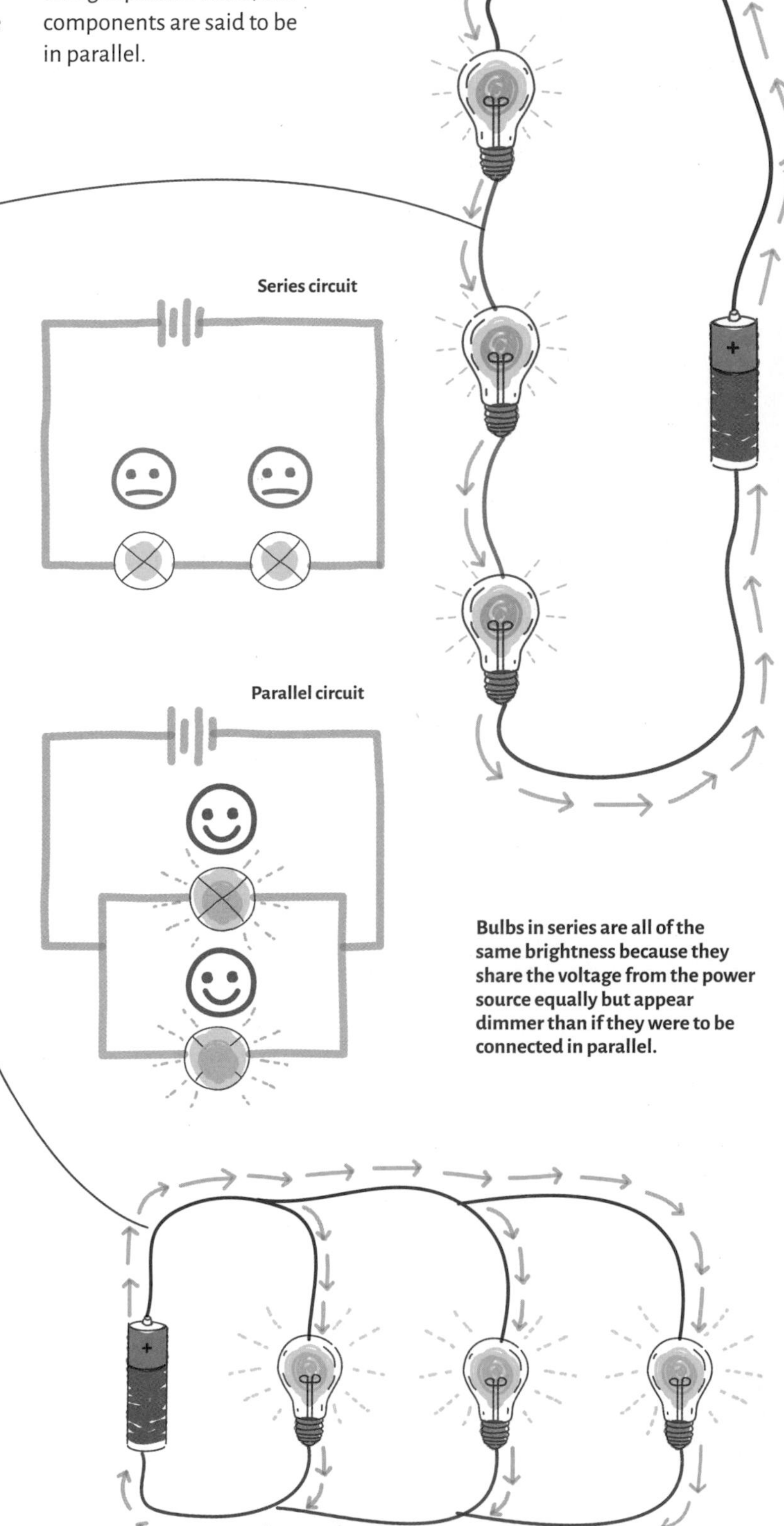

Bulbs in series are all of the same brightness because they share the voltage from the power source equally but appear dimmer than if they were to be connected in parallel.

SERIES CIRCUITS

Imagine a number of identical light bulbs. One bulb connected up to a battery of fixed emf will have a specific brightness.

If three light bulbs are connected together in a line by a conducting wire, and then to a battery, they are said to be connected in series.

If the bulbs are identical, with an equal resistance, the emf of the battery is shared equally between each of them. The total resistance, RT of all three bulbs, and hence the resistance to current flow, is the sum of their individual resistances.

PARALLEL CIRCUITS

If the three light bulbs have their own conductive wires and are connected separately to the battery, they are in parallel. The current splits up over the three paths, and the same voltage goes across each component. The total current is the sum of the currents flowing through each component.

If a component fails in a series circuit, they all fail; if a component fails in one branch of a parallel circuit, the other components continue to function as normal.

Capacitors

A **capacitor** is a device used in a circuit for storing electrical energy in an electric field. It can release the energy very quickly when required. The basic components are two metal plates (**conductors**) separated by an insulator of some kind. When a voltage is applied across the two plates, an **electric field** is created.

Capacitors come in many sizes. The amount of energy a capacitor can store relies on a number of factors. Larger metal plates will provide more electrons. Reducing the gap between the plates also increases the capacity to store energy, but there is a limit to this reduction. If the gap becomes too small, the charge will leak across to the other plate, neutralizing both plates. This charge leakage can be stopped by placing an insulating material between the plates. The material is called a **dielectric**.

If the power source is removed, providing the gap between the plates is sufficient to insulate charge flow, the charge built up on the plates will remain.

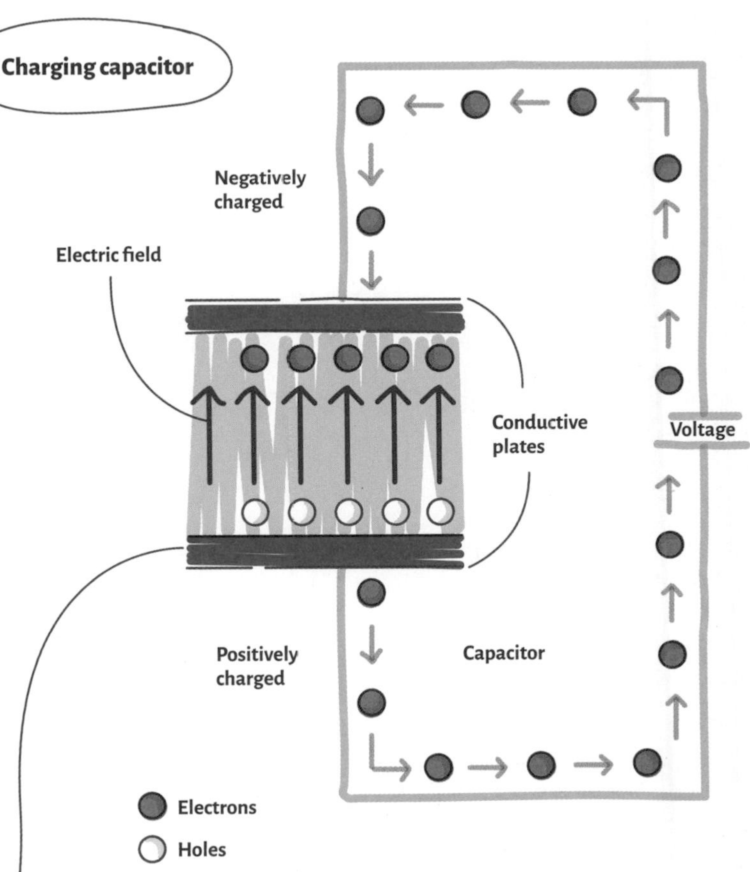

When a voltage (potential difference) is applied across a capacitor, electrons are driven from one plate (creating "holes" in the atoms where electrons existed, generating an overall positive charge) and are pushed onto the other plate (creating an excess negative charge). A parallel electric field builds across the plates as the charge difference increases.

When the potential difference is removed, the electrons cannot flow back to restabilize the charge difference due to the insulating dielectric, and so the capacitor stores this charge, which can be later used to power an electrical circuit.

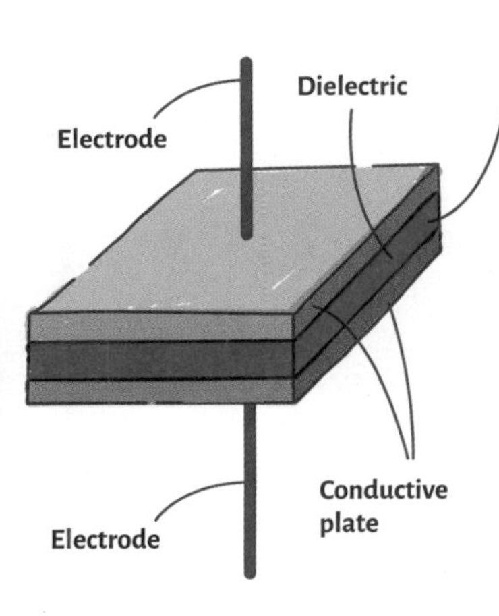

CAPACITANCE

The ability of a capacitor to store charge is called its **capacitance**, *C*, and is measured in **farads** (named for Michael Faraday, see page 87). One farad, *F*, is defined as 1 coulomb of charge stored for each volt of potential difference applied across the plates.

A farad is a huge unit and most capacitors are rated in micro-Farads, μF (one-millionth of a farad).

ELECTRICITY

CHARGE AND CHARGE TRANSFER

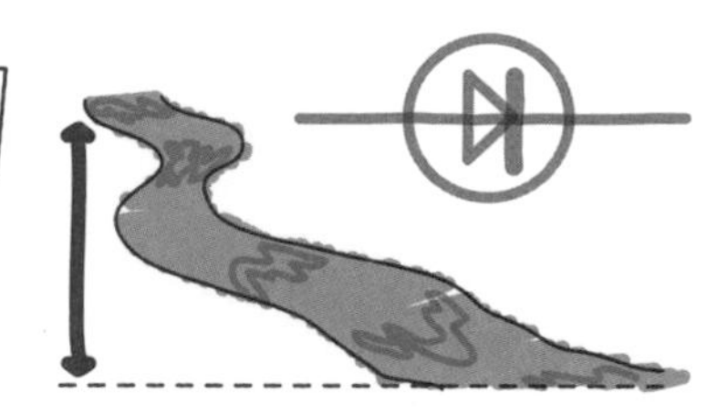

CHARGE TRANSFER

Flow of charged electrons; direction of flow depends on whether charge is positive or negative.

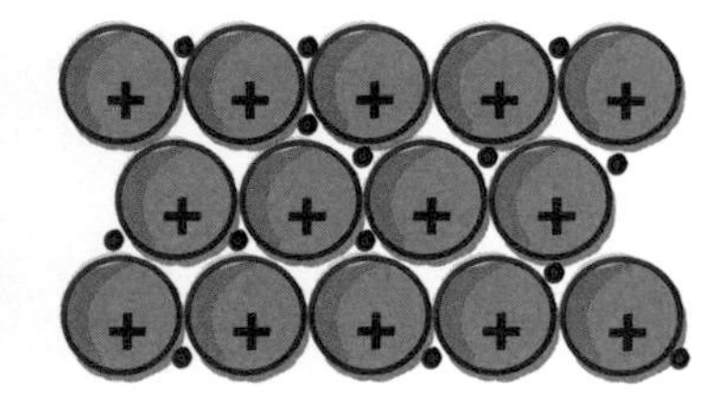

CHARGE

Measured in coulombs (C) and carried by electrons (negative) and ions (positive or negative) that have a value of $\pm 1.6 \times 10^{-19}$.

ELECTROMOTIVE FORCE

Emf; the amount of energy a battery provides to each coulomb of charge passing over it.

ELECTRICAL CIRCUITS

PARALLEL CIRCUITS

Have components placed in separate loops within the circuit.

SERIES CIRCUITS

Have all components connected within a single loop. The total resistance is the sum of each component resistance.

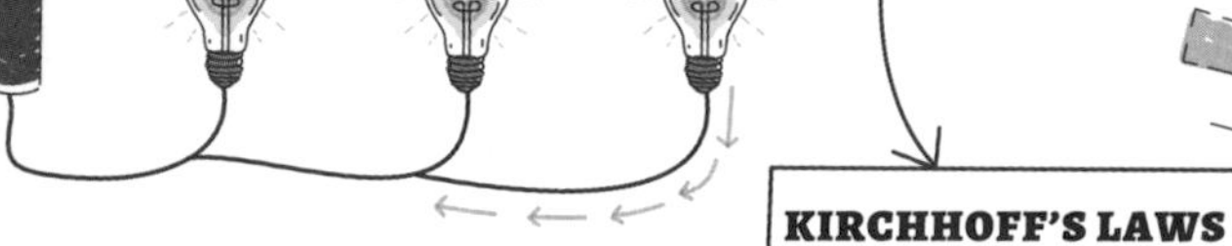

KIRCHHOFF'S LAWS

KIRCHHOFF'S FIRST LAW

Current flowing into a node (or a junction) must be equal to current flowing out of it.

KIRCHHOFF'S SECOND LAW

The sum of all voltages around any closed loop in a circuit must equal zero.

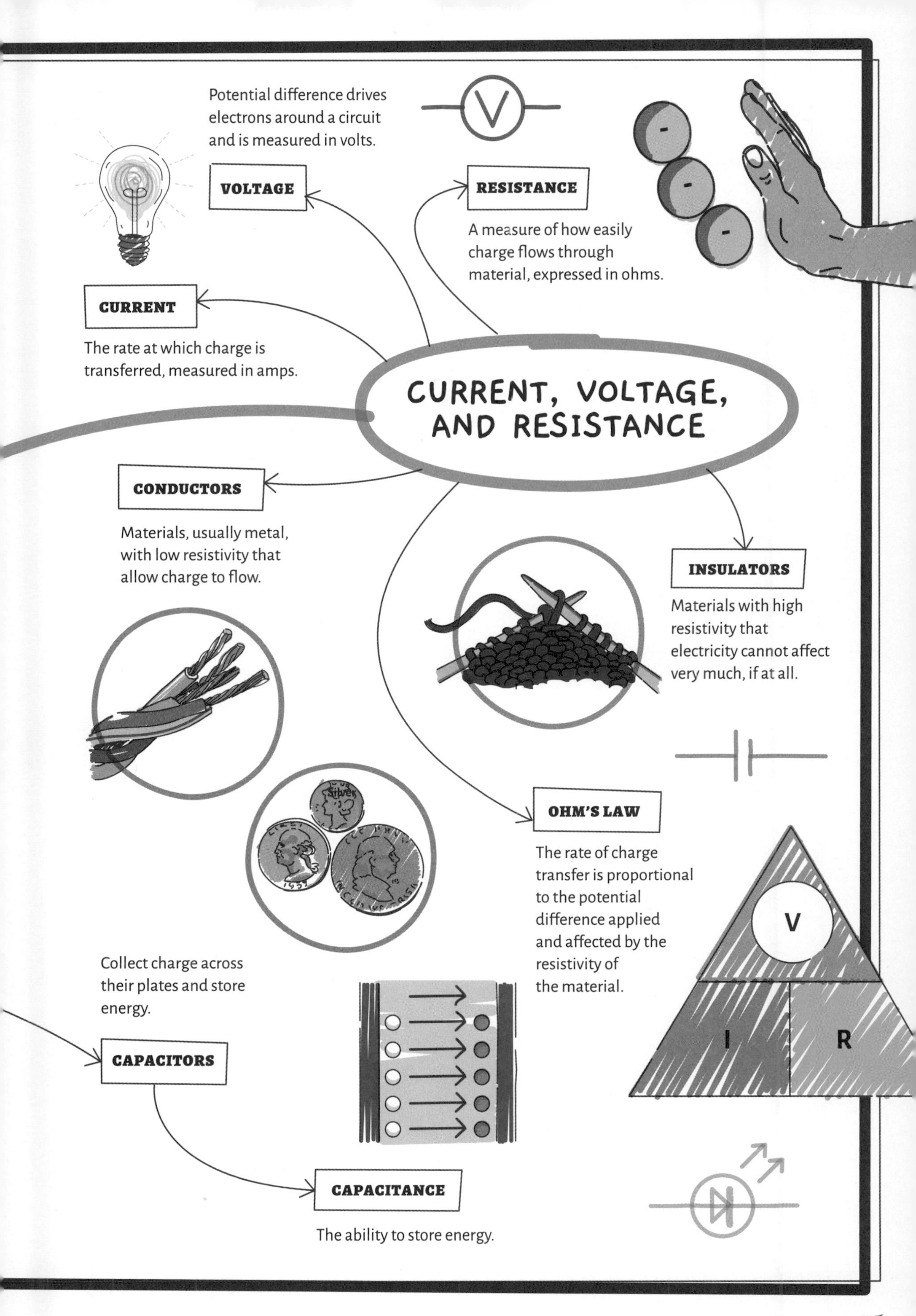
CURRENT, VOLTAGE, AND RESISTANCE
Potential difference drives electrons around a circuit and is measured in volts.
VOLTAGE
V
RESISTANCE
A measure of how easily charge flows through material, expressed in ohms.
CURRENT
The rate at which charge is transferred, measured in amps.
CONDUCTORS
Materials, usually metal, with low resistivity that allow charge to flow.
INSULATORS
Materials with high resistivity that electricity cannot affect very much, if at all.
OHM'S LAW
The rate of charge transfer is proportional to the potential difference applied and affected by the resistivity of the material.
V
I
R
Collect charge across their plates and store energy.
CAPACITORS
CAPACITANCE
The ability to store energy.

CHAPTER 6

FIELDS AND FORCES

In Chapter 1, you investigated types of forces and their effects on a body's motion. You saw that contact forces must be physically touching the body they are acting upon. In this chapter, you will be looking in more detail at noncontact forces: gravitational, electrostatic, and magnetic. All of these forces have two unifying factors: They all require a field in which to act, and their influence is greatly affected by the proximity of that field.

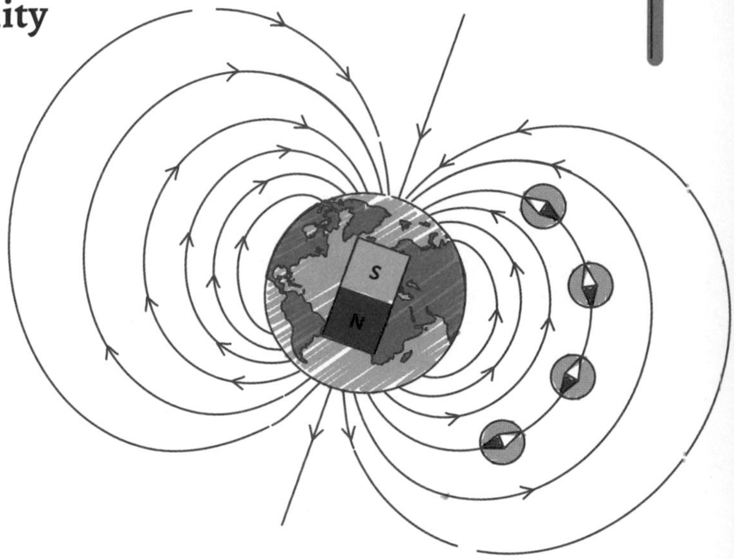

S
N

FIELDS AND THEIR EFFECTS

As discussed in Chapter 1, when a body of mass *m* is subjected to an overall force in a specific direction, it will accelerate in that direction, according to Newton's second law ($F = ma$). When the force applied to the body is a result of its presence within a field, the magnitude of that force will depend on a number of factors: the type of field, how close the body is to the cause of the field, and whether that field has the ability to affect that body.

A **gravitational field** has the capacity to affect all bodies with mass in the universe, but it is a very weak force and is only significant if the body is in close proximity to a large mass, such as a planet or star. It is always an attractive force between masses.

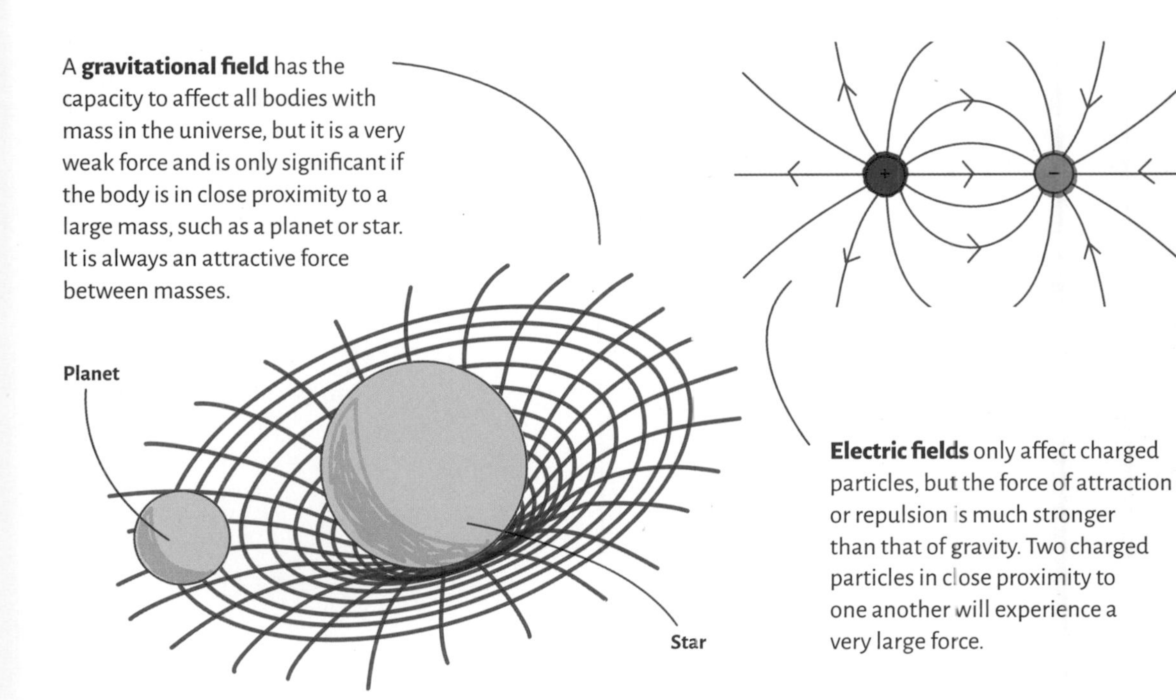

Electric fields only affect charged particles, but the force of attraction or repulsion is much stronger than that of gravity. Two charged particles in close proximity to one another will experience a very large force.

Magnetic fields affect certain materials, primarily some metals. The magnitude of the effect is determined by the nature of the magnetic material creating the field, the proximity of the body it is influencing, and the material of the body in the field.

Each type of field has its own unique properties, which we will explore in the following pages.

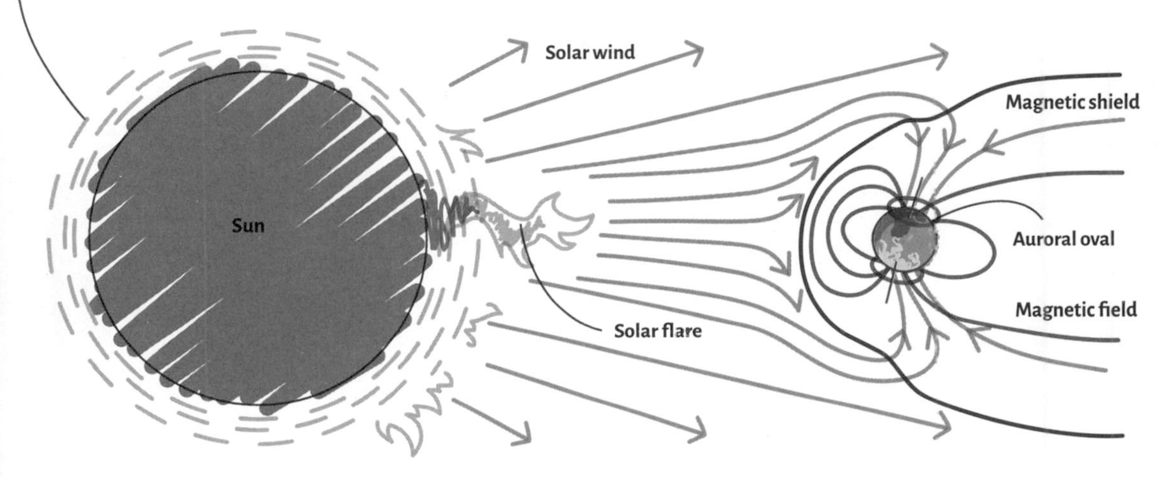

GRAVITATIONAL FIELDS

Gravitational fields occur around all objects with mass, and they exert a force on all other objects with mass. The size of the force between two masses is equal and opposite and depends directly upon the size of each mass and the distance between them.

Newton's law of gravitation

Newton defined his law of gravitation between two point masses. A **point mass** is the idea that all bodies can be considered to have their mass concentrated at a singular point in space. A point mass generates a **radial gravitational field**.

Newton stated that every particle attracts every other particle in the universe with a force that is directly proportional to the product of their masses and inversely proportional to the square of the distance between their centers.

In equation form, this statement can be written concisely as:

$$F = \frac{Gm_1m_2}{r^2}$$

where F is the force experienced, m_1 and m_2 are the masses, r is the separation between their centers, and G is the universal gravitational constant ($G = 6.67 \times 10^{-11}$).

A gravitational field is generated by a large mass, such as Earth, M, and affects a smaller mass, m, such as a person. The person is referred to as being influenced by Earth's gravitational field.

The formula then becomes:

$$F = \frac{GMm}{r^2}$$

This law can be greatly simplified. Due to the separation distances of all the particles in the universe and the strength of gravitation, only masses close enough to each other experience a force that is appreciable. Therefore, only the force between two masses is considered in order to simplify the mathematics.

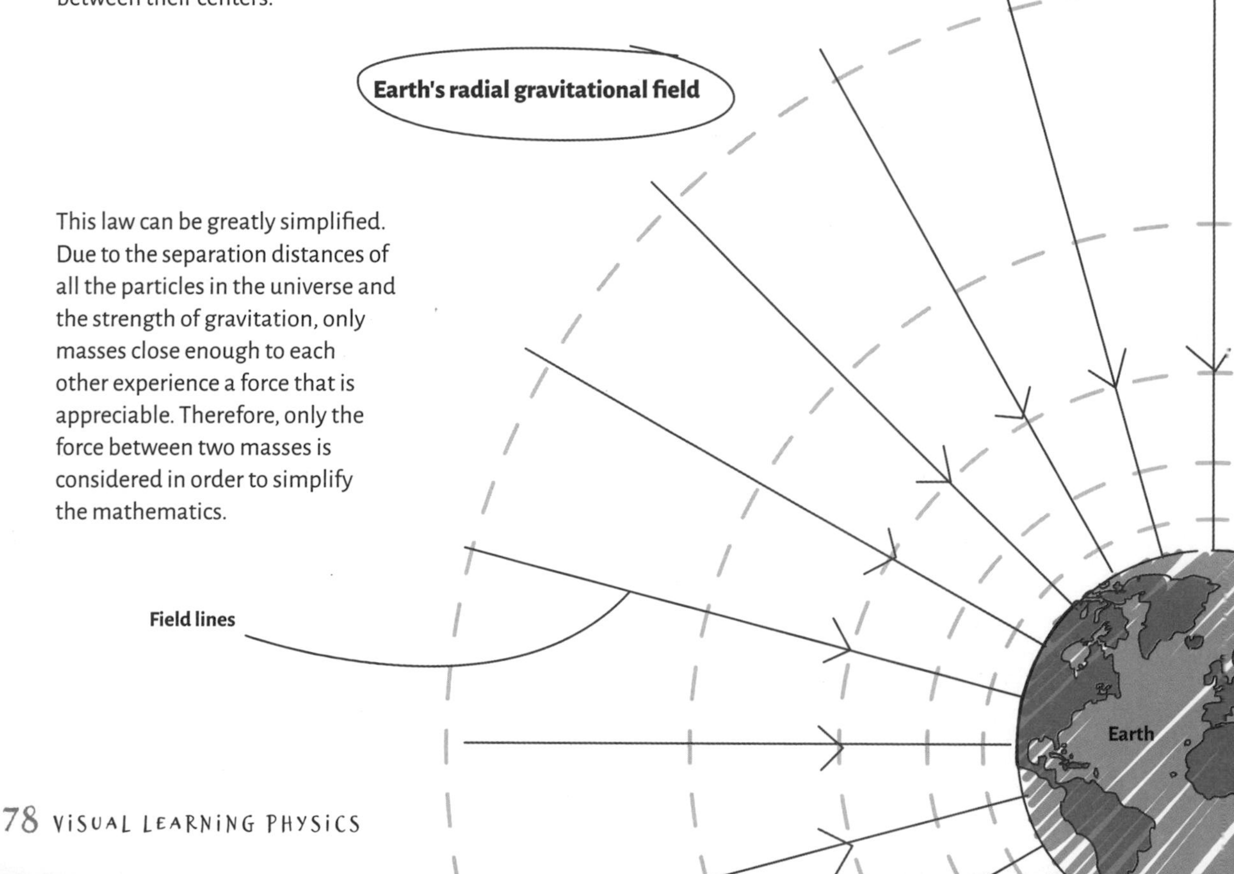

Gravitational field strength

The gravitational field strength, g, is the force experienced on Earth for each kg of mass. It is derived from the force between two masses, one of which is the mass of Earth, M:

$$F = \frac{GMm}{r^2}$$

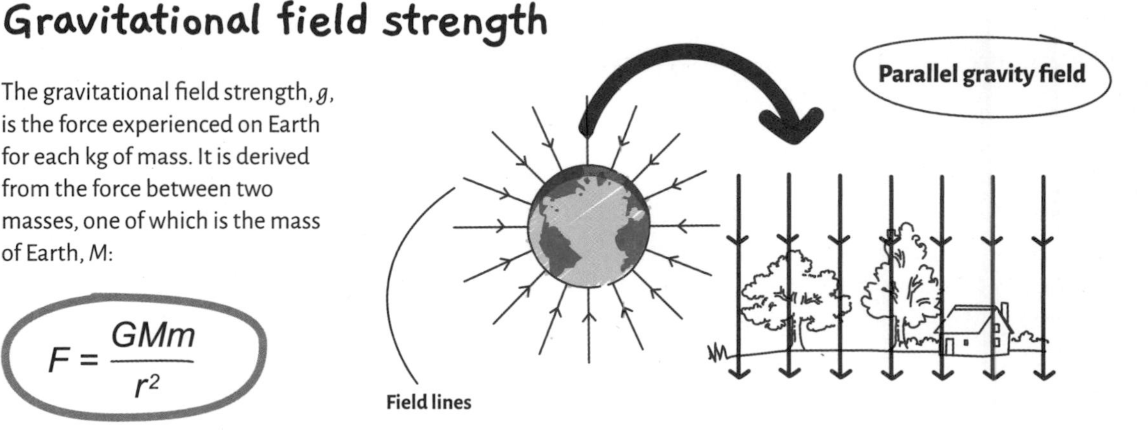

where r is the Earth's radius. Calculating this value gives $g = 9.8$ N/kg.

Multiplying g by a mass on Earth will give its weight: $W = mg$. The value of g is equivalent to the acceleration due to gravity, where $g = 9.8$ m/s^2. The units of N/kg and m/s^2 are equivalent, so gravitational field strength is the same as the acceleration due to gravity.

This value is approximately constant close to Earth's surface, as the field lines very close to Earth are very close to being parallel.

Gravitational field lines represent the direction in which the gravitational field is acting (toward the center of the mass) and the strength of field is shown by the proximity of the lines to one another.

Using this approximation, it is assumed that when moving a mass a relatively small distance upward from Earth's surface, the weight of the body is constant. From this, you know that the **potential energy** (PE) gained by the body is equal to its weight multiplied by the vertical distance moved away: PE = mgh.

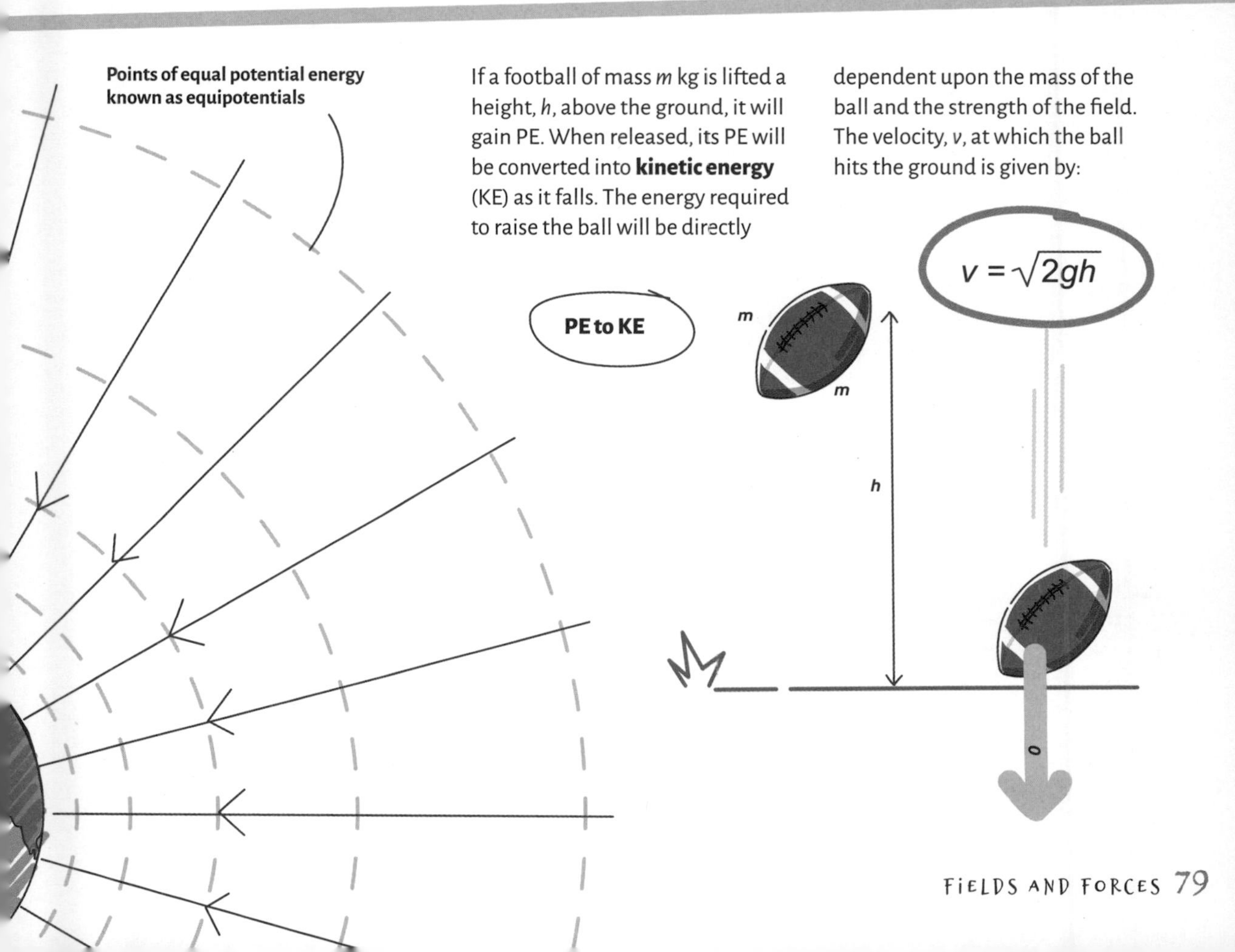

If a football of mass m kg is lifted a height, h, above the ground, it will gain PE. When released, its PE will be converted into **kinetic energy** (KE) as it falls. The energy required to raise the ball will be directly dependent upon the mass of the ball and the strength of the field. The velocity, v, at which the ball hits the ground is given by:

$$v = \sqrt{2gh}$$

MAGNETIC AND ELECTRIC FIELDS

Magnetic and electric fields are invisible and surround bodies that can create them. Moving charges, such as electric current, can produce magnetic fields, and in turn, magnetic fields can exert a force on a moving charged particle. Electricity and magnetism coexist because the presence of one affects the other.

Magnetic fields

A magnetic field influences magnetic materials, causing them to experience an attractive or repulsive force. Magnetic materials vary in how much they are influenced by magnetic fields. Metals such as iron, nickel, and cobalt are strongly affected, whereas most other metals are either weakly affected or to such a small extent that it is not appreciable.

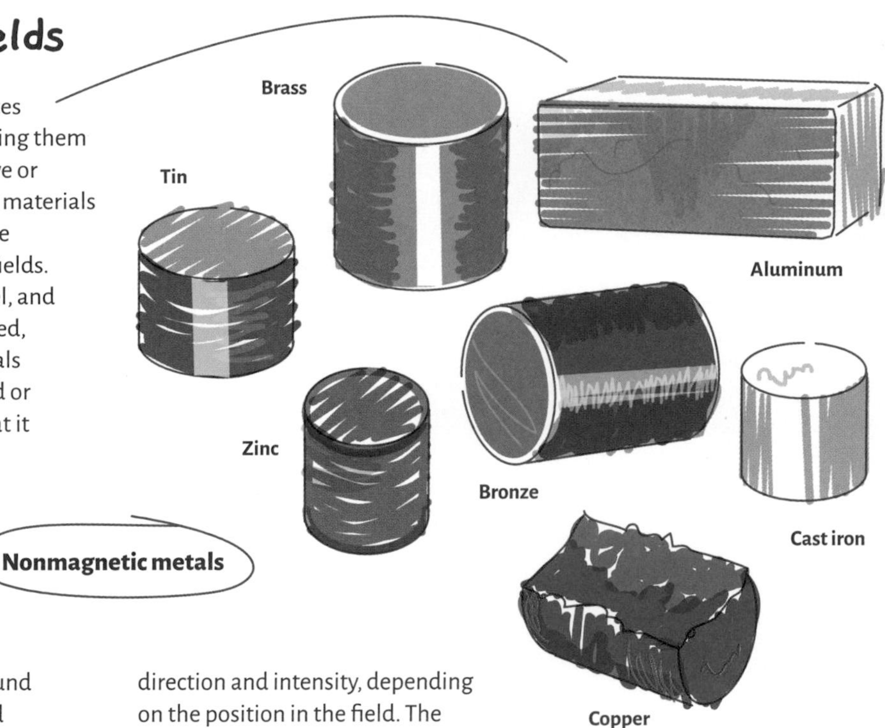

Nonmagnetic metals

Magnetic fields exist around permanently magnetized materials, such as iron, that have been exposed to a strong magnetic field. Materials that can become permanently magnetized are called **ferromagnetic**.

A bar magnet is surrounded by magnetic field lines that create a loop, running from the magnetic north to the magnetic south poles. The strength of the field surrounding a magnet is measured in **tesla** (T) and will vary in direction and intensity, depending on the position in the field. The field strength is indicated, as with gravitational fields, by the spacing of the field lines.

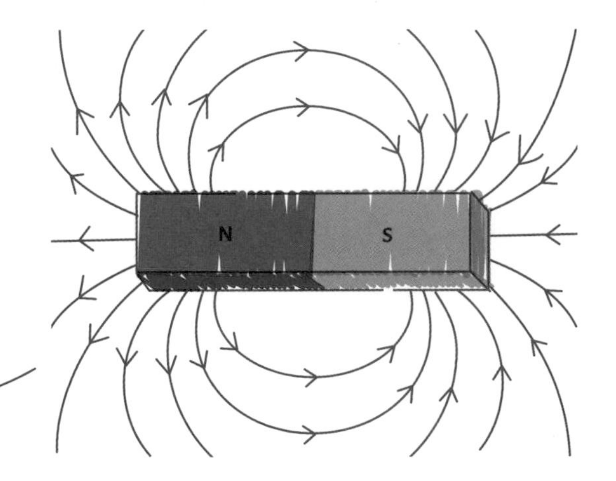

Earth is surrounded by a permanent magnetic field that looks like a giant bar magnet. The field interacts with charged solar particles, causing them to spiral down along the field lines to the magnetic poles (which happen to be close to the geographic poles). These particles strike atoms in the atmosphere, causing electrons to increase in charge states. When they drop back down, they emit **photons** of light of different colors, creating the spectacular northern and southern lights (the auroras).

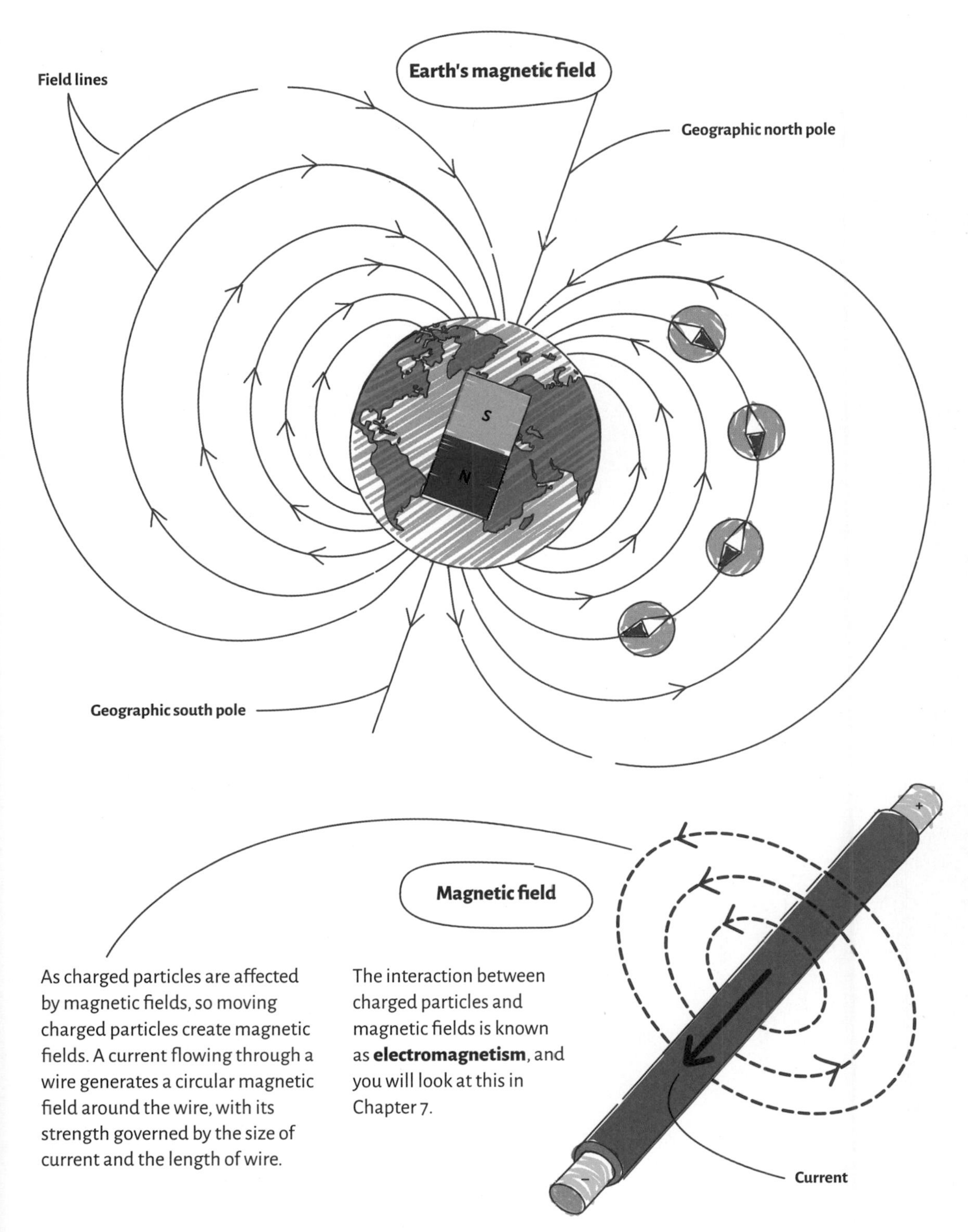

As charged particles are affected by magnetic fields, so moving charged particles create magnetic fields. A current flowing through a wire generates a circular magnetic field around the wire, with its strength governed by the size of current and the length of wire.

The interaction between charged particles and magnetic fields is known as **electromagnetism**, and you will look at this in Chapter 7.

Electric fields

Electric fields share many common attributes with gravitational fields but they only affect charged particles.

Coulomb's law states that the magnitude of the electric force between two point charges is directly proportional to the product of the charges, and inversely proportional to the square of the distance between them. A point charge creates a **radial electric field**.

The force between charged particles can be attractive (between unlike charges) or repulsive (between like charges). As you have seen in Chapter 1, it is referred to as an **electrostatic force**.

The strength of the force between two charged particles has exactly the same equation form as the force between two masses and is given by the formula:

$$F = \frac{kQ_1Q_2}{r^2}$$

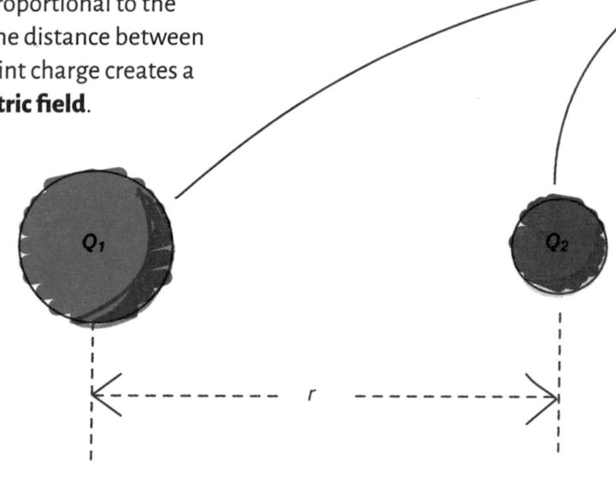

where Q_1 and Q_2 are the magnitude of the charged particles given in coulombs and k is a constant ($k = 9 \times 10^9$). Compared to the universal gravitational constant, G, this is very large indeed.

It is an electrostatic force that is responsible for binding electrons in orbit around atoms.

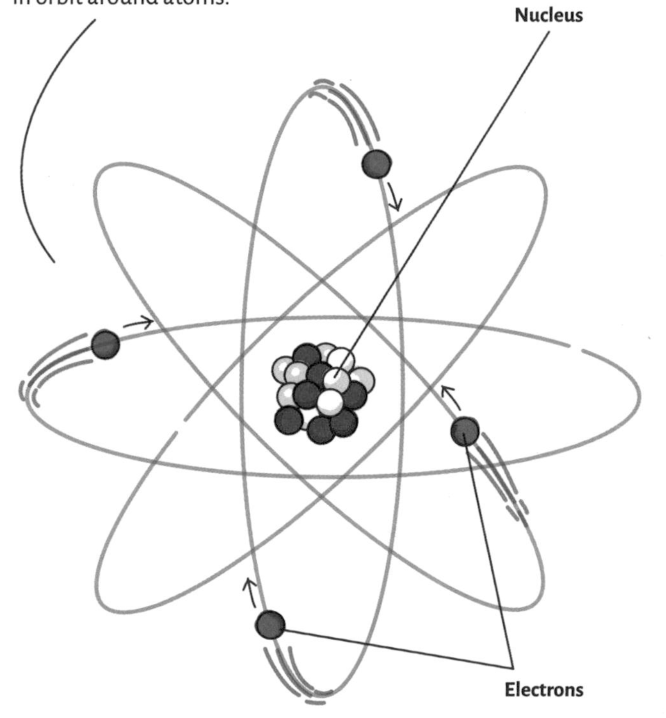

An electric field can also exist between two charged plates, as in a capacitor. In this case, the field lines are parallel, and the force on a charged particle within the field is constant, regardless of its position.

The strength of the electric field, E, between plates is given by:

$$E = \frac{V}{d}$$

where V is the potential difference and d is the plate separation.

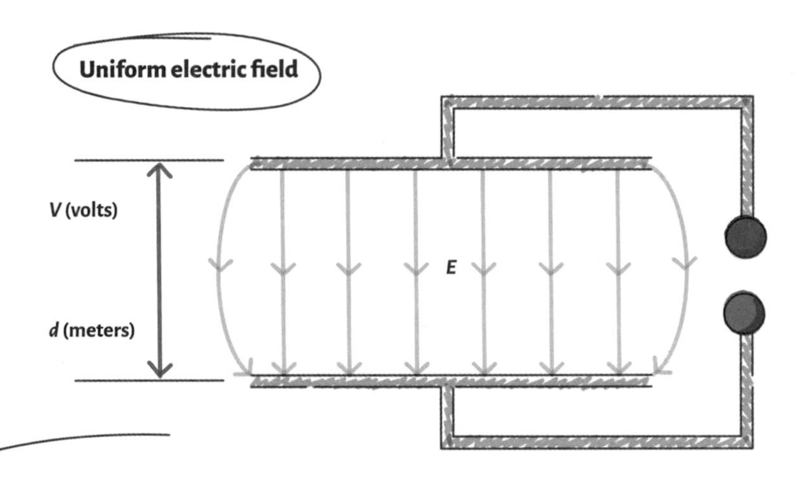

The force, F, experienced by a charge, Q, in a uniform field is:

$$F = \frac{QV}{d} = EQ$$

Cathode ray tube

A **uniform electric field** is used as the basis for a cathode ray tube (CRT). A CRT accelerates electrons within an evacuated chamber. A current heats up a metal wire, and electrons are essentially "boiled" off the surface of the wire. This is called an **electron gun**.

As the electrons are accelerated by a high potential difference, the electron stream can be diverted either by charged side plates that have a potential difference across them or by magnetic fields. This directs the beam onto specific regions of a screen coated with a layer of phosphors, which glow when struck by electrons and emit red, green, or blue light. Combinations of these primary colors displayed at differing brightnesses allow a huge range of colors. As the beam traverses across and down the screen, it strikes each pixel, and the fluorescence momentarily glows so that the entire screen is always fully illuminated.

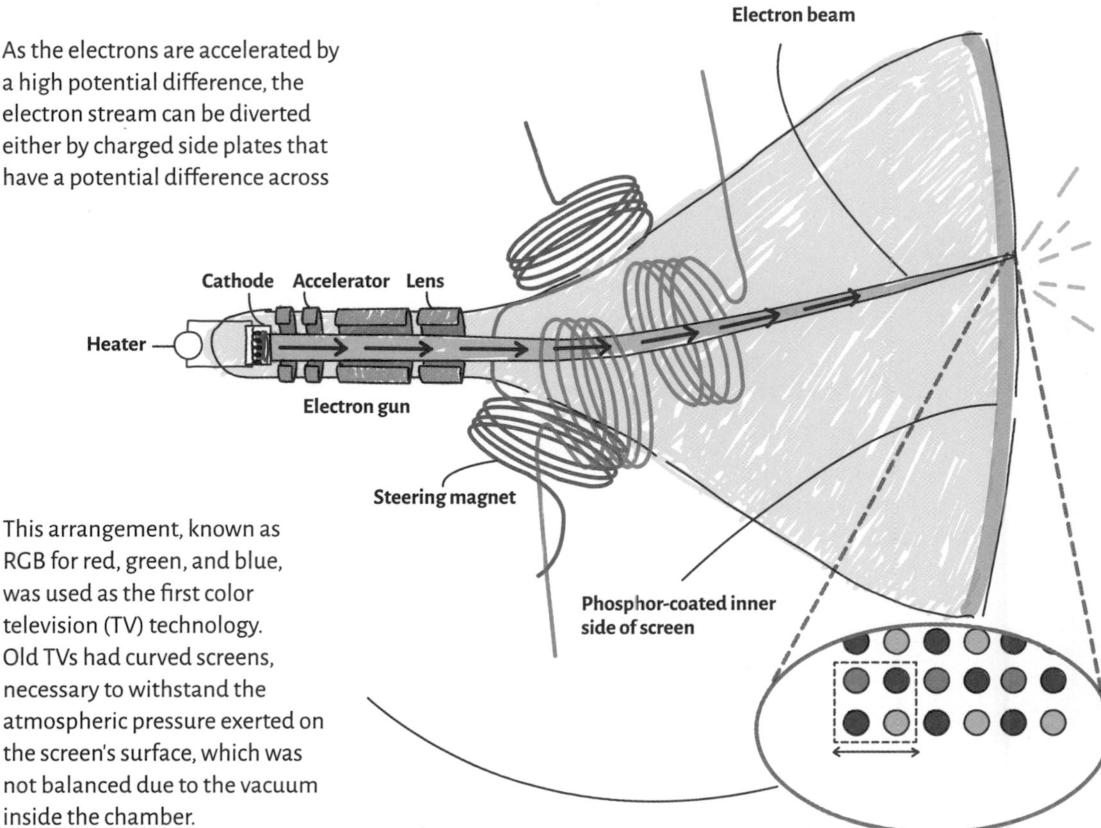

This arrangement, known as RGB for red, green, and blue, was used as the first color television (TV) technology. Old TVs had curved screens, necessary to withstand the atmospheric pressure exerted on the screen's surface, which was not balanced due to the vacuum inside the chamber.

RECAP

FIELDS AND FORCES

FIELDS AND EFFECTS

NONCONTACT FORCES

A force can be generated by a field at a distance from the object creating it.

MAGNETIC FIELD

FERROMAGNETIC METALS

Permanently affected by a magnetic field.

NONMAGNETIC METALS

Not affected by magnetic field.

MAGNETIC FIELD STRENGTH

Measured in tesla, T; strength and intensity indicated by closeness of field lines.

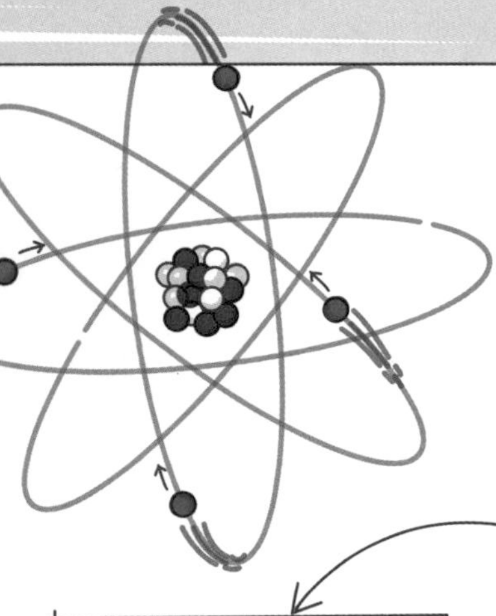

ELECTRIC FIELD

COULOMB'S LAW

$$F = \frac{kq_1q_2}{r^2}$$

ELECTROSTATIC FORCE

Like charges repel, and opposite charges attract.

STRENGTH OF FORCE

Is proportional to the product of the charges and inversely proportional to the square of the distance between them.

PARALLEL FIELD

A uniform electric field generated between two charged plates; the force on a charged particle in a parallel field is constant.

CATHODE RAY TUBE

A uniform field is the basis of the cathode ray tube.

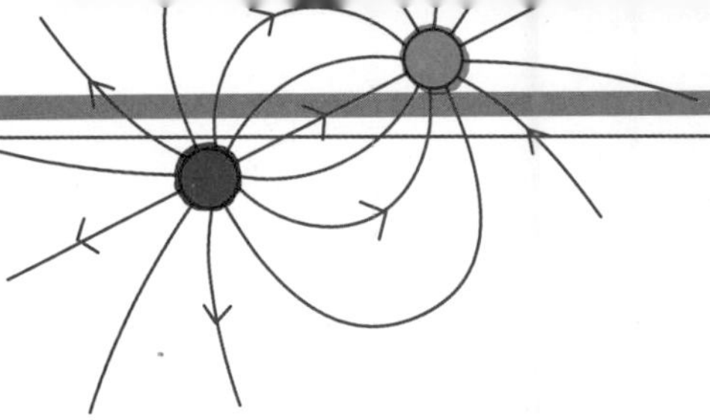

TYPES OF FIELDS

GRAVITATIONAL

Created by mass; the magnitude of the force experienced by a mass is influenced by its distance from the field.

MAGNETIC

Created by moving charged particles or magnetized material such as iron.

ELECTRIC

Created by electric charges; affects only charged particles.

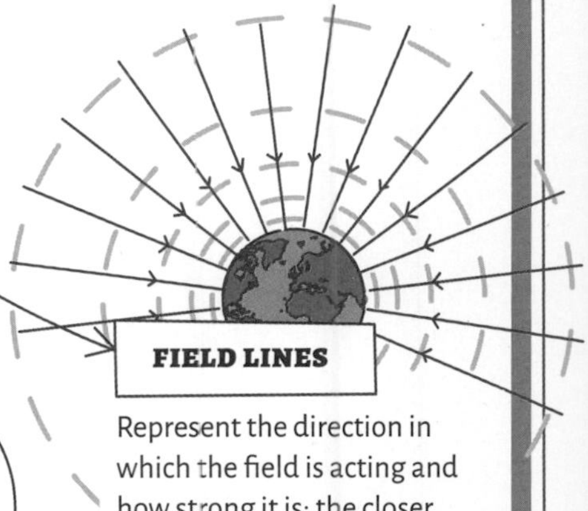

FIELD LINES

Represent the direction in which the field is acting and how strong it is; the closer the lines to each other, the stronger the field.

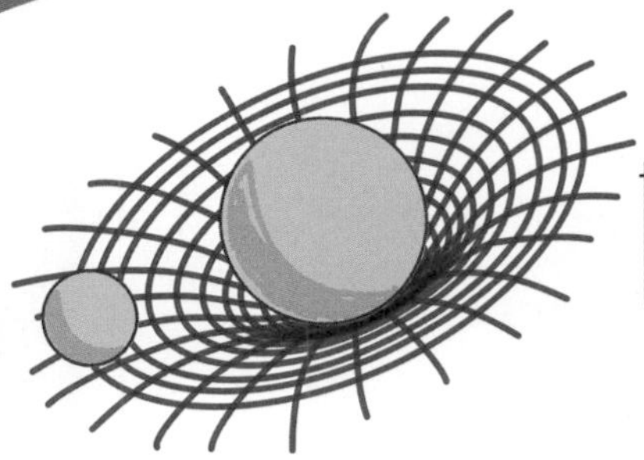

GRAVITATIONAL FIELDS

GRAVITATIONAL FIELD STRENGTH

Increases as mass increases; reduces the farther away from the mass it extends.

EARTH'S GRAVITY

The gravitational field strength, g, is the force experienced on Earth for each kg of mass.

$$g = 9.8 \text{ N/kg}$$

NEWTON'S LAW

$$F = \frac{GMm}{r^2}$$

RADIAL GRAVITATIONAL FIELD

Reduces in strength as you move away from its center.

PARALLEL GRAVITY FIELD

Close to Earth, the field lines are almost parallel.

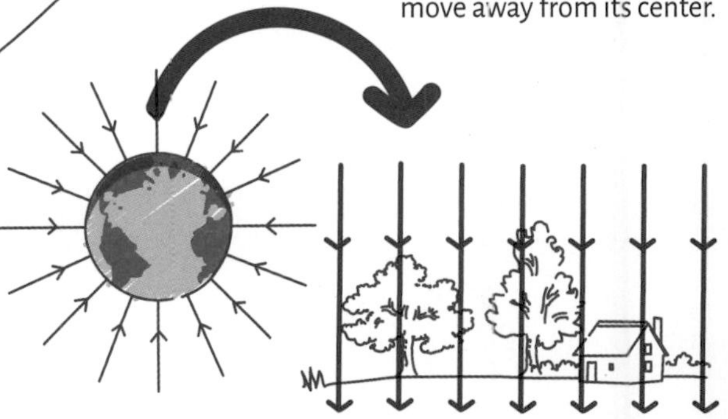

CHAPTER 7

ELECTROMAGNETISM

The interaction between electric and magnetic fields is a branch of physics known as electromagnetism. The two fields are closely linked, as both fields often coexist and interact. Magnetic fields exhibit a force on charged particles, and the movement of charged particles causes the occurrence of a magnetic field. Light is a product of a fluctuating electric and magnetic field coexisting and depending on each other. Without their mutual dependence, electricity would not exist in our homes.

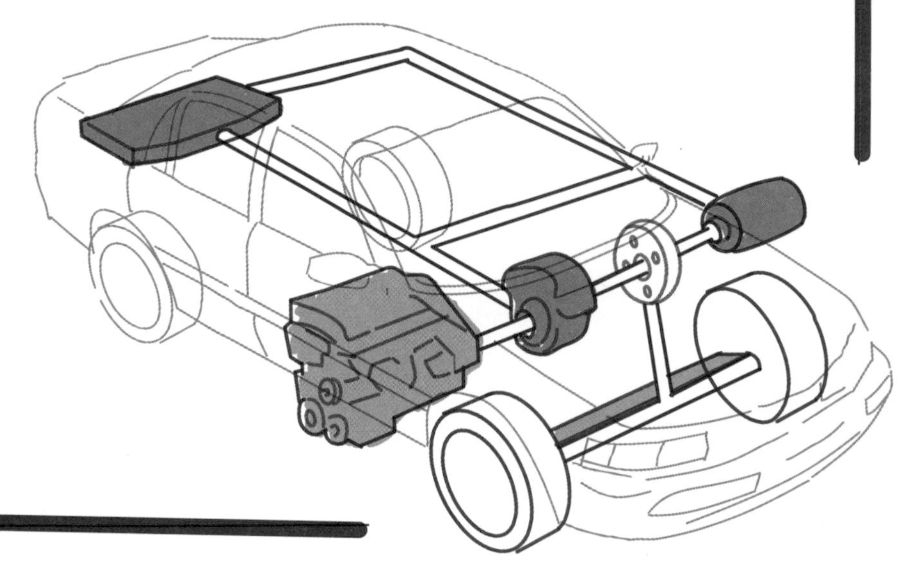

FARADAY'S LAW OF INDUCTION

The English physicist Michael Faraday (1791–1867) made a remarkable discovery in 1831: Generating a changing magnetic field in the presence of a conducting metal wire causes a current to flow through the wire. Conversely, a changing current within the wire produces a fluctuating magnetic field around the wire.

A charged particle moving relative to a magnetic field causes a force upon the particle, thus moving it. This is the very definition of electric current: a stream of moving electrons. The interaction between the three properties—magnetic fields, charged particles, and forces—is the basis of **electromagnetic induction**.

FARADAY'S LAW OF INDUCTION
An electrical conductor moving relative to a magnetic field will induce an electromotive force.

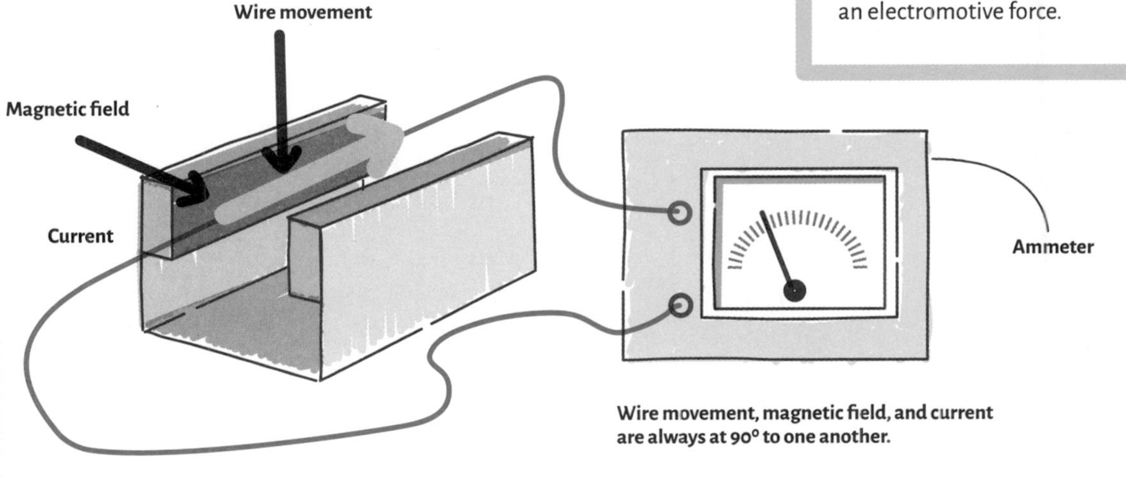

Wire movement, magnetic field, and current are always at 90° to one another.

Left-hand rule

Motion

Field

Current

Let's say you move a length of wire through a magnetic field, with field lines running at 90° to the direction of motion. The force experienced by the electrons within the wire acts at 90° to both the movement of the wire and the direction of the field. This causes a flow of electrons along the length of the wire, providing there is constant movement of the wire relative to the magnetic field.

If the direction of movement is reversed, the direction of the current is also reversed.

The direction of current flow can be determined using two fingers and your thumb in Fleming's left-hand rule, devised by English physicist Sir John Ambrose Fleming (1849–1945). By pointing fingers in the directions corresponding to known information, you can determine the direction of the unknown quantity.

ELECTROMAGNETIC INDUCTION

The phenomenon of electromagnetic induction has had a profound impact on our lives. It has two key uses: generating electricity via power stations and driving electric motors—the future of automobile technology.

Electric generators

If a conducting wire is moved in the presence of a magnetic field, a current will flow through the wire. The flow's direction relies on the direction of movement between the wire and the magnetic field. As the **angle of movement** between the wire and the field decreases from 90°, so does the strength of the current generated.

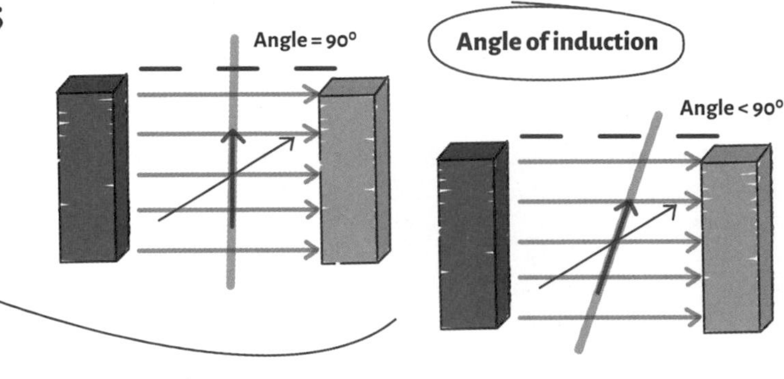

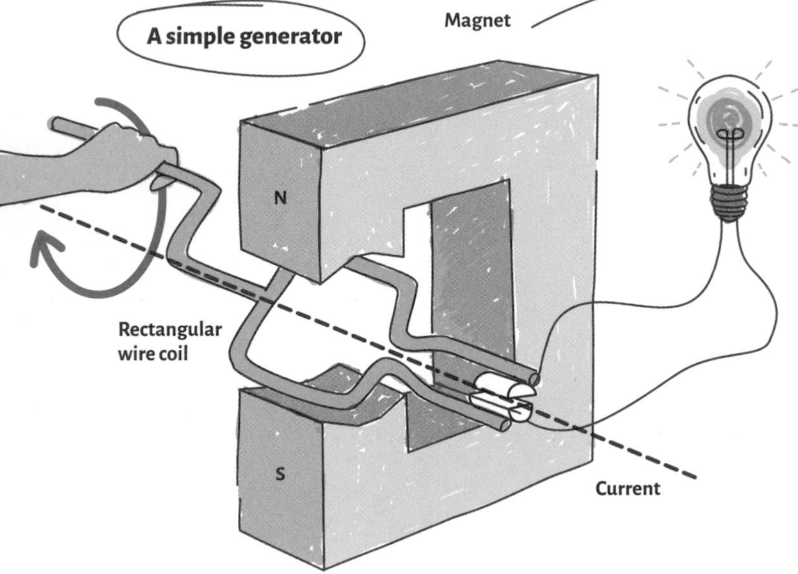

A rectangular coil of wire is connected to a bulb and placed within the strong magnetic field between two bar magnets. If the coil is rotated around an axis at 90° to the field lines, there is a relative motion between the wire and the field, so a current is induced. In reality, there are many wire loops coiled inside a **generator**, which increases the current output by a multiple of the number of loops.

As the current output changes direction and fluctuates in strength, it creates an **alternating current** (AC). If mechanical energy is used to maintain the rotational motion of the wire, then a continuous AC will be induced. This is the basic mechanism for generating electricity.

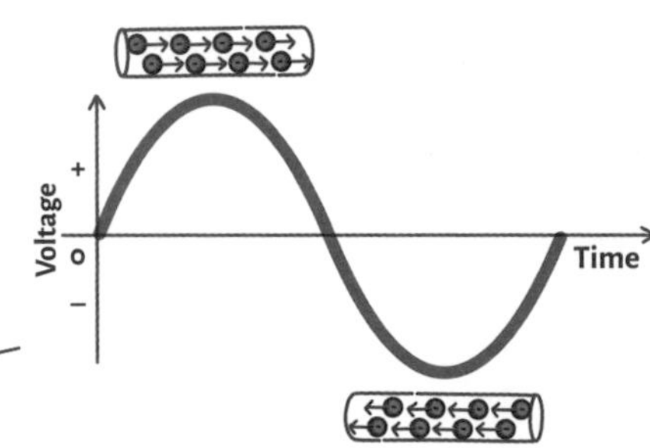

Electric motors

Electric **motors** are used to drive all sorts of electrical devices, from small radio-controlled cars, food processors, and hairdryers to full-size electric cars.

The principle behind the way they function is identical to that of an electric generator. The primary difference is that a generator converts mechanical energy into electrical energy, whereas an electric motor takes electrical energy and converts it into mechanical energy.

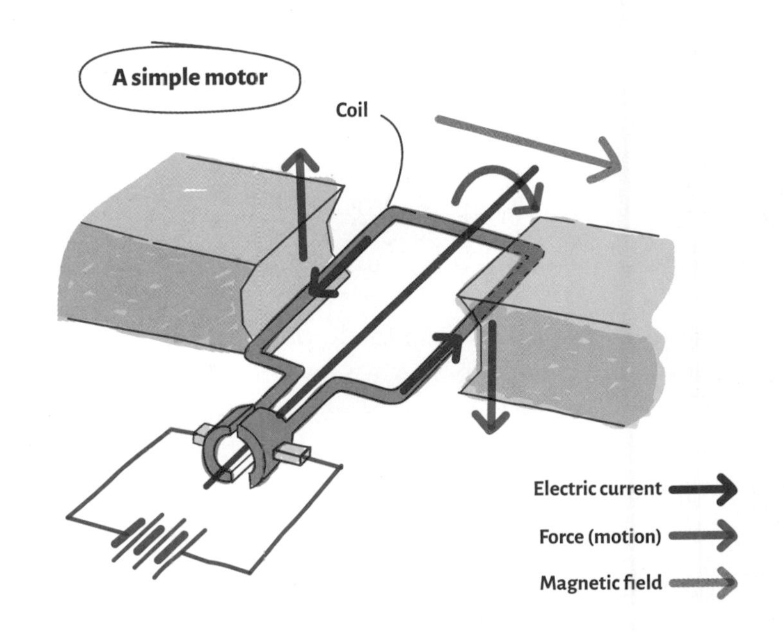

A current from a power source flows through the wires, which are within a magnetic field and free to rotate.

In this case, the electrons passing through the magnetic field experience a force at 90° to their motion. This causes a **turning moment** around the axis of rotation, so the wires rotate. This **rotation torque** is transferred via a driveshaft through the motor.

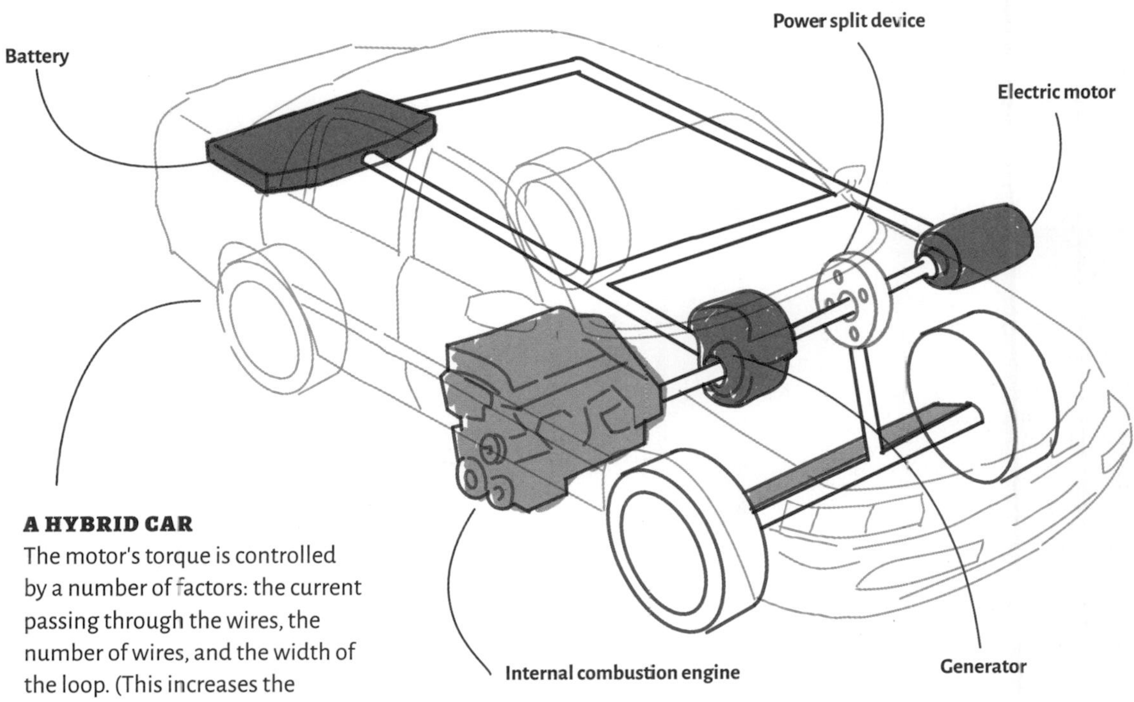

A HYBRID CAR

The motor's torque is controlled by a number of factors: the current passing through the wires, the number of wires, and the width of the loop. (This increases the distance of the force applied from the rotation axis, thus the turning moment is increased.)

If a large force is required, for example, to drive a full-size car, all of these factors need to be increased. In simple terms, this means the motor has to be big in order to provide the power needed.

In practice, a motor can be used as a generator and vice versa. A hybrid car does exactly this: It charges its battery while driving using standard fuel and then releases this stored battery energy when driving with the electric motor. These specialist devices are called **motor generators**.

ENERGY LOSS AND ENERGY TRANSFER

Electromagnetic induction has another important use in providing power to our homes. When electricity is generated in power stations by harnessing mechanical energy, it cannot be efficiently transferred through cables to our homes. It has to be "stepped up" to a much higher voltage, which minimizes power loss—this is achieved via the use of transformers.

Transformers

If a wire is wrapped many times (N_p turns) around a solid iron core and an alternating current is passed through it via an **emf** from a power source, a fluctuating magnetic field is set up around the iron core and focused through it. This is called the **primary coil**.

If a **secondary coil** is then wrapped around the opposite side of the iron core with a different number of turns (N_s), it will be exposed to the alternating magnetic field.

As the field grows and contracts in response to the alternating current in the primary coil, it passes through the secondary coil, inducing a current within it.

An emf is required to drive the current through the secondary coil. If the secondary coil has more turns, there are a larger number of electrons to motivate, thus requiring a larger emf. This is called a **step-up transformer**, and it increases the voltage in the secondary coil. A **step-down transformer** has fewer turns in the secondary coil, which reduces the voltage:

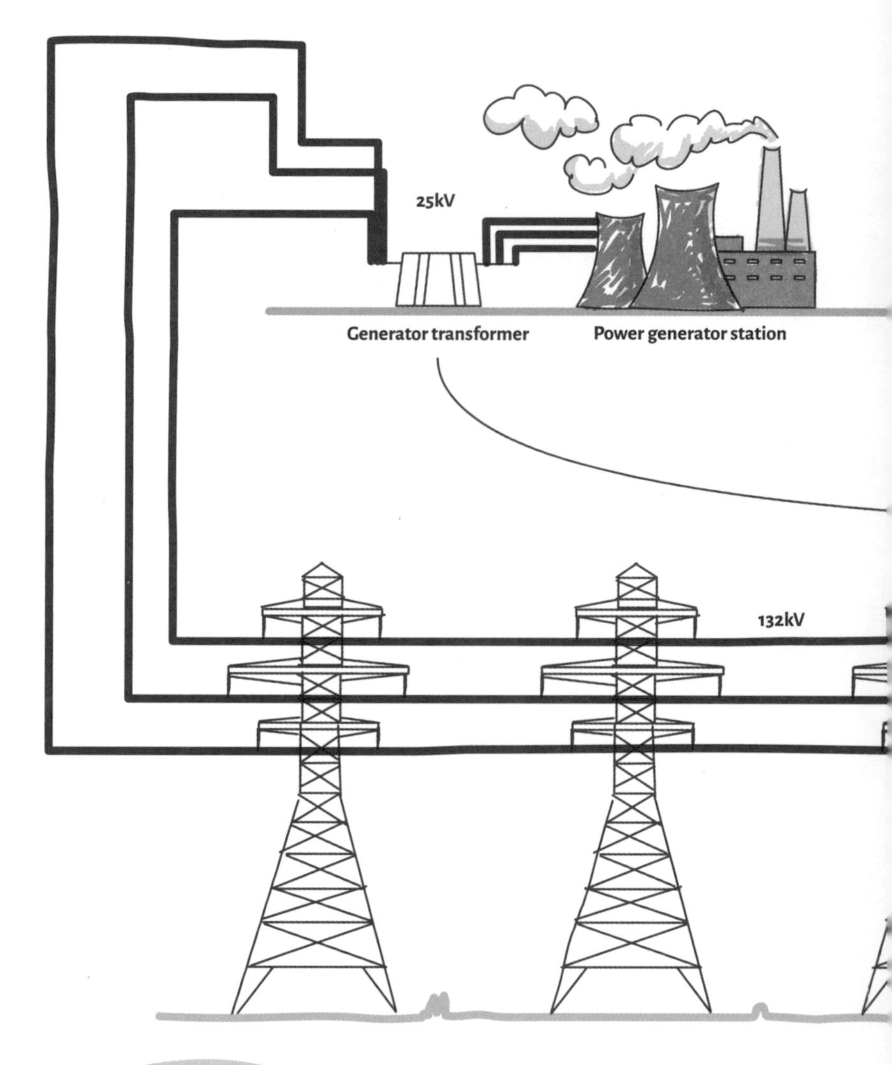

$$V_{out} = \frac{N_s}{N_p} \times V_{in}$$

The network of power stations and overhead **power cables** around the country is called the **energy grid**. Transformers are used to reduce energy losses during the movement of electricity around the energy grid by changing the current and voltage.

PVC outer sheath
Armor
Inner sheath
Insulation
Conductor

Power cables are constructed of many thick conducting wires encased in a sheath of heavy-duty insulation.

They are very good conductors, but even so, their huge length gives them a nonnegligible resistance. This resistance will cause heating and power loss if transferring high currents (more electron flow causes greater heating). Power loss (P) due to heating is highly dependent on current for a fixed resistance:

from $P = IV$ and $V = IR$, you get $P = I^2R$

It is much more efficient to transfer electricity at high voltage and low current along power cables at vast distances. Multiple transformers are used to step the voltage back down in stages as it is supplied to heavy industries, light industries, and finally, small businesses and homes. Changing the voltage within transformers is highly efficient, but there are still some losses from heat and sound within the iron core.

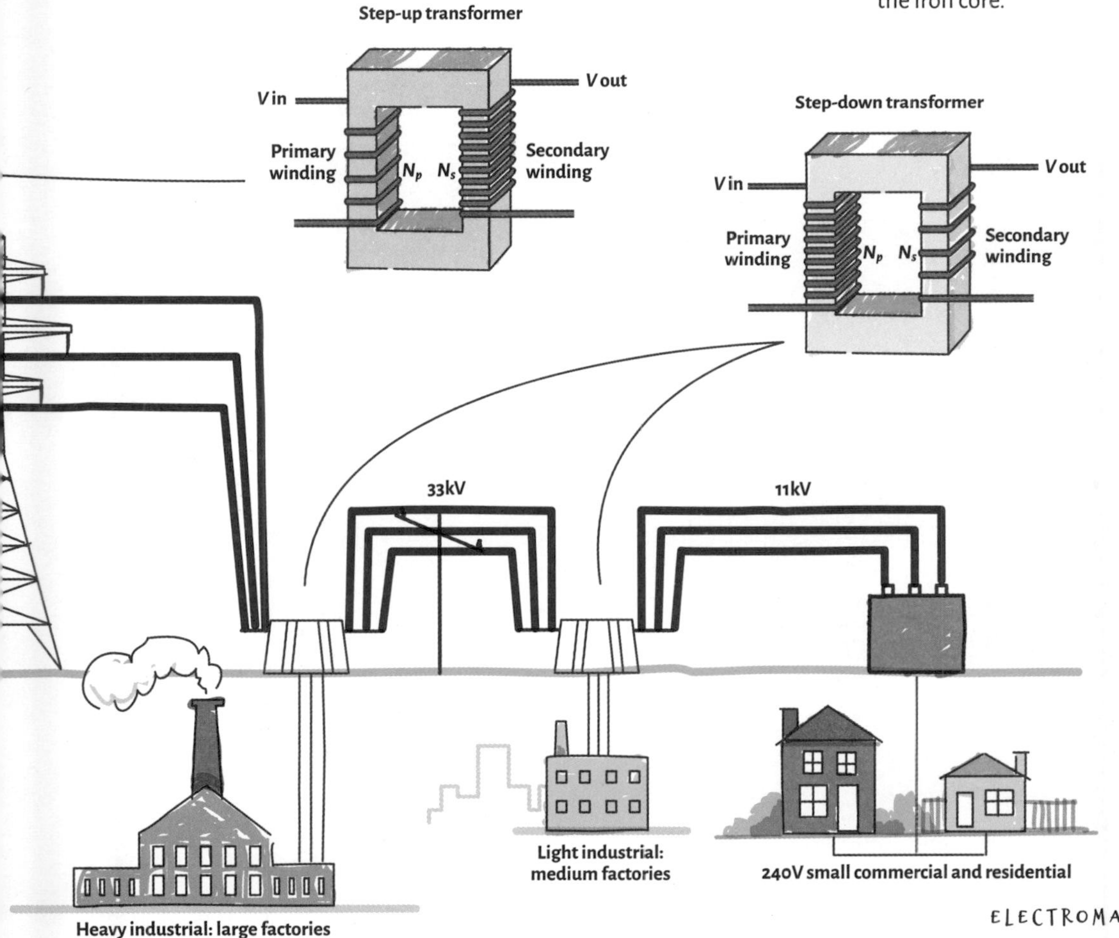

ELECTROMAGNETIC RADIATION AND SPECTRUM

Light waves consist of an oscillating magnetic and electric field that cross each other at a 90° angle. It is known as the electromagnetic spectrum and consists of all frequencies of light, from low-energy radio waves up to extremely high-energy gamma rays.

Electromagnetic radiation

Electromagnetic radiation transfers energy through space or a transparent medium, such as air or glass. The speed of this radiation varies according to the properties of the medium. Light travels at approximately 3×10^8 m/s in a vacuum and about 2×10^8 m/s in glass.

Light was first thought to be made up of tiny particles that carried kinetic energy. This idea was first explored by Newton, but certain properties of light were difficult to explain using this particle theory.

In 1678, the Dutch physicist Christiaan Huygens (1629–1695) theorized that light was made up of waves vibrating at 90° to the direction of motion—this became known as **Huygens' principle**.

In fact, there were problems with both theories, with certain observations being difficult to explain entirely using either idea. Later, Einstein proposed that light was made up of packets of energy known as **photons**, where each photon has a very specific amount (or quantum) of energy. A photon's energy changes with **frequency** (number of oscillations per second).

Photon energy by wavelength

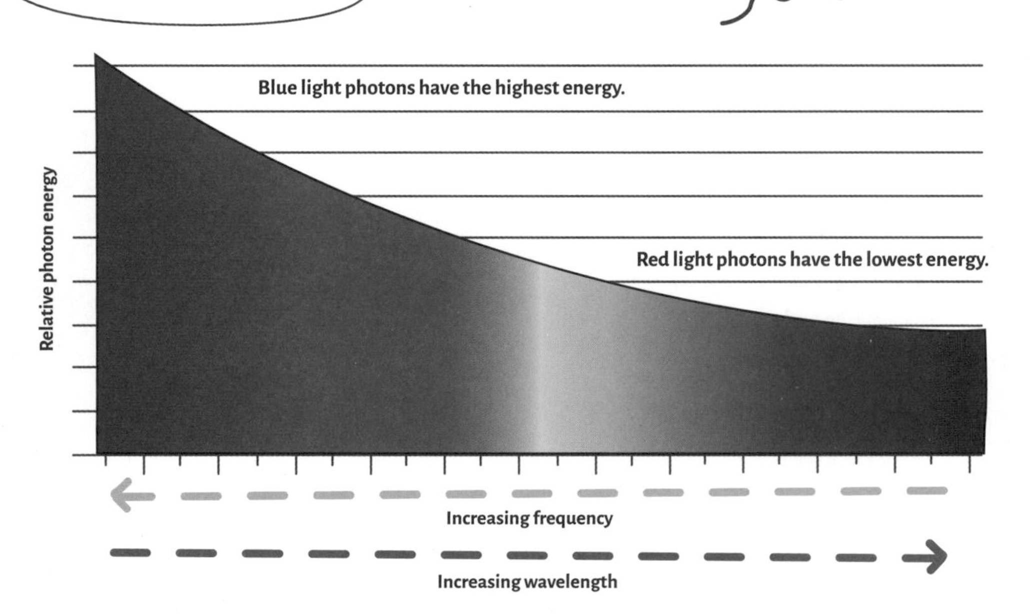

Other light sources

Electromagnetic radiation is caused by many mechanisms. Most light that you see every day is generated by **nuclear fusion** taking place in the sun and radiates from its surface to Earth, where it is reflected by the surfaces you observe. But light can also be emitted by chemical and nuclear processes on Earth as well as warm or hot objects.

A metal, for example, will glow when heated. As its temperature increases, so does the energy of the photons it emits, and its color changes.

Bioluminescent animals, such as glowworms or anglerfish, generate their own light. It's produced by chemical reaction, either by the animals alone or by bioluminescent bacteria.

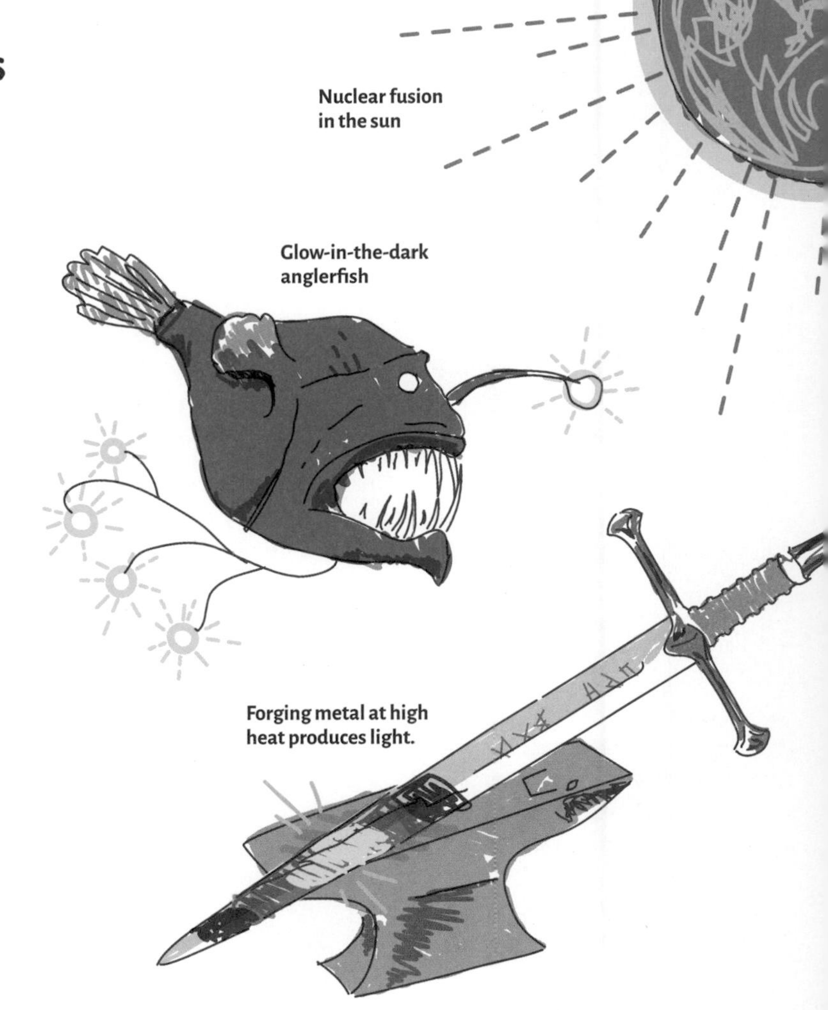

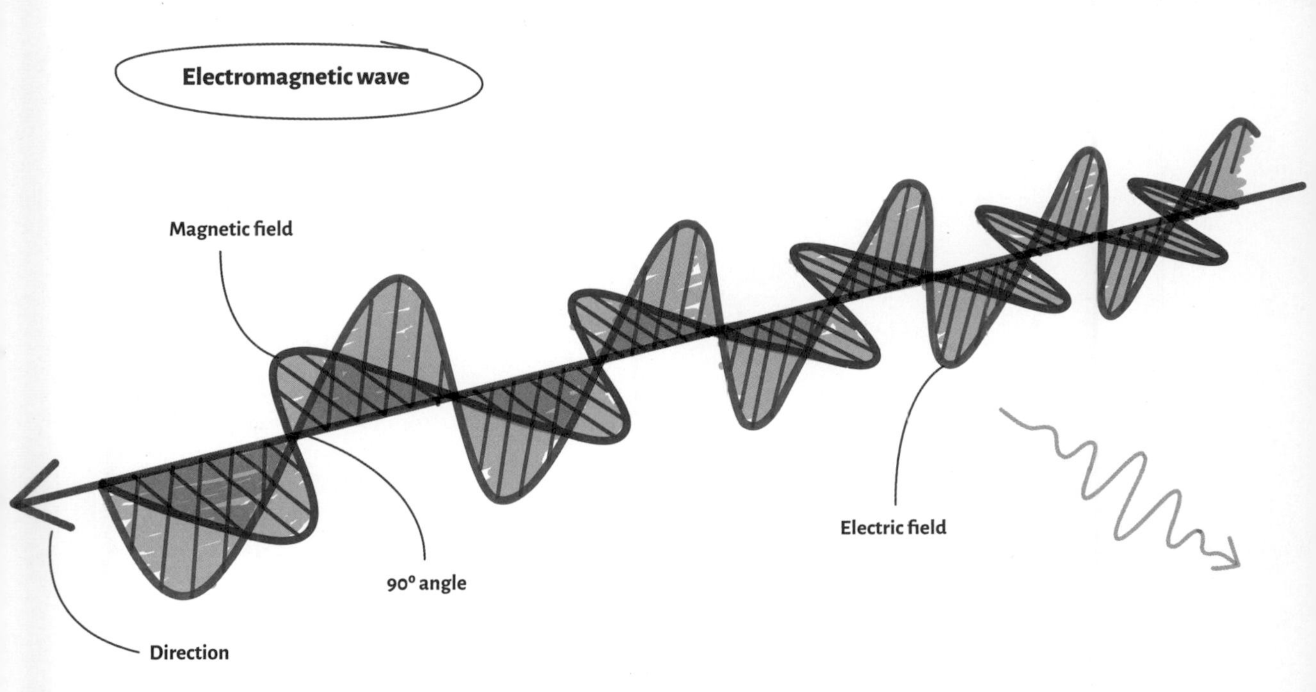

THE ELECTROMAGNETIC SPECTRUM

The electromagnetic spectrum consists of all energies of light. A photon's energy is governed by its frequency, f (measured in hertz), and wavelength, λ (measured in meters). Higher-frequency waves have smaller wavelengths and higher energies.

Gamma rays have the highest energy of all photons. These are created in nuclear reactions and very high-energy events in space, such as **supernova** explosions.

X-rays are emitted by some very hot objects in the universe, such as the hot gas around **black holes**. We use X-rays to image bones because they absorb much of the radiation, whereas skin does not.

Ultraviolet (UV) light has the next highest energy after blue light and is not visible. It is emitted by the sun at quite high levels, although some of the higher-energy UV is absorbed by the atmosphere. UV light is dangerous and can cause skin damage and cancer.

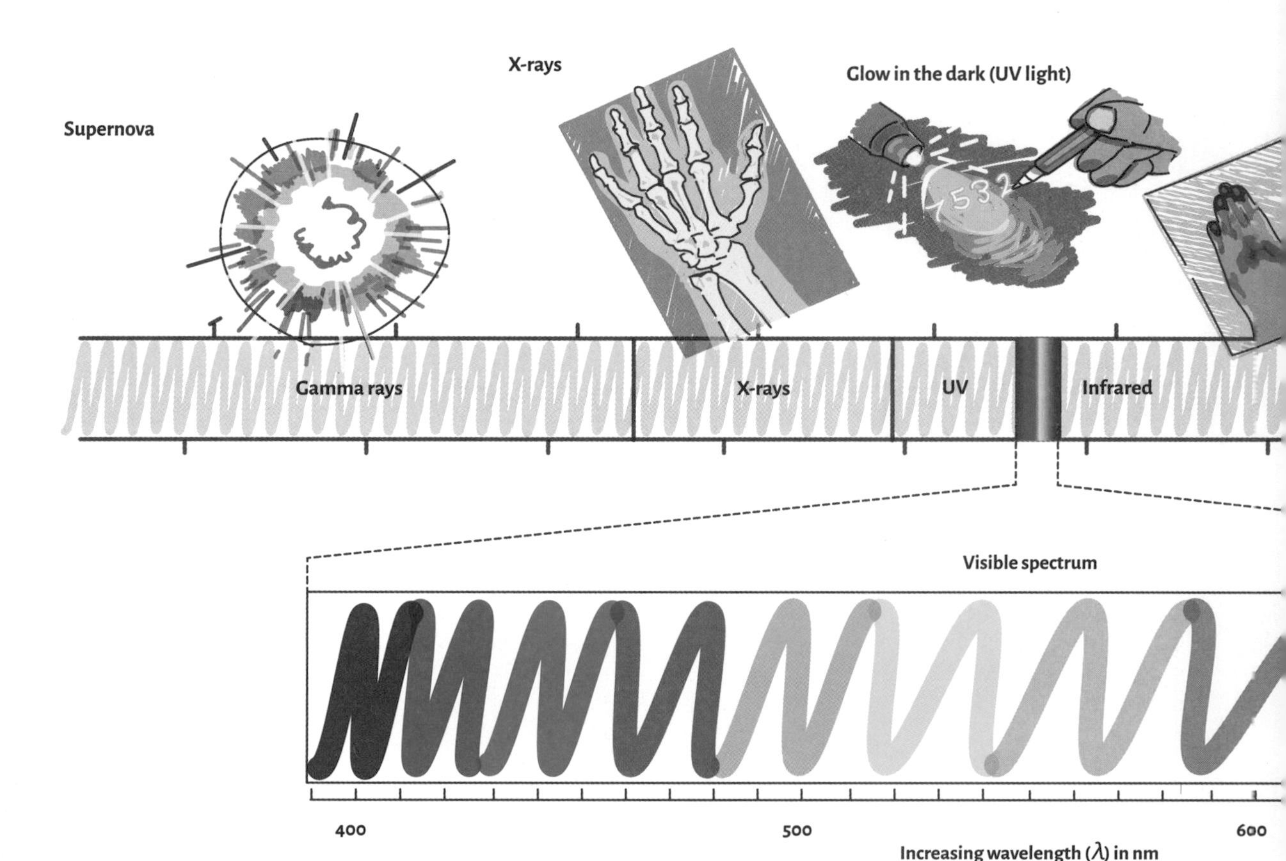

Everyday electromagnetism

The electromagnetic (EM) spectrum has been harnessed by humankind in many areas of everyday life, science, and medicine. From X-rays in hospitals through infrared (IR) scanners used to detect warm bodies at night to radio waves and microwaves for cell phone technologies and radio broadcasting, we have all benefited from the huge range of frequencies present in the EM spectrum.

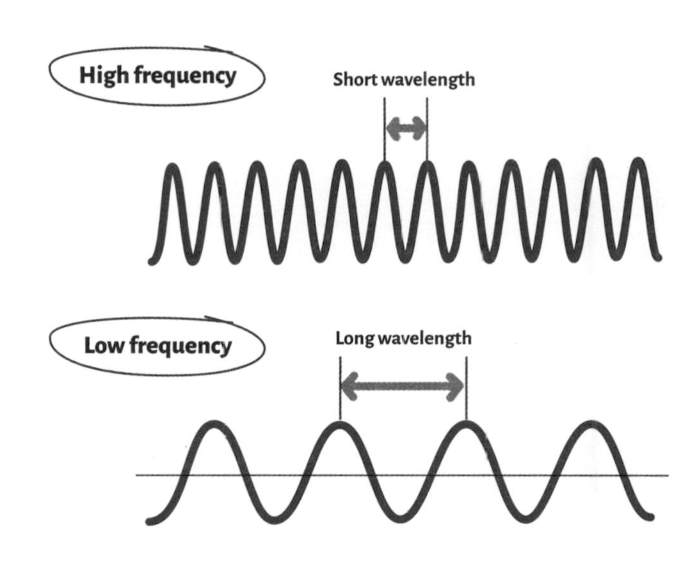

Infrared (IR) is the part of the spectrum that is emitted by warm bodies, such as animals, and can be used to detect warm objects when it is dark.

Microwaves are the next region of the spectrum. These are used in cell phones and at certain frequencies have the ability to heat water.

Radio waves come at the lowest-frequency end and have the least energy. These make up a large section of the electromagnetic spectrum and are used widely for communication and broadcasting.

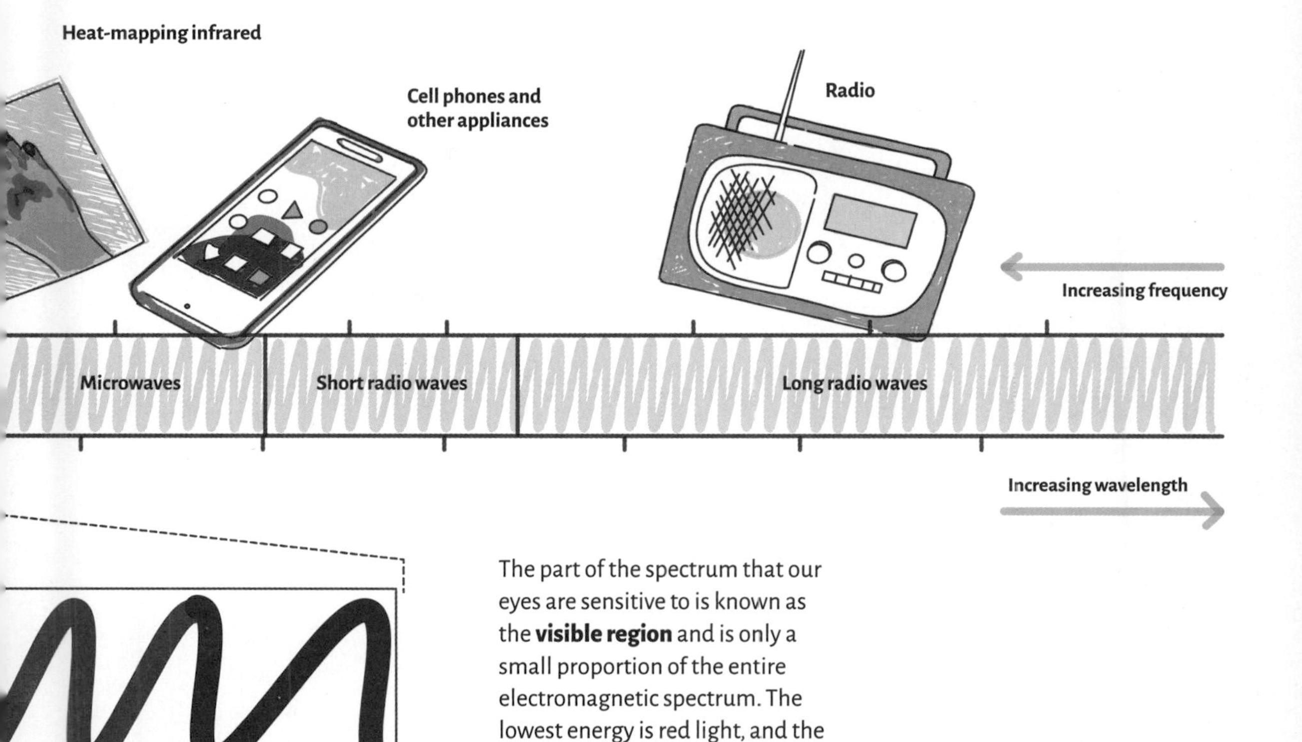

The part of the spectrum that our eyes are sensitive to is known as the **visible region** and is only a small proportion of the entire electromagnetic spectrum. The lowest energy is red light, and the highest energy appears blue.

ELECTROMAGNETISM

INDUCTION

FLEMING'S LEFT-HAND RULE

Point first finger in direction of magnetic field, second finger in direction of current, and thumb will indicate direction of force or motion.

FARADAY'S LAW

An electrical conductor moving relative to a magnetic field will induce an electromotive force.

EM INDUCTION

The three direction vectors—wire movement, magnetic field, and current—are always at 90° to one another.

ELECTROMAGNETIC SPECTRUM

THE SPECTRUM

Consists of all energies of light.

- **GAMMA RADIATION**
- **X-RAYS**
- **UV LIGHT**
- **VISIBLE SPECTRUM**
- **INFRARED LIGHT**
- **MICROWAVES**
- **RADIO WAVES**

WAVELENGTH

Length between waves, measured in meters; the longer the wavelength, the lower the frequency.

FREQUENCY

How often a wave repeats, measured in hertz.

ELECTROMAGNETIC RADIATION

ELECTROMAGNETIC WAVES

Light waves are the result of oscillating magnetic and electric fields crossing each other at 90°.

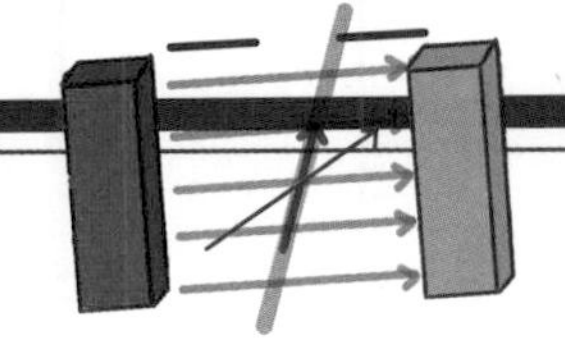

ELECTROMAGNETIC INDUCTION

ANGLE OF INDUCTION

Angle between the wire and the magnetic field; current strength decreases as the angle decreases from 90°.

ELECTRICAL GENERATOR

Converts mechanical energy into electrical energy.

MOTOR GENERATOR

Device that can both generate power from mechanical energy and convert it back, such as a hybrid car.

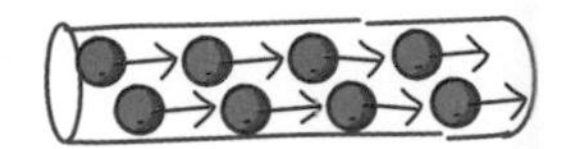

ALTERNATING CURRENT

Changes direction and magnitude periodically over time.

ELECTRIC MOTOR

Converts electrical energy into mechanical energy.

ENERGY TRANSFER

ENERGY GRID

Network of power cables and generators that can carry power over large distances using alternating current.

TRANSFORMER

Electrical device that transfers electrical energy from one circuit to another.

STEP-UP TRANSFORMER

Increases voltages between circuits.

STEP-DOWN TRANSFORMER

Decreases voltages between circuits.

POWER CABLE

Heavy-duty, highly insulated cables that carry electric current overhead or buried in the ground.

ENERGY LOSS

As the transformers change the voltage along a circuit, some energy is lost through heat and sound.

OTHER LIGHT EMITTERS

BIOLUMINESCENCE

Some fish, insects, and plants can create their own light source.

HEAT

Heated metal and burning wood or coal produce light-emitting heat.

CHAPTER 8

WAVES

Waves carry energy from one place to another through a medium. Electromagnetic waves can transfer the energy created by nuclear fusion within stars across the vacuum of space and through our atmosphere. Water waves carry energy from storms across our oceans as the vertical oscillations of water molecules. Sound is a type of wave and requires a fluid, such as air or water, through which to move (called propagation) via vibrations. Regardless of the type of wave, there are many common factors: oscillations, transfer of energy, and wave properties.

AMPLITUDE, FREQUENCY, AND PERIOD

Before you explore different types of waves and their properties, it is important to define some key physical properties shared by all types of waves. A wave is fundamentally an oscillation or cyclical movement of some form of energy. A single full oscillation takes place over a fixed period of time for a specific wave—this is the time period, *T*, for the oscillation.

The time period is the shortest time elapsed at which the wave looks identical as it propagates. The number of times the wave looks the same per second as it passes is called the **wave frequency**, *f* (measured in hertz). Time period and frequency are connected by the following relationship:

$$T = \frac{1}{f}$$

The time period of a wave is measured by noting the time it takes for a peak of the wave to be replaced by another peak. The distance between successive peaks is called the **wavelength** and is represented by the Greek letter λ (lambda), measured in meters.

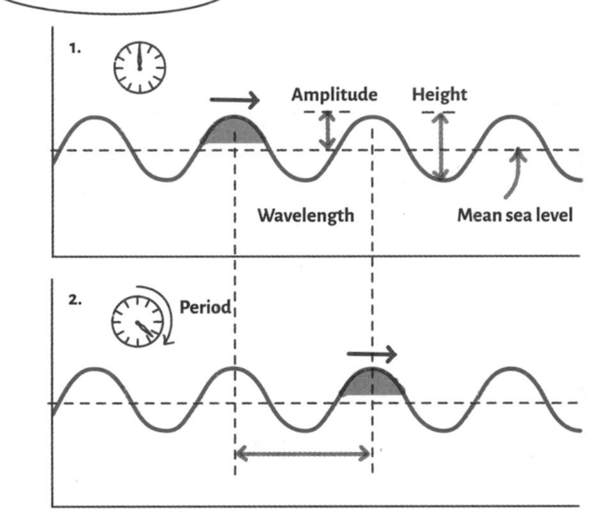

All waves have an **equilibrium point** (an undisturbed point), which is defined as the average displacement of each particle. The maximum distance from this equilibrium position is called the **amplitude**, *A*, of the wave. The maximum amplitude is called a **peak** (or crest), whereas the minimum amplitude is known as a **trough**.

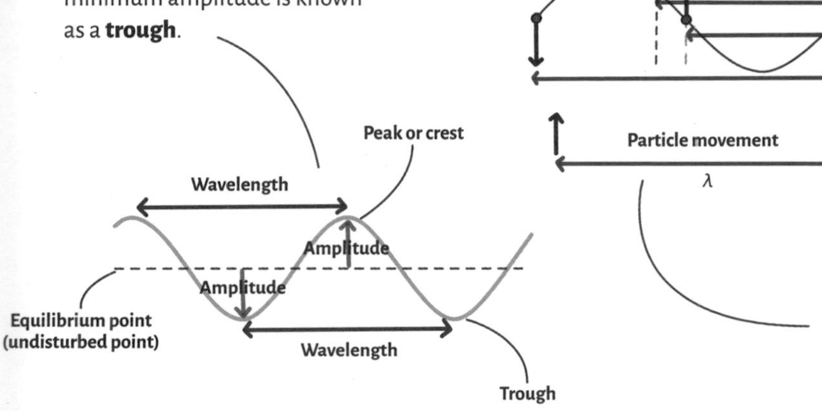

The distance between any two like points on a wave is also the wavelength. At any two like points, the oscillation of a particle is identical: Direction and speed of movement are the same.

SIMPLE HARMONIC OSCILLATIONS

The fluctuations of a physical property that repeat over time are called oscillations. Examples include an object's displacement measured from a fixed central point or the variations in the electric and magnetic field strengths of an electromagnetic wave.

Oscillations are cyclical in nature, and the variation is repeated over fixed time intervals, but the maximum **magnitude** (amplitude) of the variation may change over time as the system loses (or gains) energy. A **simple harmonic oscillator** will experience a restoring force directed back to its equilibrium position, the magnitude of which directly depends on the size of that displacement.

Imagine a ball on a string moving in a circular path. If a light is shone at the ball and its resultant shadow is projected onto a screen, the shadow will appear to move back and forth around the equilibrium position.

The shadow will be moving fastest at this point and will slow as it reaches the ends of its motion (when the ball's motion is parallel to the light beam). The motion of the shadow is described as a simple harmonic oscillator (SHO).

The displacement, x, of the shadow from its equilibrium is either positive or negative, and if plotted on a displacement-time graph (see page 32) will describe a perfect sine (or cosine) wave, depending on its start position.

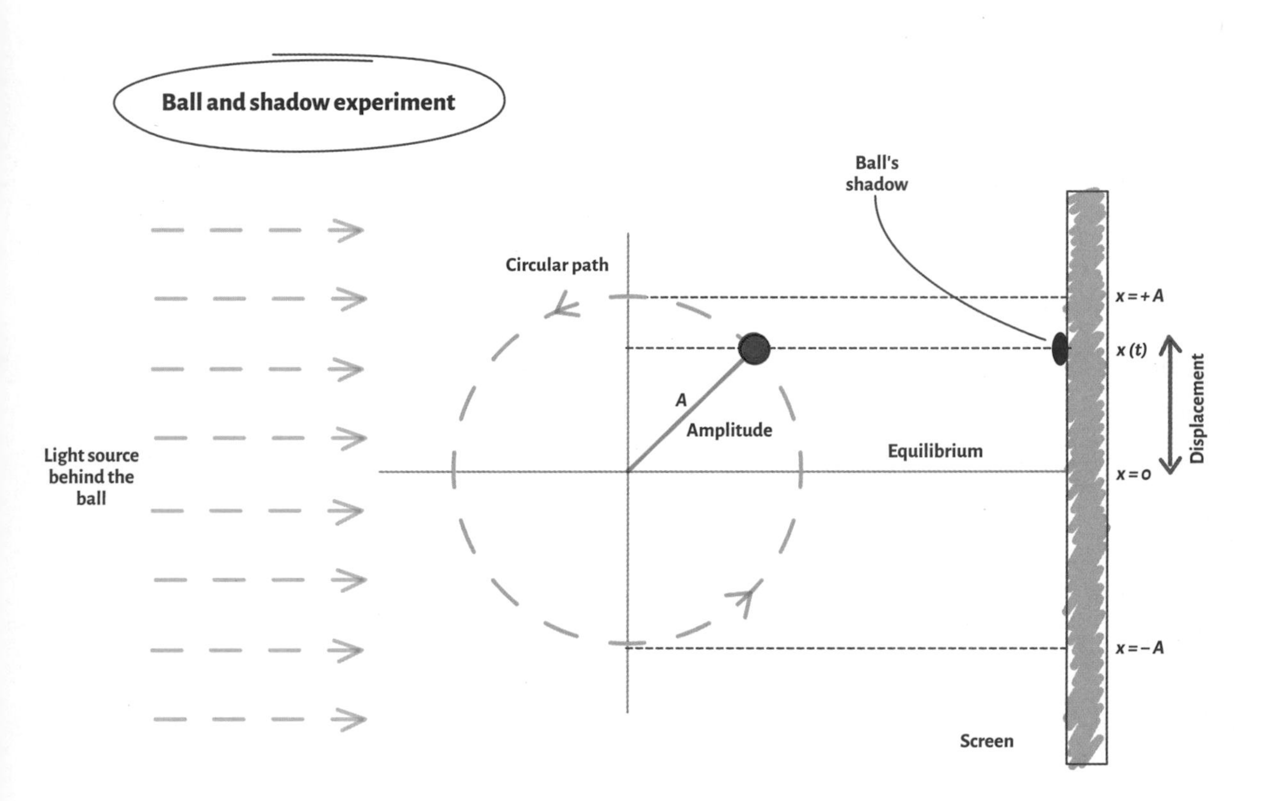

The properties of a simple harmonic oscillator

A simple harmonic oscillator moves back and forth with varying velocity depending on its position relative to the equilibrium position. As it moves, it is always accelerating toward the center of its motion: Its speed is increasing toward the midpoint and decreasing as it passes the midpoint.

The acceleration of the body is directly proportional to its distance from the midpoint, *x*, and directed toward it.

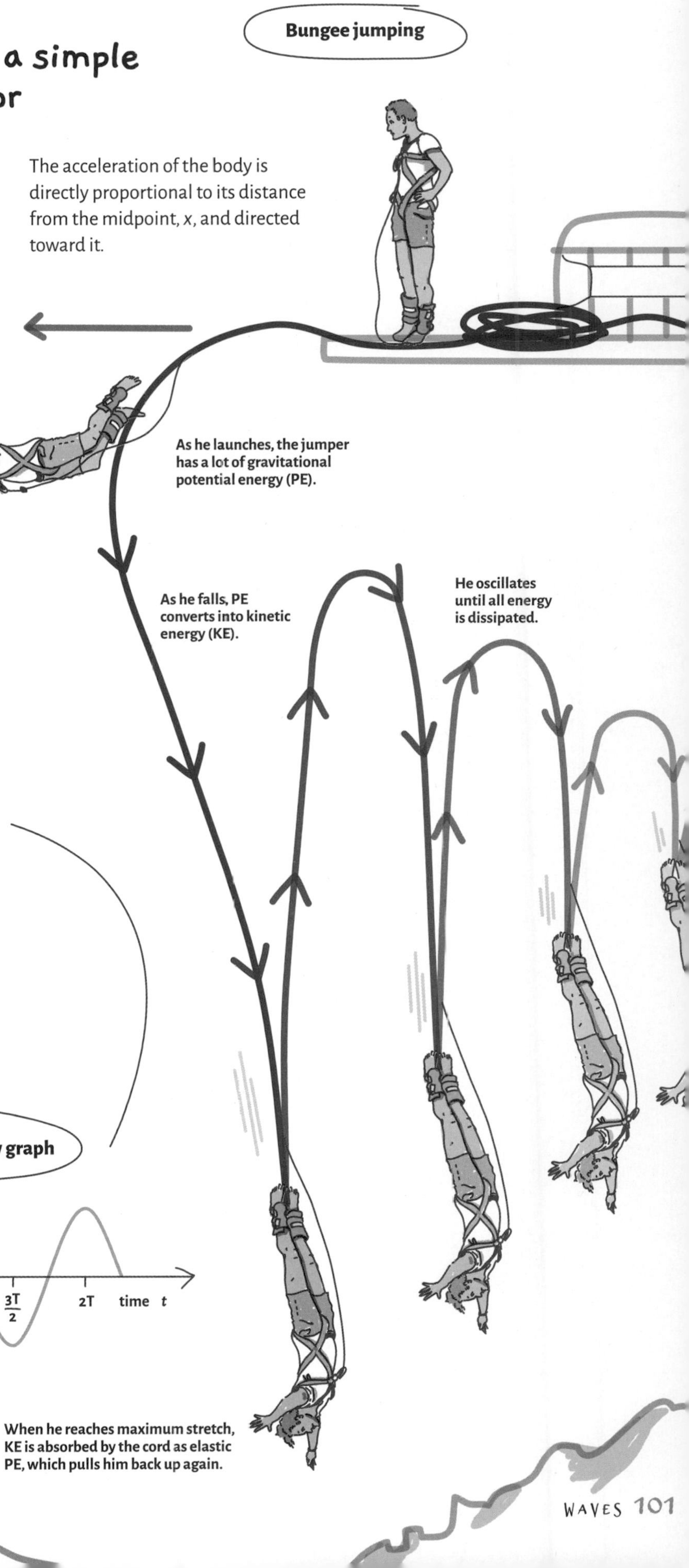

Bungee jumping

As he launches, the jumper has a lot of gravitational potential energy (PE).

As he falls, PE converts into kinetic energy (KE).

He oscillates until all energy is dissipated.

When he reaches maximum stretch, KE is absorbed by the cord as elastic PE, which pulls him back up again.

This is the definition of a simple harmonic oscillator, and there are many examples that display these characteristics in the real world, such as bungee jumping.

The variation of each property of a simple harmonic oscillator—displacement, velocity, and acceleration—are all sinusoidal in nature. **Sinusoidal** means that these are continuous waves, like a sine wave. As displacement increases, acceleration toward the center increases (slowing down the mass) and velocity decreases. The graph below, describing the ball and shadow experiment opposite, summarizes the details of this motion.

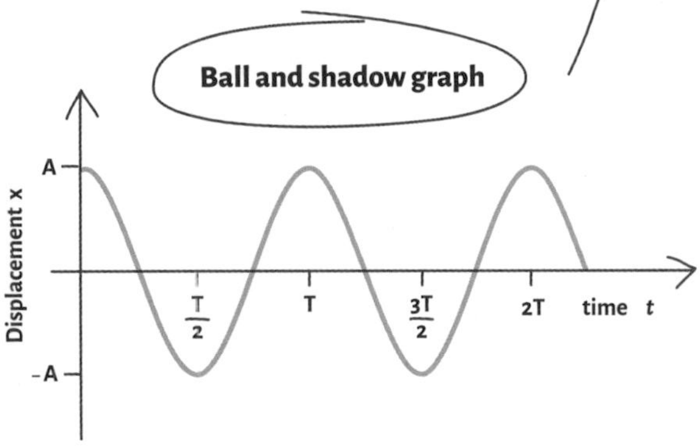

Ball and shadow graph

Displacement x
A
-A
$\frac{T}{2}$
T
$\frac{3T}{2}$
2T
time t

Simple pendulums

A mass (called a **bob**) connected to a string and free to oscillate back and forth with the string fixed at one end behaves like a simple harmonic oscillator, providing that the amplitude, *A*, of oscillation is small compared to the string length, *l*, and the mass of the string is small compared to that of the bob.

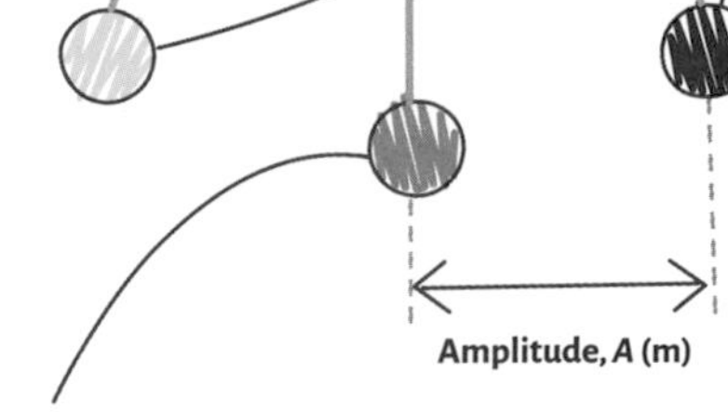

This setup is called a **simple pendulum** and is the basis for how an old-style grandfather clock works. The period of motion, *T*, for one complete cycle depends on the length of the string but is not affected by the bob's mass and its amplitude (provided it is small).

A simple pendulum can be used to measure the value of *g* on Earth. Using a very long pendulum string and timing many oscillations (this increases the accuracy of the measurement), the formula can be rearranged to find the value of *g*. Indeed, this method is used to do exactly that and is so accurate that it can detect small variations in the value of *g* at different altitudes on mountains on Earth!

The relationship between time period and string length, *l*, is:

$$T = 2\pi\sqrt{\left(\frac{l}{g}\right)}$$

where *g* is the acceleration due to gravity.

The period of motion for a grandfather clock is exactly two seconds, so that each swing takes exactly one second. The corresponding length of the pendulum is almost exactly 1 meter. This makes this type of clock very big indeed.

Mass-spring systems

A **mass-spring system** uses the principle of stored spring energy exchanging the KE and PE of a mass on a spring. A mass, m, is connected to a spring with a spring constant (spring stiffness), k. It is pulled down a small displacement from its equilibrium position, A, and released. The mass will then oscillate as energy is converted between the spring energy and KE of the mass.

Unlike a simple pendulum, a mass spring oscillates vertically instead of from side to side. As you have seen from Hooke's law (see page 12), the restoring force of the spring is directly proportional to the displacement of the mass from equilibrium ($F = kx$).

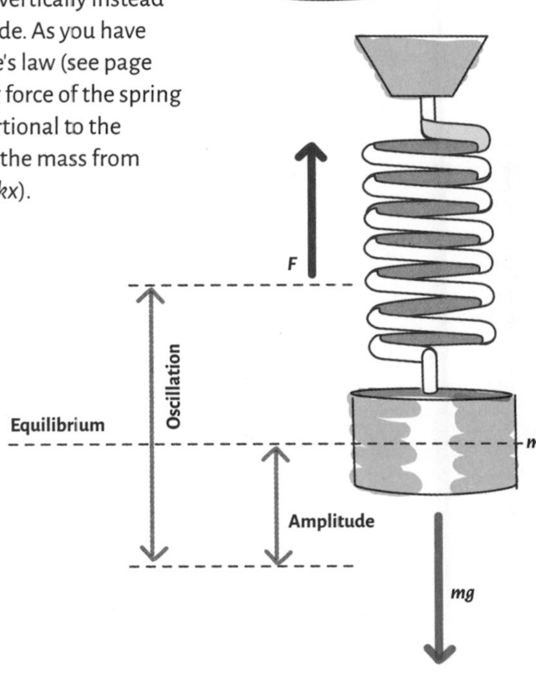

From Newton's second law ($F = ma$), the acceleration, a, of the mass is given by

$$a = \frac{F}{m}$$

Combining the two, you get:

$$a = \frac{k}{m}x$$

Since k and m are constant for a given spring, you have the requirements for a simple harmonic oscillator. In this case, the acceleration of the body depends directly on its mass, m, and the stiffness of the spring, k.

It can be shown that the time period:

$$T = 2\pi\sqrt{\left(\frac{m}{k}\right)}$$

This relationship can be used to find the value of k for any spring.

CAR SUSPENSION

A mass-spring system has many practical uses, such as car suspension. To create a smooth ride, the oscillations of a car must be minimized. Springs with very high spring constants connected to each wheel have a very short time period of oscillation, which reduces the up and down movement of the vehicle.

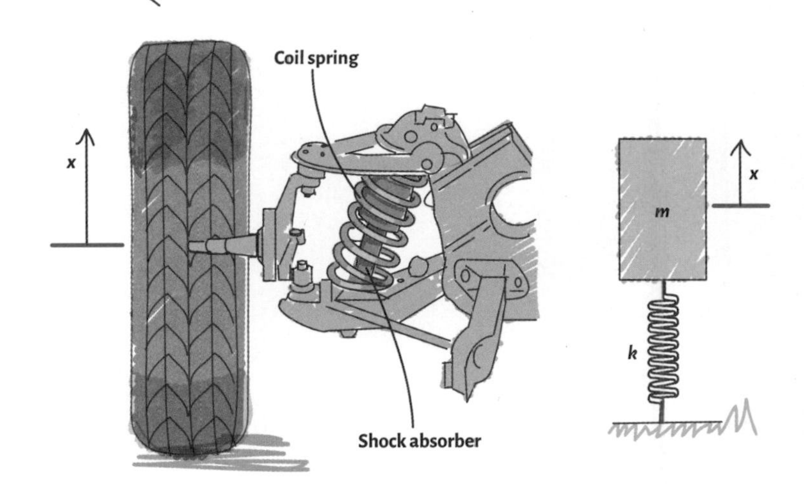

TRAVELING WAVES

Traveling waves oscillate in the same way as simple harmonic oscillators, while moving energy from one place to another in the direction of propagation. The transportation of energy occurs when particles in the wave oscillate up and down (transverse waves) or back and forth (longitudinal waves).

Transverse and longitudinal waves

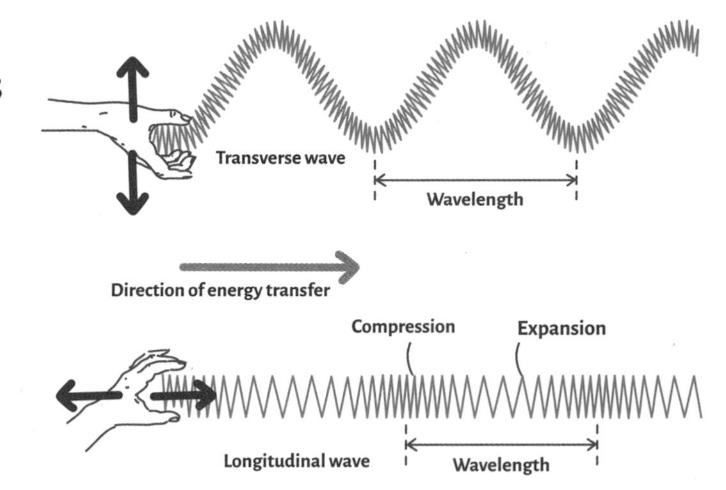

Traveling waves can be categorized into two types: transverse and longitudinal waves.

Transverse waves are classified by the oscillations of their particles at 90° to the transfer of energy.

The particles in **longitudinal waves** oscillate back and forth in the same direction as the energy transfer or propagation. Examples of this are sound and shock waves.

Water waves displace particles vertically up and down (imagine a boat on the ocean) as the wave travels forward.

Light rays travel in straight lines across a vacuum or through air or glass. You have seen that their electric and magnetic fields fluctuate at 90° to the direction of motion (see page 80).

Both types of wave consist of oscillations and transfer energy. It is the direction of oscillation relative to the transfer of energy that defines the wave type.

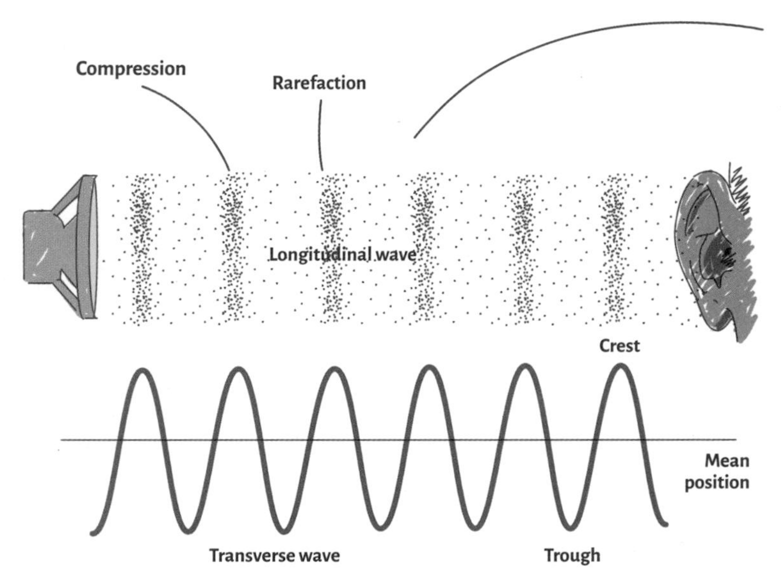

SOUND WAVE

Sound travels through fluids, such as air or water, and is a result of a sudden disturbance of the fluid. The particles are given energy by this disturbance (such as a stone landing in a pond) and oscillate back and forth. This energy is then transferred to neighboring particles, sending it outward from the source.

The position of each particle from its equilibrium position can be represented on a graph that looks like a sine wave.

Wavelength and speed

Wavelength and wave speed are interrelated: If one changes, so does the other. Wave speed, v, is determined by the medium through which the wave is traveling and is fixed for that system.

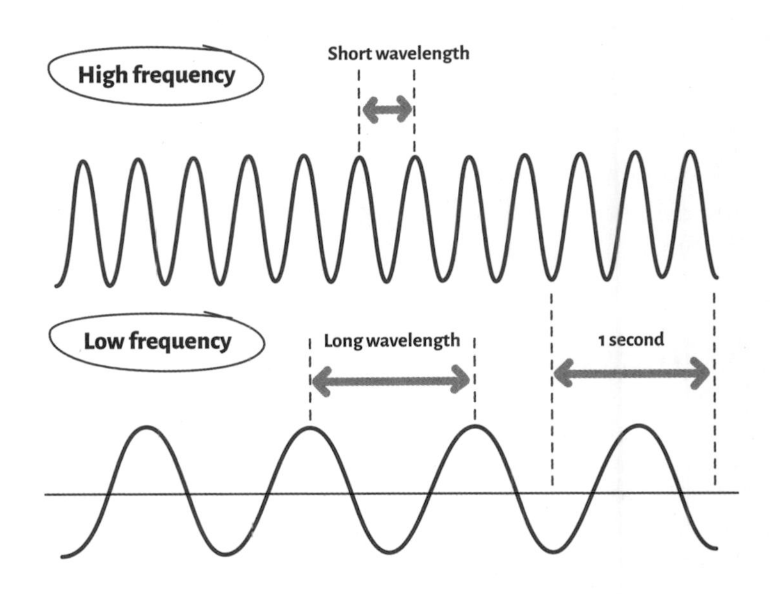

The **time period** of a wave is defined as the shortest time interval at which the wave looks identical as it passes or the time the wave takes to travel exactly one wavelength, λ. The shorter the wavelength, the smaller this time period and the more oscillations there are per second—this is the **frequency**, f.

Since v is fixed in a given medium, as wavelength, λ, decreases, the frequency increases and vice versa. This leads to the relationship:

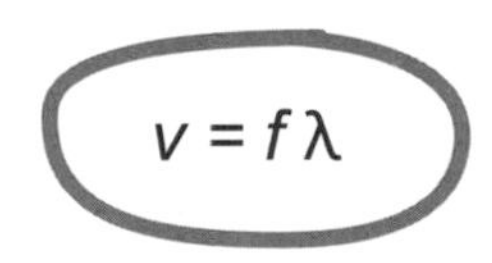

$$v = f\lambda$$

This relationship allows the velocity of waves to be accurately calculated if both the wavelength and frequency are known.

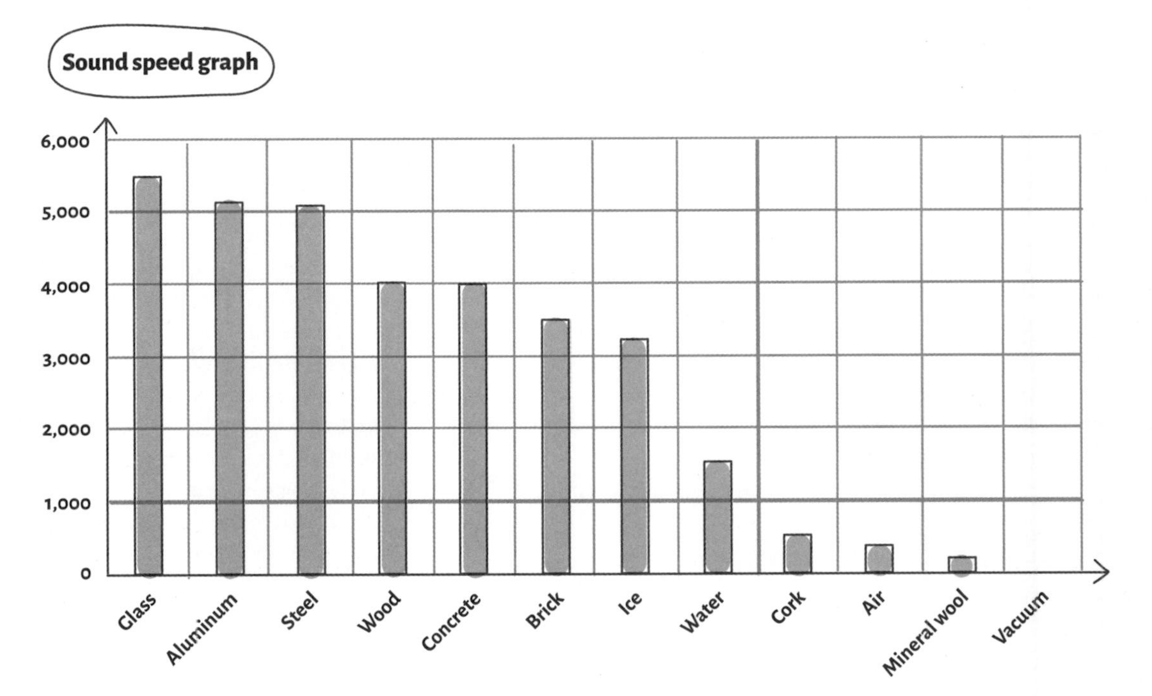

Sound speed varies dramatically with the medium through which it is passing. A simple experiment can be constructed to determine the speed of sound through air. This speed will vary slightly depending on the temperature of the air but is more or less constant on Earth at sea level.

If a speaker is attached to a signal generator that creates a known frequency input, the wavelength of the sound wave can be measured using a microphone. This relies on reflecting the sound wave and creating **wave interference**.

WAVE PROPERTIES

All types of waves share a number of properties that affect their movement through a medium. All waves can be reflected, refracted, and diffracted. Each of these properties changes the speed of the wave, its direction of travel, or both.

Reflection

All waves can be reflected by various barriers or surfaces depending on the wave type. Light is reflected by shiny (reflective) surfaces such as glass or metals, sound can be reflected by a rigid surface (creating an echo), and water waves reflect off solid barriers above the water level (such as large rocks).

A wave moving toward the barrier is called the **incident wave**, and a wave reflected by the barrier is called the **reflected wave**.

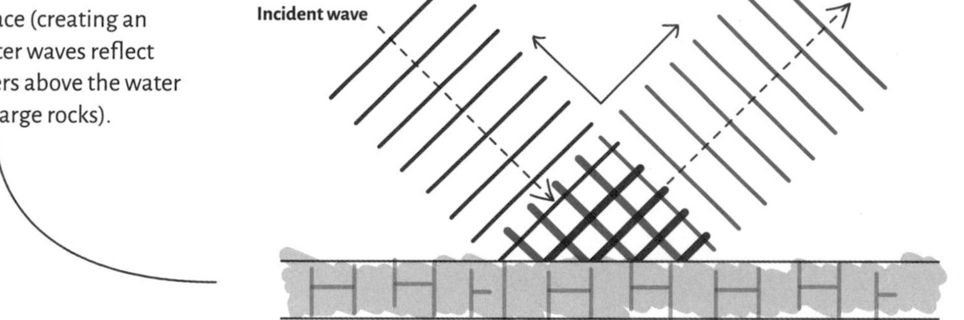

A wave that is reflected strikes the surface relative to a line at 90° to that barrier (called the **normal**) and is reflected at the same angle, θ. These are called the **angle of incidence** and **angle of reflection**, respectively.

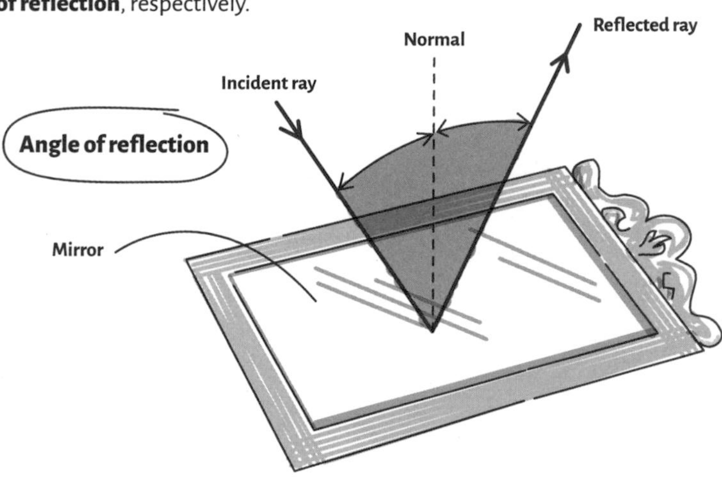

If it weren't for the wave property of reflection, you would not be able to see most things on Earth. Light from the sun travels to Earth, passes through our atmosphere, and hits the surfaces of all the objects around us. Some or all of the light is then reflected in all directions, depending on the nature of the surface.

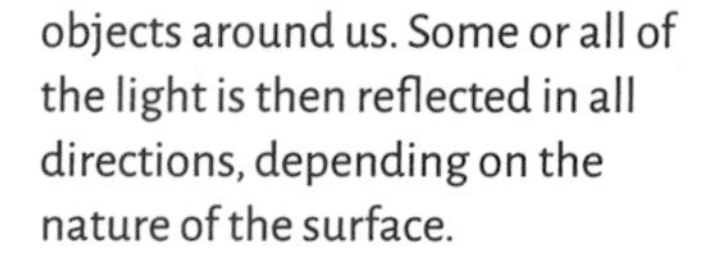

Sunlight is made up of all the different colors in the visible spectrum. Black objects reflect no light, whereas white objects reflect all frequencies of light equally. There is an infinite number of reflective surfaces, all appearing to us as different colors.

Refraction

Refraction is a property of light that changes both the speed of the wave and its direction of travel. Refraction primarily slows the propagation of a wave front, which has the result of changing the direction in which it is moving.

For light waves, refraction occurs when the wave front passes from one transparent medium (such as air) into another (such as water). At the barrier, the wave front slows down as the medium becomes more optically dense—this reduces the wavelength.

The angle that the incident wave forms is called the angle of incidence, θ_i, and the angle at which the wave is refracted is called the **angle of refraction**, θ_r.

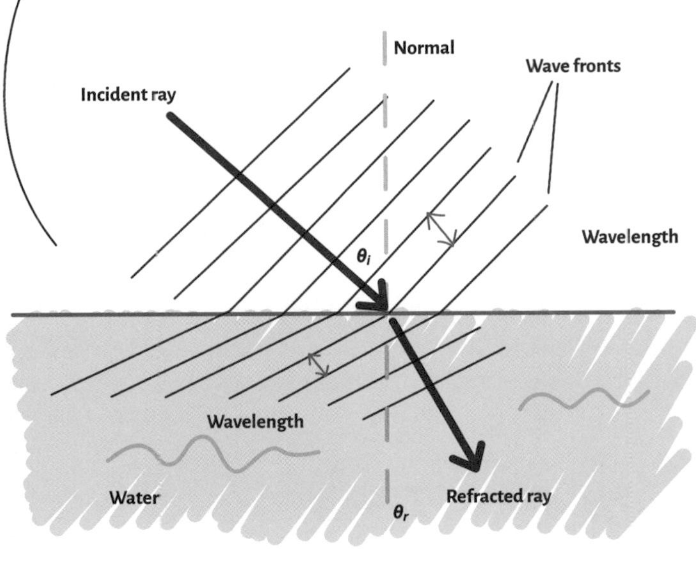

At the boundary between air and water, there is a small amount of reflection called **partial reflection**.

A reason for rainbows

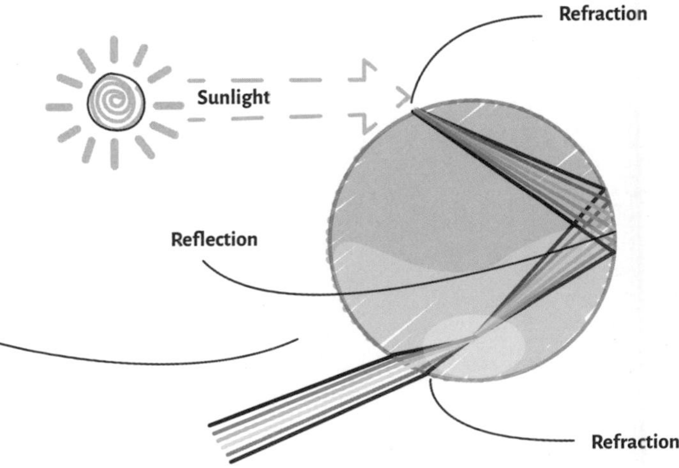

Refraction depends on two factors: the difference in the optical density of each medium and the frequency of light passing across the boundary.

Light dispersion

Glass prism
White light
Red
Orange
Yellow
Green
Blue
Indigo
Violet
Screen

Frequency of light affects the angle of refraction: Red light is refracted least, whereas blue light is refracted most. This causes what is known as **light dispersion**, a phenomenon demonstrated by light passing through a glass prism. It is the reason for rainbows, as sunlight is refracted at the boundary between air and water droplets.

Diffraction

Diffraction (spreading or bending) changes the direction and speed of a wave front. It is a result of the wave passing through a gap in a barrier or around an obstacle without a change in medium. As a straight wave front passes through a gap, it fans out and bends outward, causing a circular wave pattern.

Diffraction through gaps

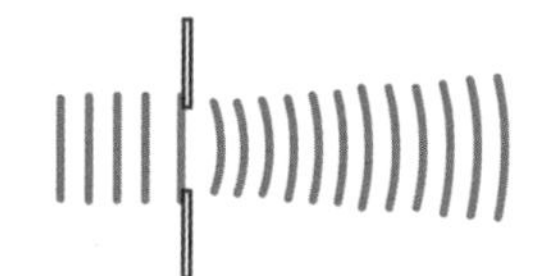

Wide gap – small diffraction effect

Narrow gap – large diffraction effect

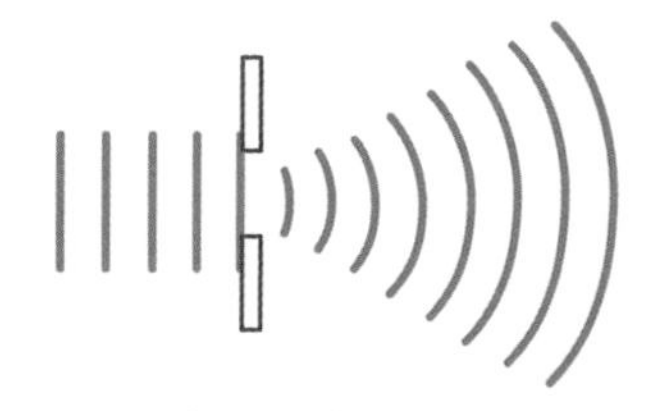

Large wavelength – large diffraction effect

The size of the effect is dependent on the relative sizes of the wavelength of the wave compared to the gap through which it is passing. If the two are of comparable size, the effect due to diffraction is maximized.

Light waves, water waves, and sound waves can all be diffracted.

A low-frequency sound travels more readily around a large object because its wavelength is a similar size to that of the object around which it is moving. This produces a larger dispersion of low-frequency sound, making it easier to hear around objects than high-frequency sound.

Diffraction around objects

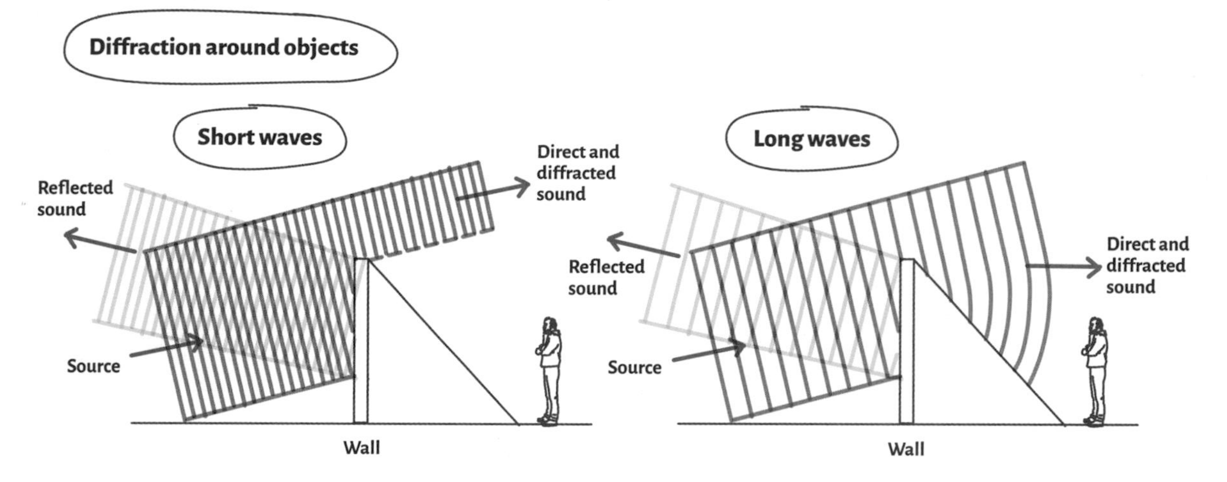

Communication via radio waves is hugely affected by diffraction. Radio transmissions occur at many different radio frequencies.

Short wavelengths are blocked by hills and other large objects, whereas long wavelengths can easily diffract around them. The region behind a hill, a transmitter, and a receiver is called the **wave shadow**. This shadow barely affects the quality of a long-wave radio transmission but completely blocks short-wave transmissions.

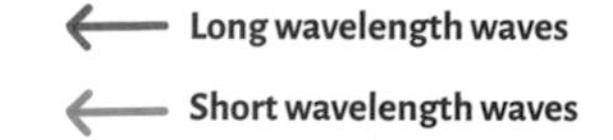

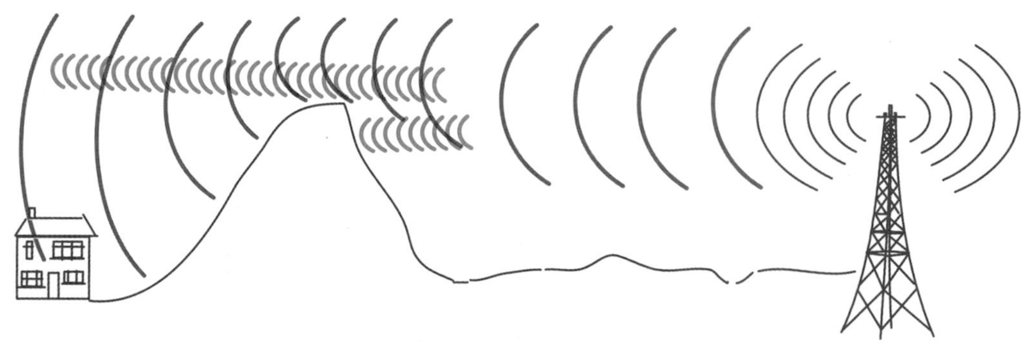

INTERFERENCE AND STANDING WAVES

When two separate traveling waves interact with each other, they either combine to form a large peak or a large trough, or they cancel each other out partially or completely. This is called interference and is the mechanism by which standing waves are produced.

Interference

Two waves of the same wavelength and frequency that coincide with each other exactly are called **coherent waves**. Coinciding waves of different wavelengths and frequencies are called incoherent.

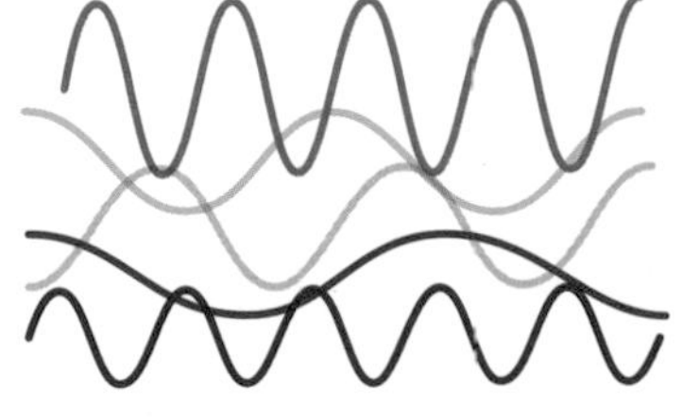
Incoherent

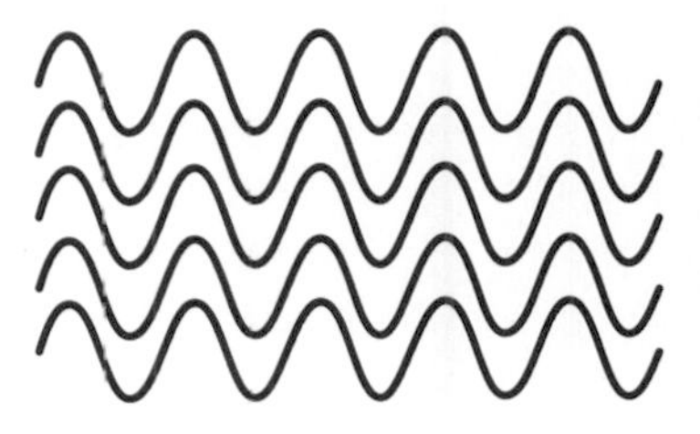
Coherent

If two coherent wave peaks or two coherent wave troughs coincide, while traveling in the same direction or opposite directions, they reinforce each other, producing a much larger peak or trough. This is called **constructive interference**.

If a trough of one wave interacts with a peak of another with the same amplitude, the result is a canceling out of each, producing a flat region. This is called **destructive interference**.

If two fishermen cast their lines in the water at the same time at different points, **ripples** will fan out in circles. As they interact, large **troughs** and **peaks** will form during constructive interference, whereas regions of flat water will result where a trough meets a peak, causing destructive interference.

Waves that match up exactly peak to peak are said to be **in phase** and will constructively interfere. As they travel past each other, the peaks will no longer be in alignment. At the point where a peak meets a trough, the waves are now one half of a wavelength out of alignment—they are said to be 180° **out of phase**.

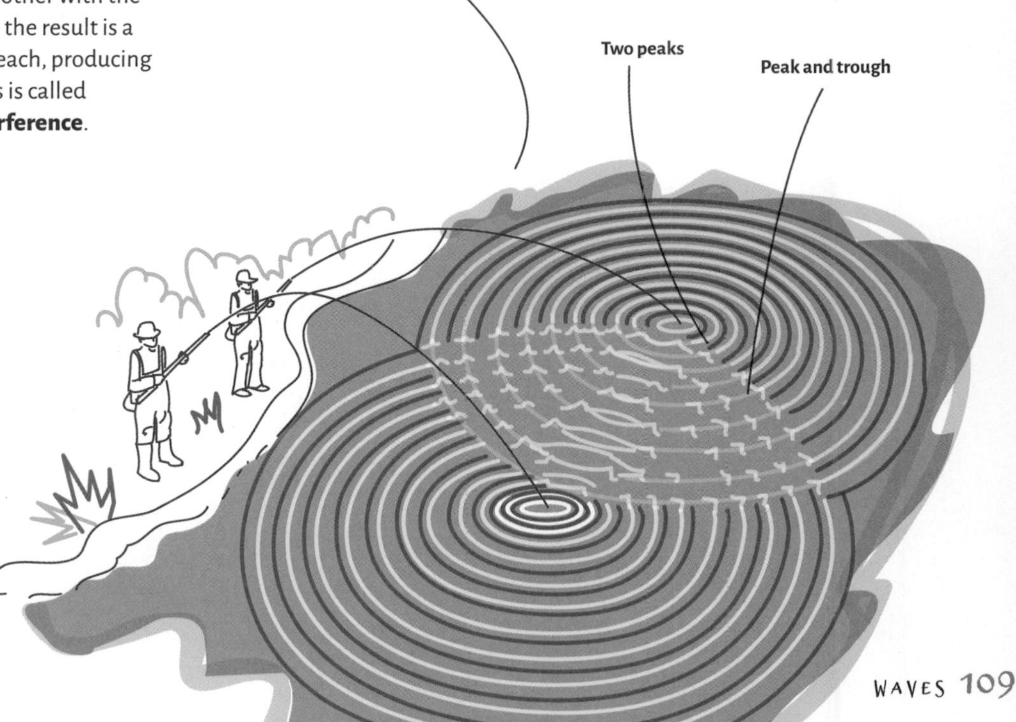

Standing waves

A continuous wave that travels toward a boundary and reflects will meet an identical wave traveling in the opposite direction.

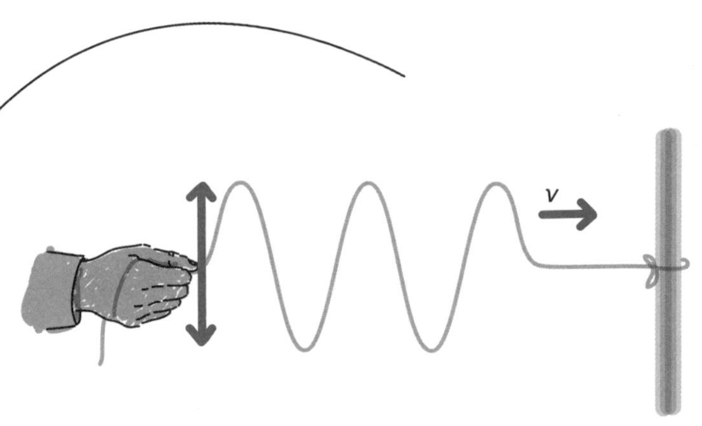

Consider a string of length, *L*, which is free to move at one end and fixed at the other. A **transverse wave** is generated at the free end at a constant frequency that travels toward the fixed end, acting as a reflective boundary. The wave will reflect off the boundary and travel in the opposite direction, meeting the incoming waves. The resultant pattern will be a combination of constructive and destructive interference.

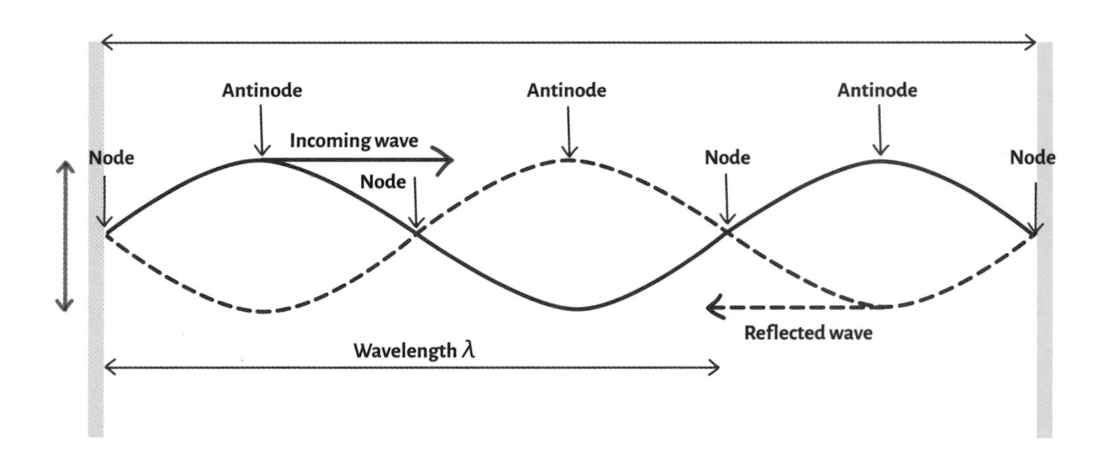

At certain frequencies, the **reflected wave** will meet the **incoming waves** exactly in phase. If the string length is an exact multiple of half of the wavelength of the traveling wave, the reflected wave will coincide with the incoming wave.

The result is destructive interference at a fixed point on the string (called a **node**) and constructive interference at a different point on the string (called an **antinode**). The antinode will oscillate up and down at a certain frequency.

Using a strobe light of the same frequency will illuminate the wave, which will appear to stand still—a **standing wave** (often referred to as a stationary wave).

HARMONIC

A guitar string that is plucked at a point that is exactly a fraction of its length will form a standing wave. This is called a **harmonic**. If a guitar string is plucked in the middle, it is called the first harmonic. If plucked at one-quarter of its length, it is the second harmonic, and so on.

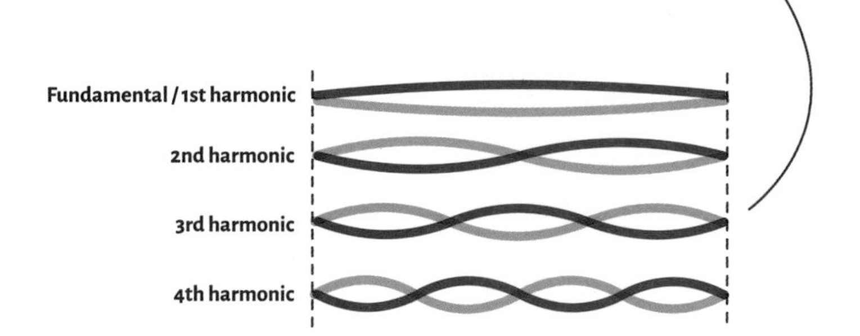

DOPPLER SHIFT

When a wave source moves through space, the frequency of its emitted wave—sound or electromagnetic—changes, either compressing the wavelength or extending it. This is called the Doppler shift.

Sound travels through air at a fixed speed, 1,235 kph (767 mph). Its frequency determines the pitch of the sound. A higher frequency will have a higher pitch.

If an object such as an ambulance broadcasting a siren is moving toward you, the wavelength will be compressed and appear to have a higher frequency or pitch than that at which it was broadcast.

As the ambulance passes you, its speed moving away from you will appear to stretch the sound waves, making them **longer in wavelength** and therefore a lower frequency. The change in pitch is caused by a decrease in frequency—the sound appears to change. This is called **Doppler shift**.

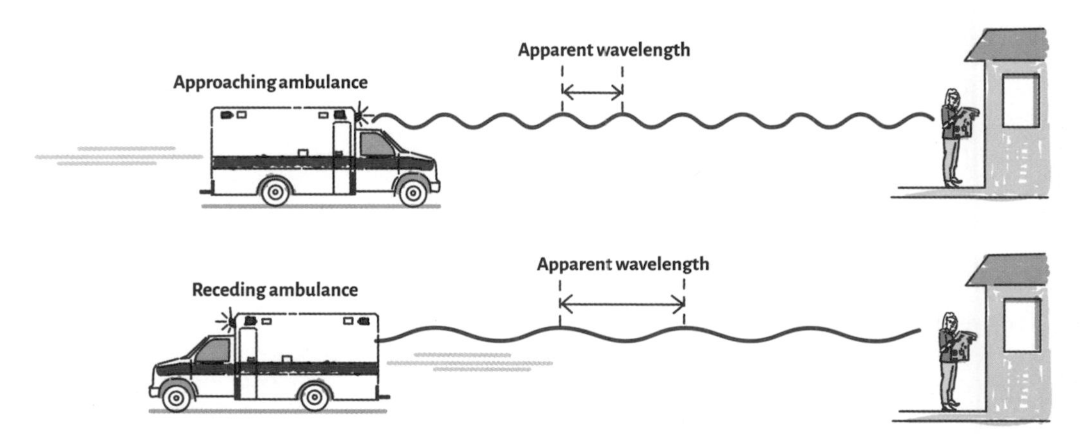

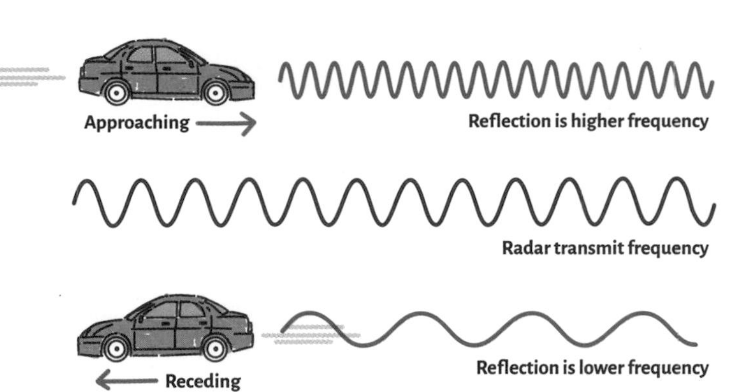

MEASURING SPEED

Doppler shift is used to measure the speeds of cars. A speed camera transmits a high-frequency radio wave (**radar**) of known frequency, which is reflected from the car and compressed by its forward speed. The change in frequency observed by the camera is used to calculate the speed of the vehicle.

The difference in frequencies of the emitted and received signals will give an accurate value of the speed of the vehicle, based upon the equation:

$$f_o = \frac{v}{v - v_e} f_e$$

where, f_e is the frequency of signal emitted by the speed camera, f_o is the observed frequency reflected by the car, v_e is the speed of the car, and v is the speed of sound.

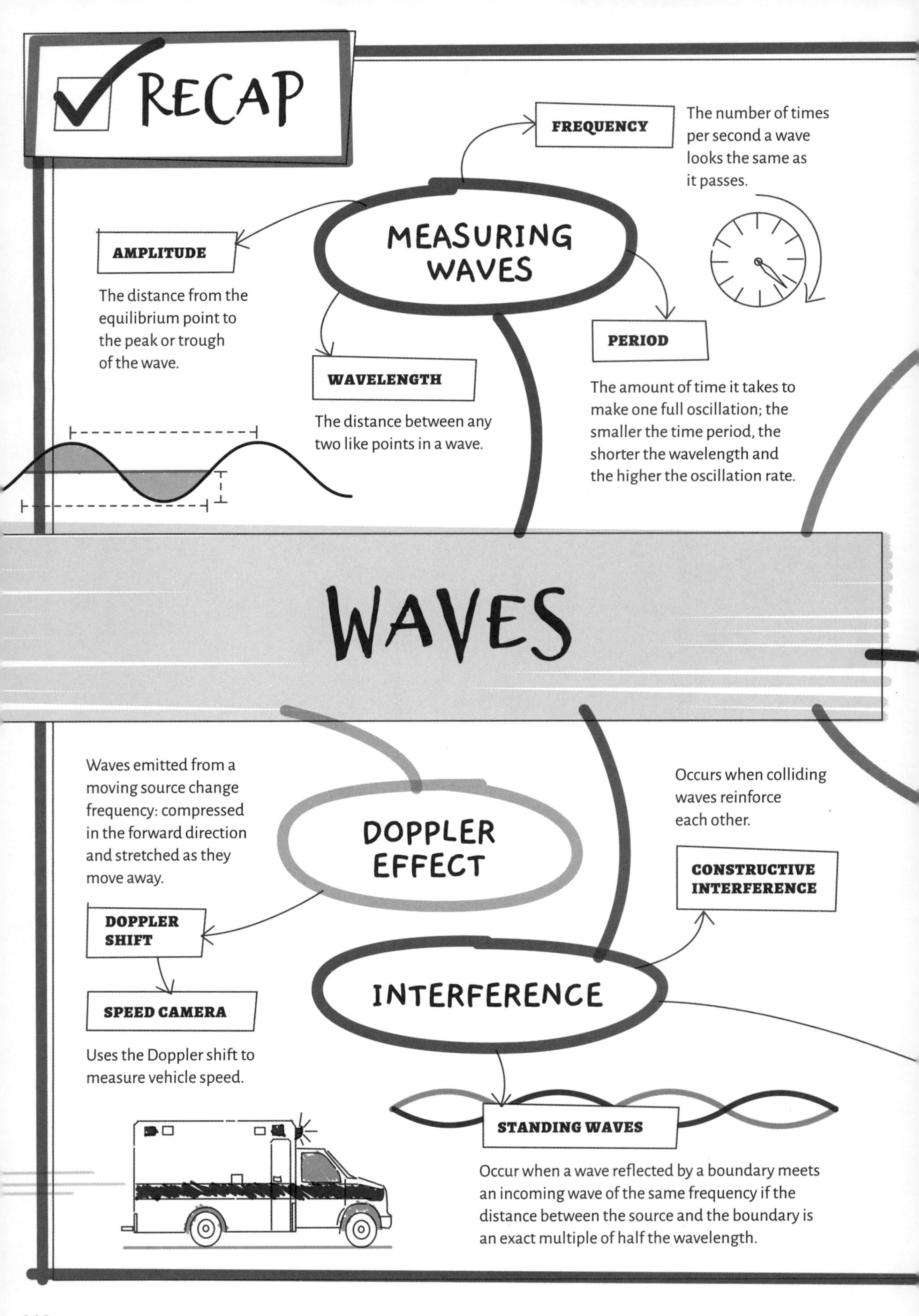
RECAP
MEASURING WAVES
FREQUENCY
The number of times per second a wave looks the same as it passes.
AMPLITUDE
The distance from the equilibrium point to the peak or trough of the wave.
WAVELENGTH
The distance between any two like points in a wave.
PERIOD
The amount of time it takes to make one full oscillation; the smaller the time period, the shorter the wavelength and the higher the oscillation rate.
WAVES
DOPPLER EFFECT
Waves emitted from a moving source change frequency: compressed in the forward direction and stretched as they move away.
DOPPLER SHIFT
SPEED CAMERA
Uses the Doppler shift to measure vehicle speed.
INTERFERENCE
Occurs when colliding waves reinforce each other.
CONSTRUCTIVE INTERFERENCE
STANDING WAVES
Occur when a wave reflected by a boundary meets an incoming wave of the same frequency if the distance between the source and the boundary is an exact multiple of half the wavelength.

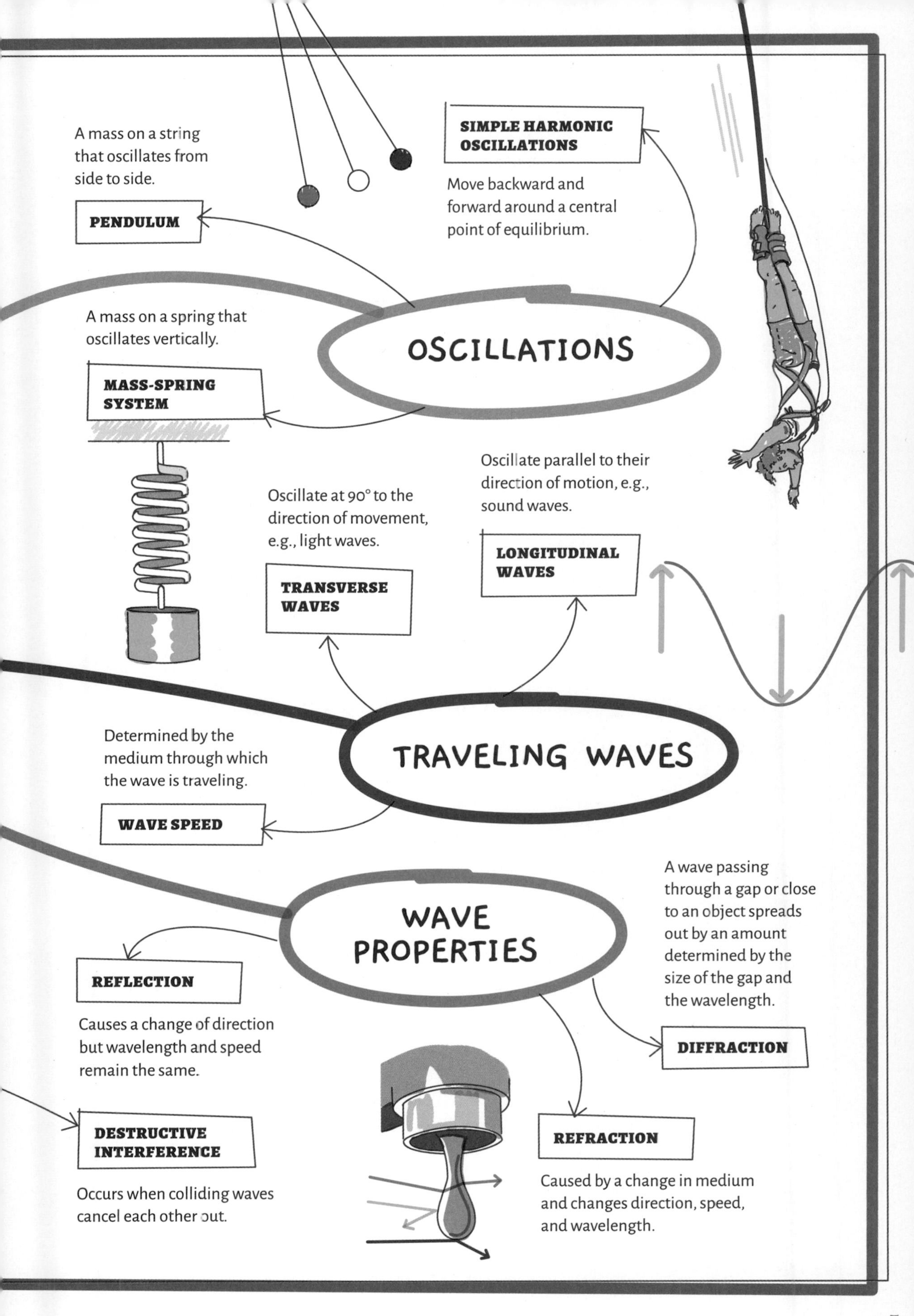
A mass on a string that oscillates from side to side.
PENDULUM
SIMPLE HARMONIC OSCILLATIONS
Move backward and forward around a central point of equilibrium.
A mass on a spring that oscillates vertically.
OSCILLATIONS
MASS-SPRING SYSTEM
Oscillate parallel to their direction of motion, e.g., sound waves.
Oscillate at 90° to the direction of movement, e.g., light waves.
LONGITUDINAL WAVES
TRANSVERSE WAVES
Determined by the medium through which the wave is traveling.
TRAVELING WAVES
WAVE SPEED
A wave passing through a gap or close to an object spreads out by an amount determined by the size of the gap and the wavelength.
WAVE PROPERTIES
REFLECTION
Causes a change of direction but wavelength and speed remain the same.
DIFFRACTION
DESTRUCTIVE INTERFERENCE
Occurs when colliding waves cancel each other out.
REFRACTION
Caused by a change in medium and changes direction, speed, and wavelength.

CHAPTER 9

OPTICS

Optics is the branch of physics that explores the behavior of light. Light is an electromagnetic wave and is controlled by the same laws as all waves: reflection, refraction, diffraction, and interference. These properties can be exploited in a wide range of practical applications, including mirrors, lenses, and relatively new ideas that have had a massive impact on our lives, such as fiber-optic cables for superfast broadband. Even 3D movie theaters make use of the polarization of light through 3D glasses.

THE LAWS OF REFLECTION

All waves can be reflected off a surface. All light waves (not just visible light) are reflected, and the amount of reflection depends on the nature of the surface. Most materials absorb certain frequencies of light, but some, such as mirrors, are highly reflective.

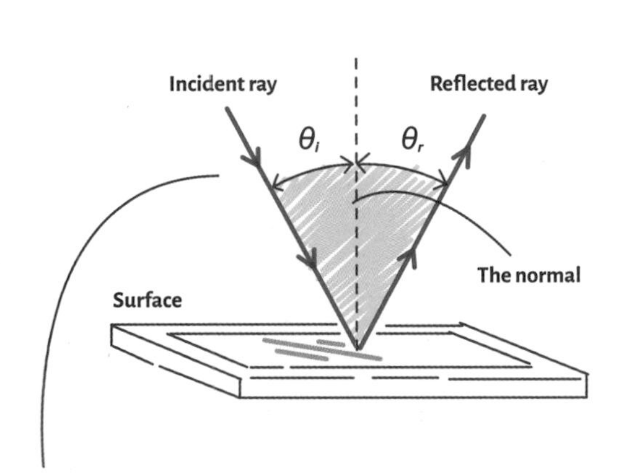

THE TWO BASIC LAWS OF REFLECTION

1. The incident ray, the reflected ray, and the normal are all in the same plane or surface.
2. The angle between the incident ray θ_i and the normal is equal to the angle between the reflected ray θ_r and the normal.

All electromagnetic radiation can be reflected, but the nature of the surface and the frequency of the incident ray will determine the amount of reflection.

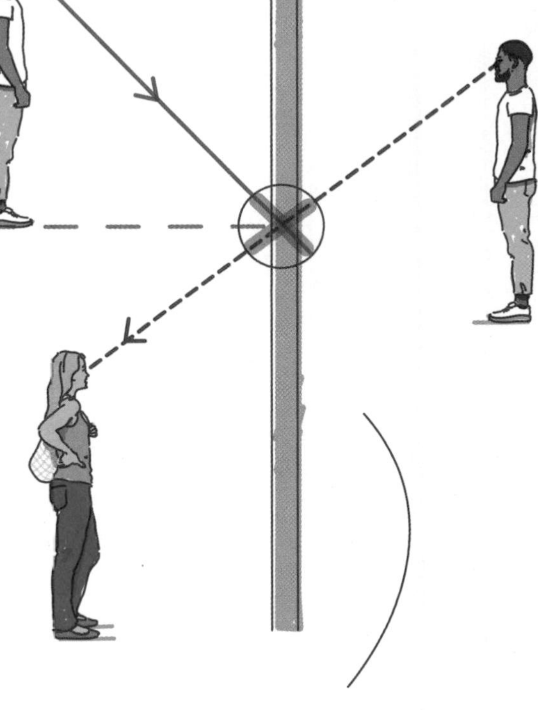

Strong sunlight is readily absorbed by black surfaces and reflected by white or metallic ones. Sunlight is also strongly reflected by ice and snow but absorbed by water. This effect is called **albedo**.

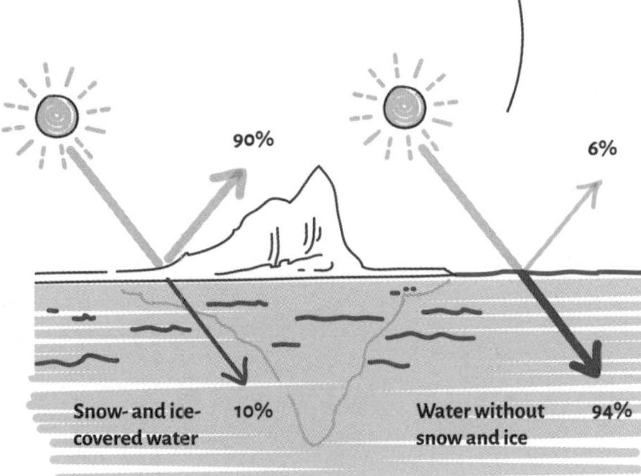

Mirrors are highly reflective. You can see someone who is standing a distance away from you in a mirror if the angle at which you are observing the mirror is the same as the angle at which they are observing your reflection.

REFRACTION, SNELL'S LAW, AND TOTAL INTERNAL REFLECTION

As you have seen, when light travels from one transparent medium to another at an angle, the wave is refracted; it changes in wavelength, speed, and direction. A small amount of light is also reflected. This is called partial reflection.

Snell's law

The change in wavelength, speed, and direction are all determined by the nature of the transparent material: its **optical density**.

A beam slows down upon moving into a more optically dense material, whereas a beam speeds up as it moves into a less optically dense material.

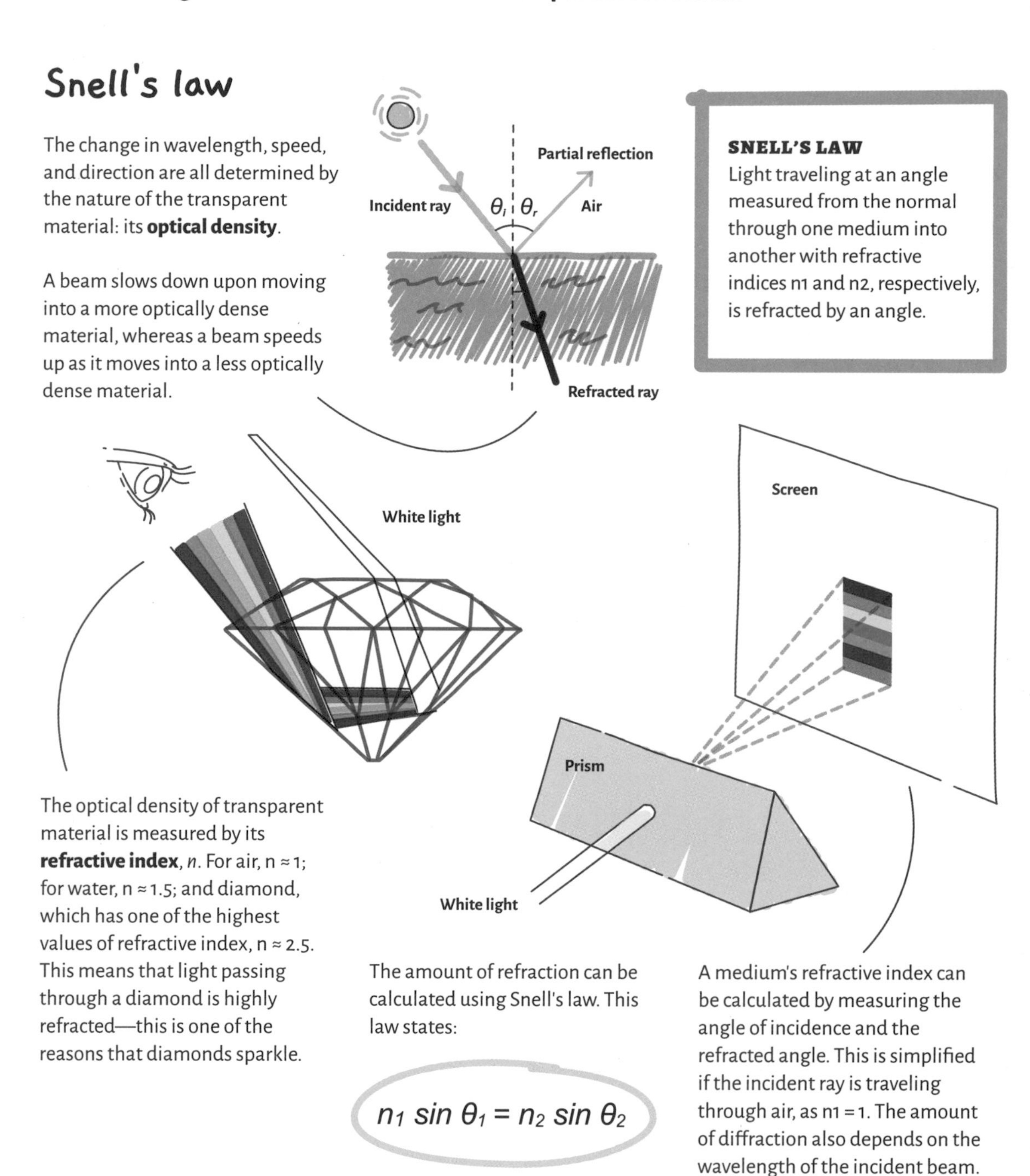

SNELL'S LAW
Light traveling at an angle measured from the normal through one medium into another with refractive indices n1 and n2, respectively, is refracted by an angle.

The optical density of transparent material is measured by its **refractive index**, *n*. For air, n ≈ 1; for water, n ≈ 1.5; and diamond, which has one of the highest values of refractive index, n ≈ 2.5. This means that light passing through a diamond is highly refracted—this is one of the reasons that diamonds sparkle.

The amount of refraction can be calculated using Snell's law. This law states:

$$n_1 \sin \theta_1 = n_2 \sin \theta_2$$

A medium's refractive index can be calculated by measuring the angle of incidence and the refracted angle. This is simplified if the incident ray is traveling through air, as n1 = 1. The amount of diffraction also depends on the wavelength of the incident beam.

Total internal reflection

If a light ray is passing from glass into air, its angle of refraction θ_r is greater than its angle of incidence θ_i. If the angle of incidence is large enough, the refracted ray is at an angle of 90° to the normal. This angle of incidence is referred to as the critical angle, c.

Using Snell's law and substituting 90° as the angle of refraction and n1 = 1.5 for glass and n2 = 1 for air, the equation becomes:

$$1.5 \sin \theta_c = 1$$

Giving a value of

$$\theta_c = \sin^{-1} \frac{1}{1.5} \approx 41.8^0$$

At this angle of incidence, no light emerges from the glass as the refracted ray travels along the boundary. For any angles larger than this critical angle, the light ray is entirely reflected back into the glass. This is called **total internal reflection**.

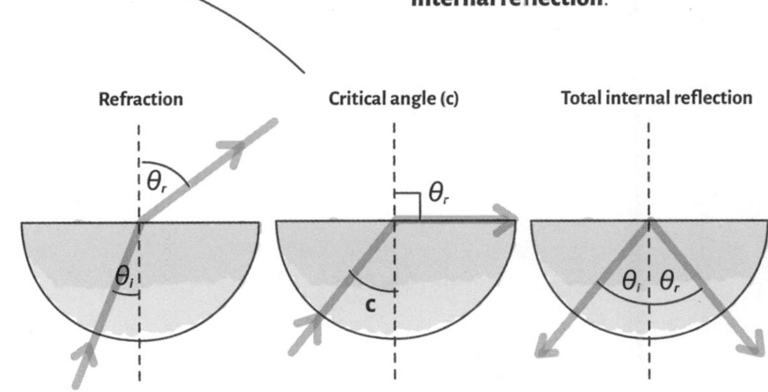

FIBER OPTICS

Total internal reflection is a highly useful property of light within a thin glass tube or capillary. It is the basic mechanism behind fiber optics and, until recently, was used primarily for creating pretty lighting. Now it drives the Internet.

The use of **fiber-optic cables** has revolutionized communications and the Internet, meaning that superfast broadband signals can be transferred across large distances.

A fiber-optic cable is formed of a number of glass fibers, each protected by a plastic sleeve, that are grouped together in a protective plastic layer.

These cables connect our world across vast distances as never before, allowing information to be shared almost instantaneously with almost no loss of signal from one end to the other.

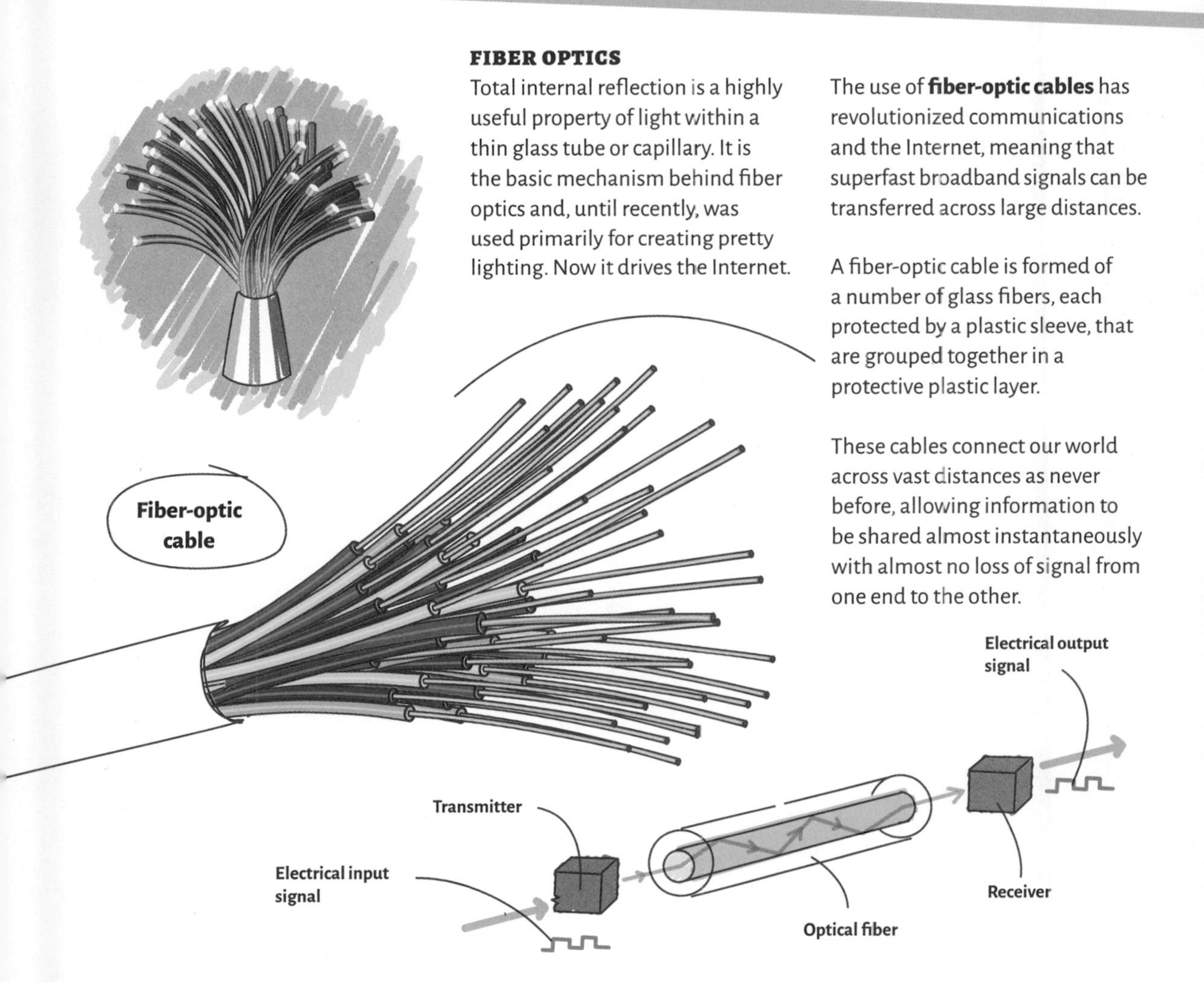

THE SCIENCE OF OPTICS

The science of optics is a powerful tool. It has allowed us to create microscopes to observe the microscopic world and to build reflecting optical telescopes constructed of mirrors up to 10 m (more than 30 ft) across to observe distant objects in the universe.

Lenses and mirrors

Lenses use the property of refraction to focus light beams into a point (convex lenses) or disperse them (concave lenses). Lenses can be used to enlarge an object's image, as with a microscope, or they can be used to correct blurred eyesight, by helping the eye's natural lenses to properly focus the light entering them.

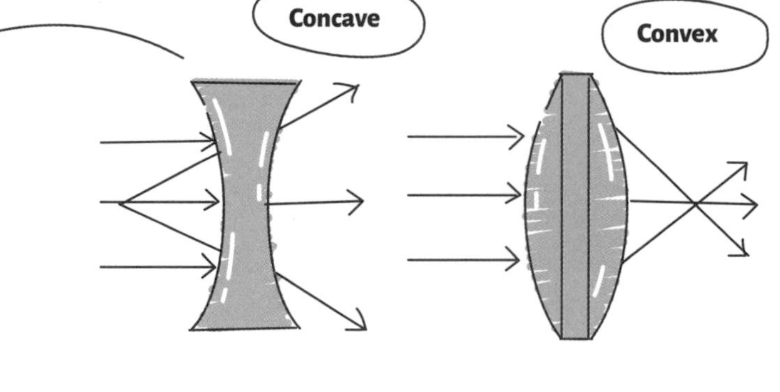

There are two main types of lens: **concave** and **convex**. A concave lens takes a parallel beam of light and disperses it outward, whereas a convex lens focuses a parallel beam into a point. A concave lens widens a parallel beam of light and has the effect of making an image smaller than the object. A convex lens will do the opposite, focusing the light beam into a point, making the image appear larger than the actual object. For example, a fly seen through each type of lens will appear a different size.

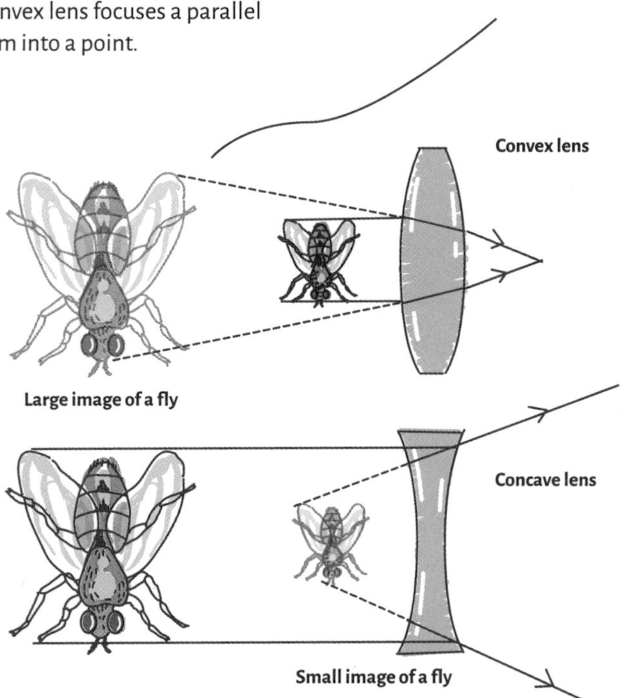

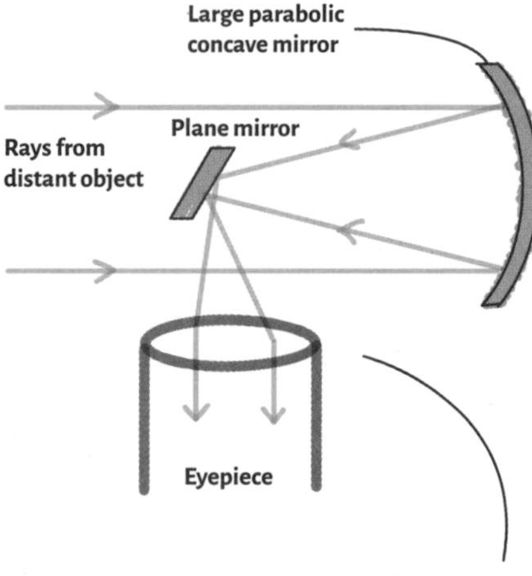

Mirrors can be used in the same way. A **reflecting telescope** uses a curved mirror to enlarge the image of distant objects, such as stars and galaxies. A simple telescope will have one large concave mirror and a small plane mirror that focuses the light into an eyepiece, enlarging the image.

Real and virtual images

Images made by lenses and mirrors are classified as **real** or **virtual**.

A real image is where the light passing through a lens is focused into a point and can be projected onto a screen. An example of a real image is that of a light from a projector in a movie theater. A flashlight shining through tracing paper with an image drawn on it, and focused by a concave lens onto a screen, demonstrates the basic mechanism by which a projector works. The image projected onto the screen is real and **inverted** (upside down).

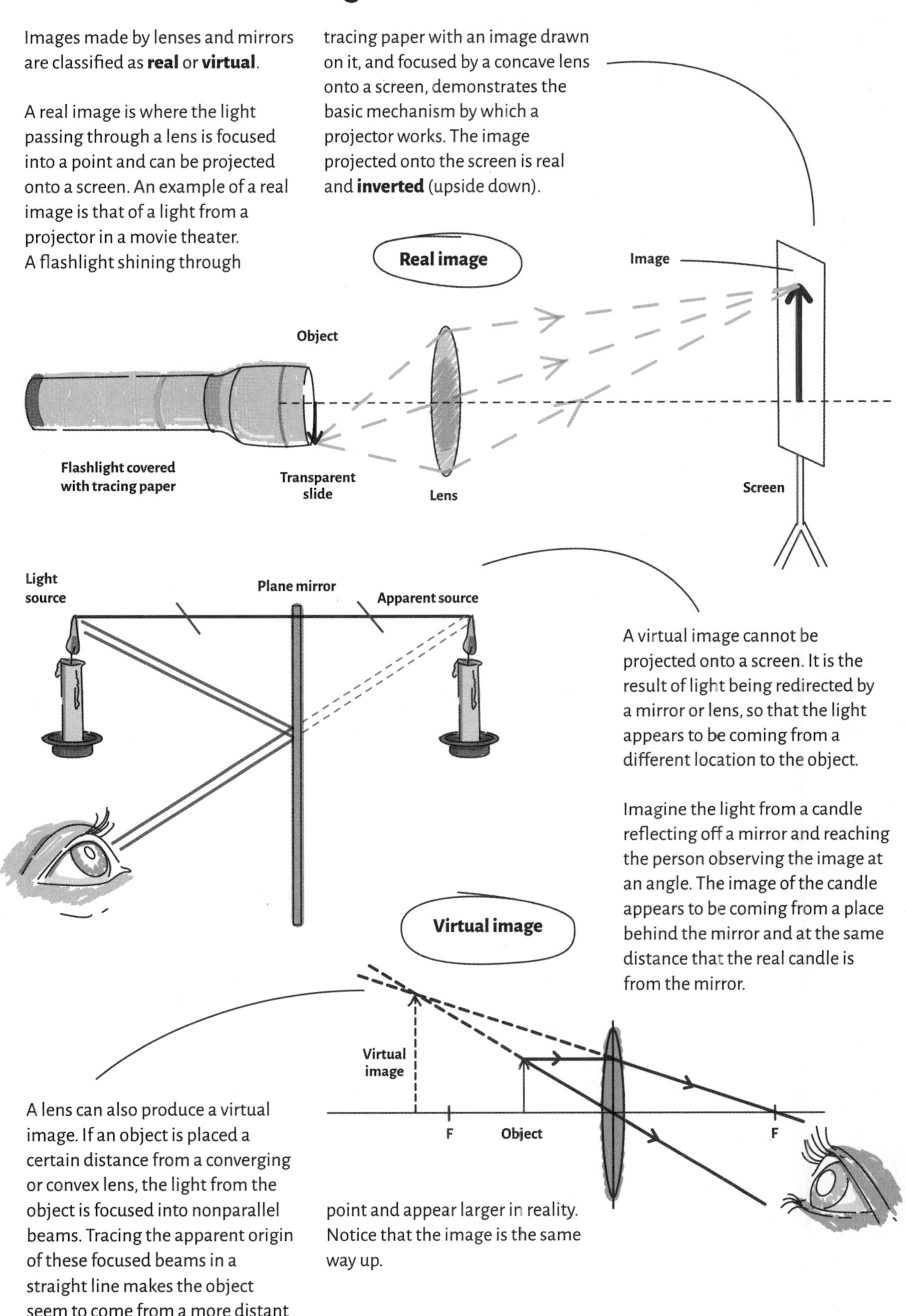

A virtual image cannot be projected onto a screen. It is the result of light being redirected by a mirror or lens, so that the light appears to be coming from a different location to the object.

Imagine the light from a candle reflecting off a mirror and reaching the person observing the image at an angle. The image of the candle appears to be coming from a place behind the mirror and at the same distance that the real candle is from the mirror.

A lens can also produce a virtual image. If an object is placed a certain distance from a converging or convex lens, the light from the object is focused into nonparallel beams. Tracing the apparent origin of these focused beams in a straight line makes the object seem to come from a more distant point and appear larger in reality. Notice that the image is the same way up.

THE BEHAVIOR OF LIGHT

Light from the sun is a series of different frequencies of electromagnetic photons, which consist of crossed electric and magnetic fields. Each photon's electric and magnetic fields will be oriented in a different way as they travel through space. When they enter our atmosphere, many photons are scattered, and their orientation changes.

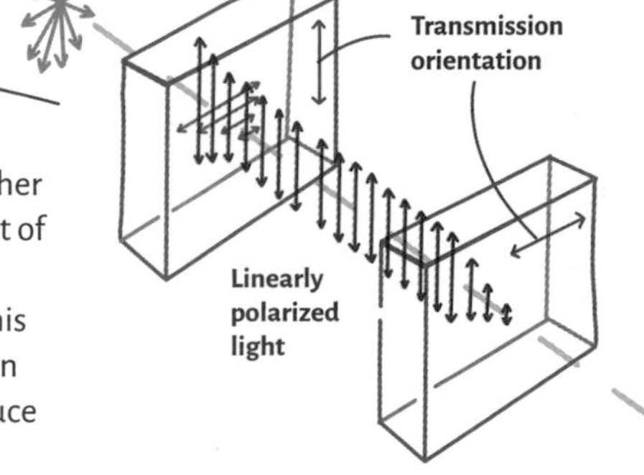

Polarization

Polarization of light is the name given to the process that singles out specific photons that have their electric and magnetic fields aligned in a certain direction. A **polaroid lens** (or polarizer) only allows light of a certain orientation to pass through, blocking all other photons. This also has the effect of reducing the amount of light passing through the lens. For this reason, polaroid lenses are often used in sunglasses, as they reduce the amount of glare.

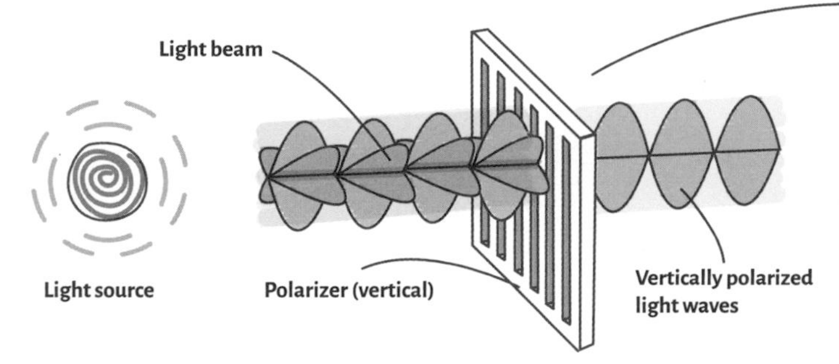

If a combination of two polarizers is used, with one aligned to transmit vertically polarized light and the other to transmit horizontally polarized light, there will be no transmission of light through the second polarizer.

An example is 3D glasses for movie theaters. To us, images appear 3D because each eye sees the world from a slightly different viewpoint. 3D movie theaters synthesize this effect by filtering out vertically and horizontally polarized images, allowing each eye to see the screen from a different viewpoint. One eye will see one image, the other will see a slightly different image, just like in real life.

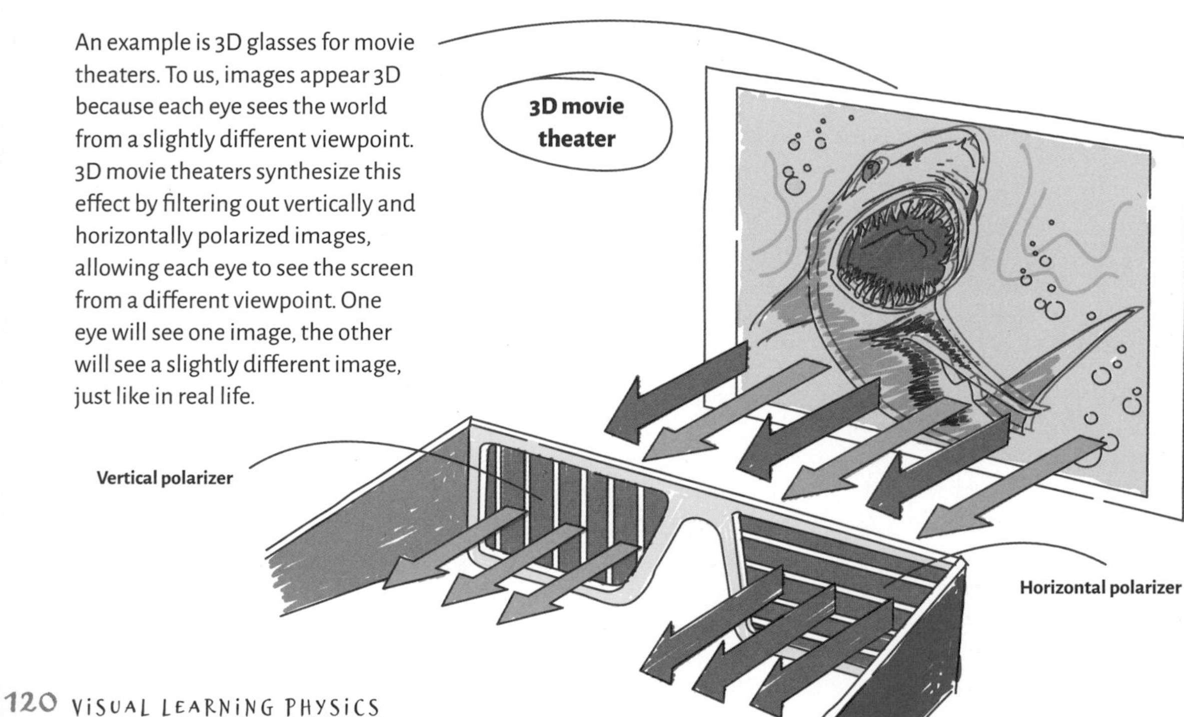

Scattering

Water vapor in clouds absorbs and scatters light in all directions. From a plane window, clouds appear white and fluffy, their bubble-like structure reflecting most wavelengths back into the upper atmosphere. When viewed from the ground, the same clouds have effectively stopped most of the light from the sun. On a cloudy day, some still reaches the Earth's surface, but much has been scattered back into space by reflection.

White light is scattered in all directions.

Some light penetrates to cloud base.

Sunlight frequencies

Many of the frequencies of sunlight are absorbed by different elements and molecules in our atmosphere, which are often redirected at different wavelengths, such as infrared, as the particles warm up. Our atmosphere protects us against much of the harmful radiation, such as UV and X-rays from other cosmic events.

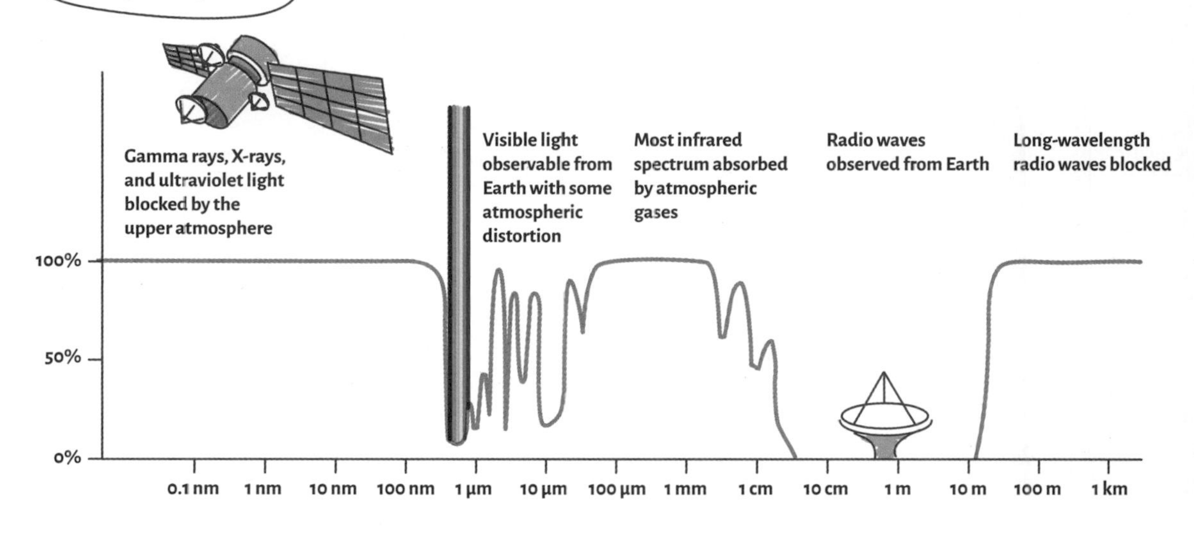

Greenhouse gases

Light from the sun heats the land and oceans and is stored as heat, which is then redirected back into the atmosphere as **infrared radiation (IR)**. The wavelength of this infrared radiation means that certain gases (such as carbon dioxide) prevent it from escaping into space as they heat up. These are called greenhouse gases, and the amount of these gases in our atmosphere controls how much heat escapes via radiation.

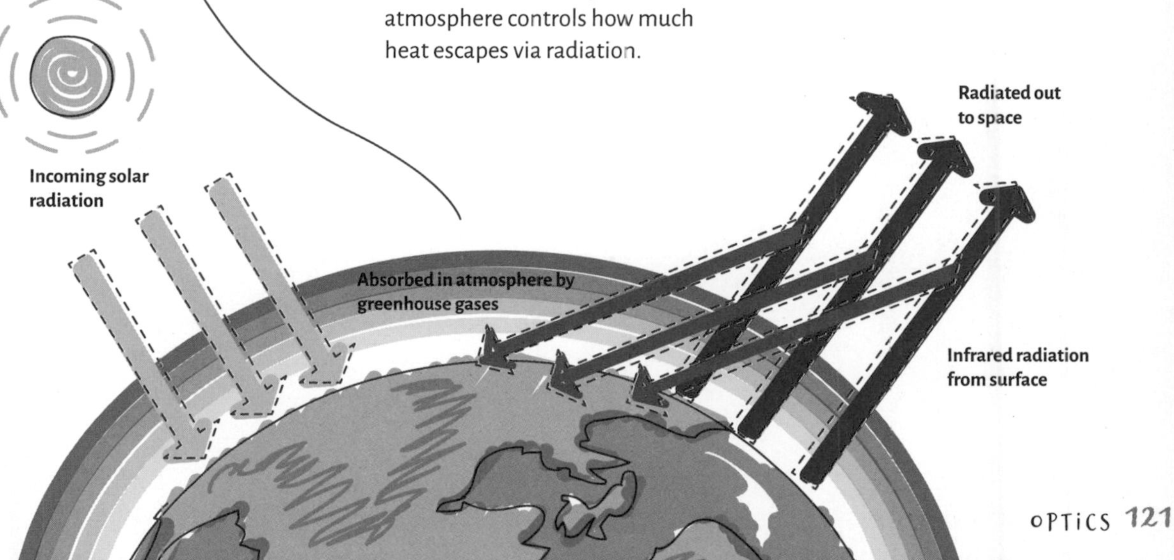

The colors of the sky

As seen in the previous chapter, scattering of solar radiation is the reason you see different colors. White light from the sun, made up of all colors, strikes an object's surface, and certain frequencies are absorbed and others reflected.

The sky appears blue as a consequence of the scattering of sunlight in the atmosphere. Due to the makeup of the atmosphere, blue light is scattered while all other visible frequencies (of different colors) pass straight through, and so the sky appears blue.

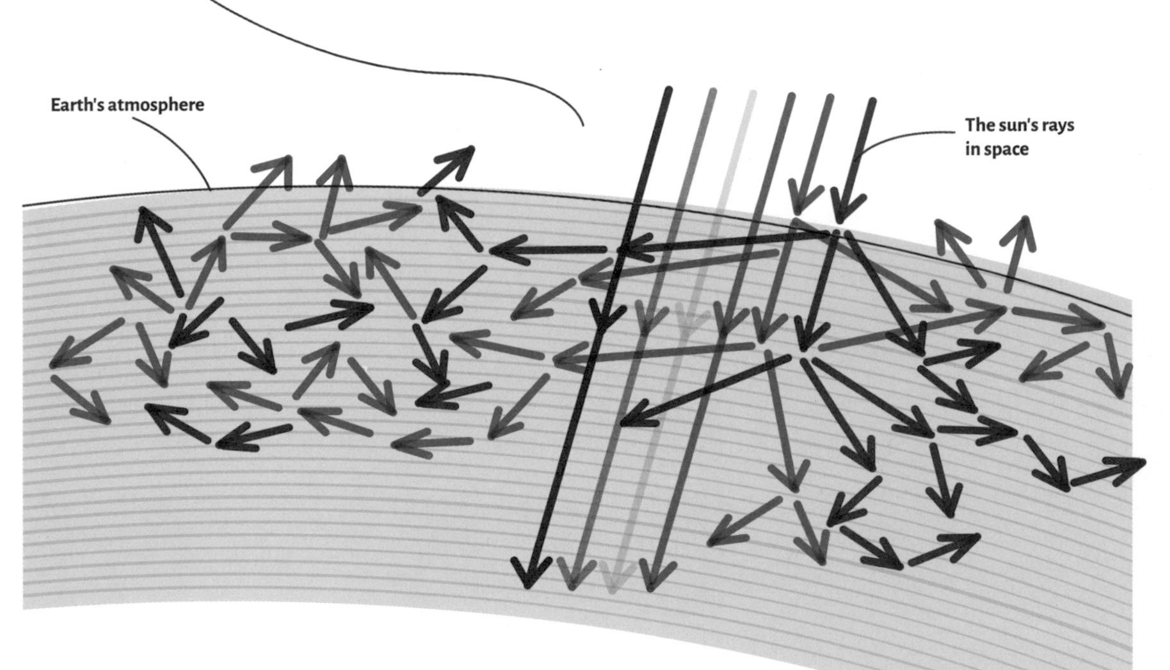

RED SKY AT NIGHT

A red sky appears when tiny dust molecules are present in the atmosphere (usually due to high pressure). These scatter all wavelengths of light except red, which passes straight through. This usually occurs at sunrise or sunset, as the dust molecules lie closer to the ground and the sunlight travels a larger distance through the atmosphere. When the sun is low in the sky, the light interacts with a larger number of dust particles, therefore scattering more light of wavelengths other than red—the observed light is made up of predominantly lower red frequencies, causing a beautiful sunset.

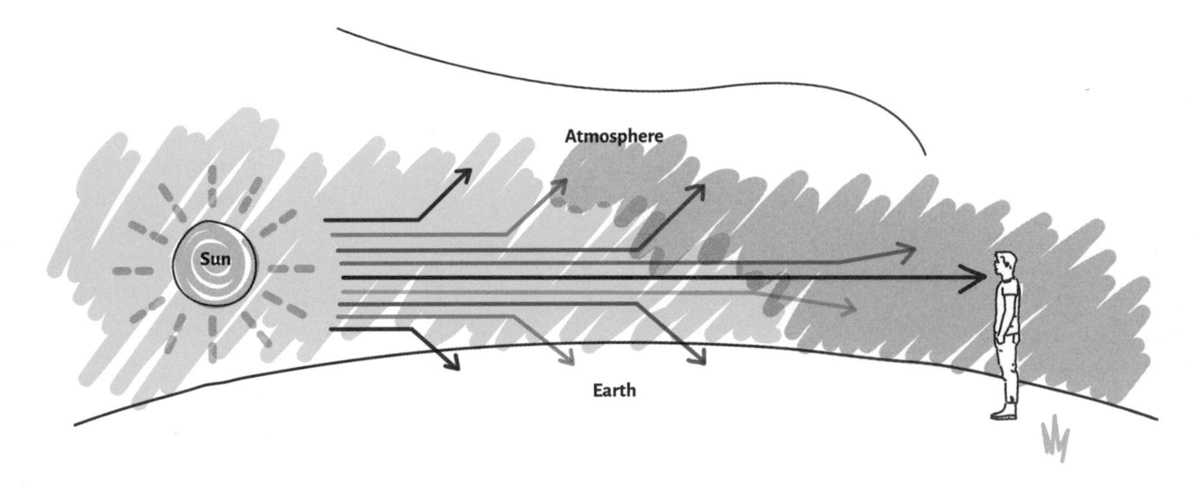

INTERFERENCE AND INTERFEROMETRY

The phenomenon of wave interference can produce different observable effects. It can be used to determine the wavelength of light or even utilized to detect tiny movements, using the Michelson interferometer.

Thin-film interference

If a surface is covered by a thin transparent film, such as detergent or oil, it can produce a variety of colors, which may appear to shift and change. The process by which this happens is called **thin-film interference**.

As the thickness of the film varies slightly, the constructive interference changes with the wavelength and color of light. This produces a rainbow-like pattern across the surface of the water.

Imagine the surface of a pond on which there has been an oil spill. The oil on the water's surface will spread out to produce a very thin film. When light strikes this surface, some will pass through it and reach the opposite side of the film, where it is reflected, whereas the remainder of the photons are immediately reflected from the top surface of the oil.

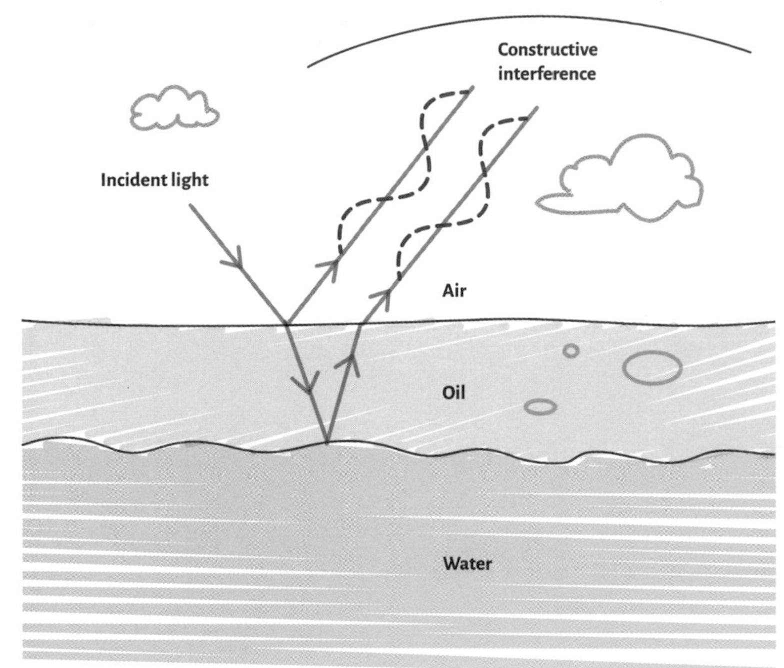

Under the right conditions, the two waves will leave the oil's surface exactly in phase, causing **constructive interference** and therefore producing a larger amplitude of the observed wave appearing as brighter colors.

The thickness of the oil film is critical here, with specific thicknesses enhancing different colors of different wavelengths. The thickness of the oil layer must be an exact multiple of the wavelength of the enhanced color. This ensures the two waves (the wave reflected from the top surface and the wave reflected from the bottom surface) are an exact whole number of wavelengths apart.

The Michelson interferometer

Albert Abraham Michelson (1852–1931) was a German-born American physicist who established that the speed of light was a fundamental constant in a vacuum. He invented the **Michelson interferometer**, which uses the principle of wave interference to measure very small changes in length.

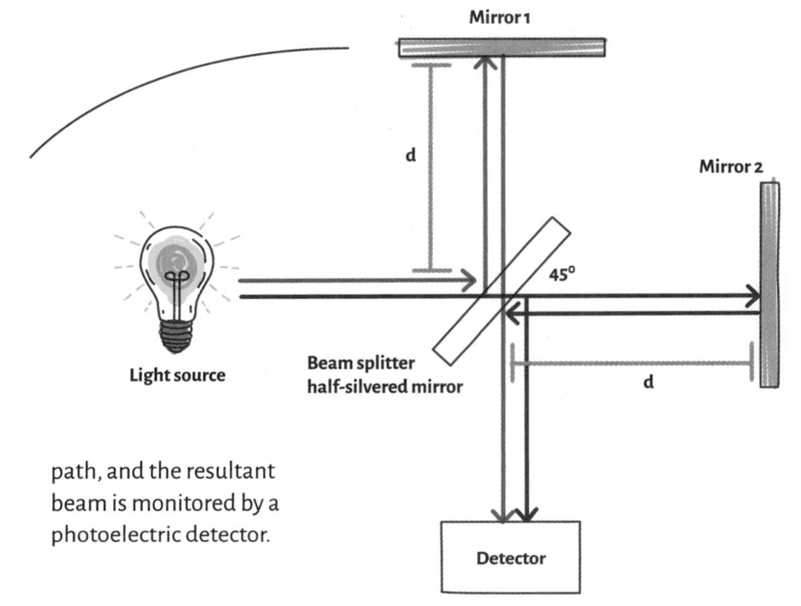

A **monochromatic** (one-color) laser beam can be split in two by a semitransparent mirror to travel along two separate arms with a mirror at each end. Each beam is reflected back along its original path, and the resultant beam is monitored by a photoelectric detector.

THE WAVE DETECTOR

One arm is movable and can change in length. If the arm lengths are identical, the beams arrive in phase and produce constructive interference. If the distance of one arm changes, the detector will record a change in light intensity as the two beams arrive out of phase. Knowing the wavelength of the laser, the tiny change in length can be calculated by the light intensity. The sensitivity of the Michelson interferometer is in the order of nanometers (10^{-9} m).

A very large Michelson interferometer is used at Caltech's **LIGO** (Laser Interferometer Gravitational-Wave Observatory) to detect tiny variations in space-time due to passing gravitational waves caused by the merging of black holes. The detector is sensitive to one ten-thousandth of the diameter of a proton.

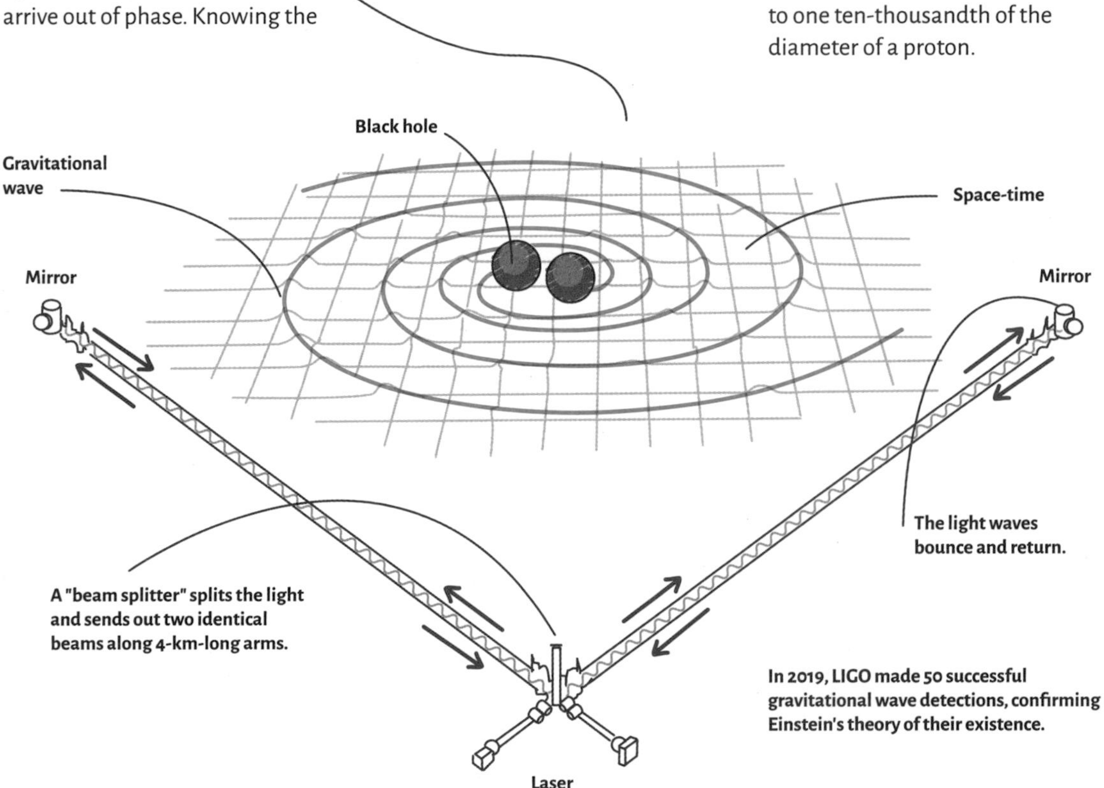

Two-slit interference

The **double-slit experiment** was first conducted by the British scientist Thomas Young (1773–1829) in the early nineteenth century to calculate light wavelengths.

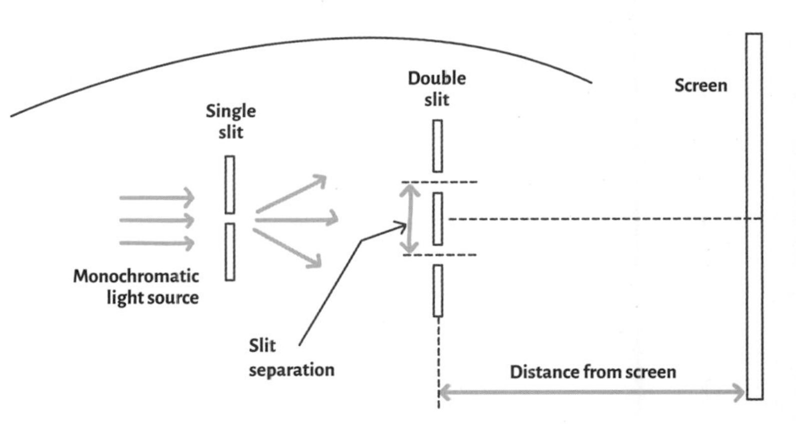

It uses a monochromatic light source that is shone through a single vertical slit, creating a straight light beam. This beam is incident upon two vertical slits separated by a small distance, *d*. Two coherent light beams emerge from the slits, which are diffracted through each small gap, creating two circular wave fronts.

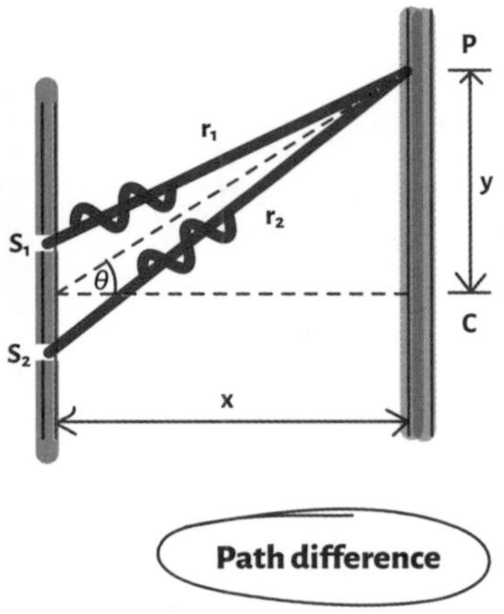

Path difference

As these wave fronts coincide, they interfere constructively and destructively, creating regions of brightness and darkness, respectively. The resultant pattern is then projected onto a screen, revealing a series of bright and dark stripes known as an **interference pattern**.

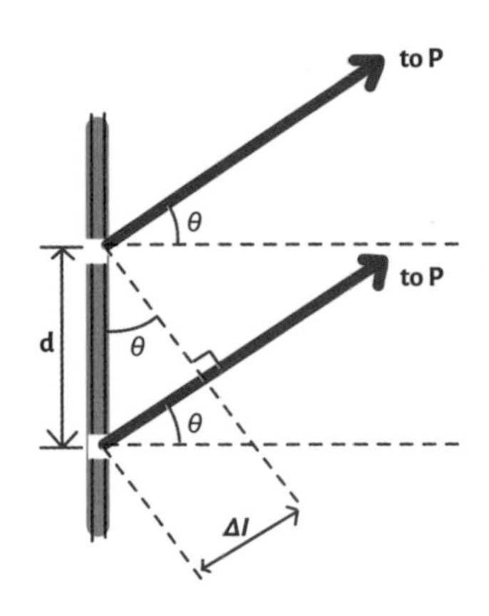

The two beams emerging from each slit will arrive in phase with each other at the screen (labeled P), causing a bright bar if the difference in path lengths (Δl) of the beams is an exact whole number of wavelengths.

Screen output

This will occur at different points along the screen, with each bright bar decreasing in intensity the farther it is from the central point.

The first bright bar at either side of the central maximum coincides with exactly one wavelength path difference. At the second bright region, the waves arrive two wavelengths apart, and so on. These are called *n*=1, *n*=2, *n*=3, etc., indicating by how many wavelengths' separation the waves arrive at the screen.

Mathematically, it can be shown that:

$$d\sin\theta = n\lambda$$

where *d* is the known slit separation, *n* is the number of the bright bar from the central point on the screen, and θ is the angle measure from the center line to the light path producing the bright bar. Using this formula and measuring the value of θ, the wavelength of the light beam can be calculated.

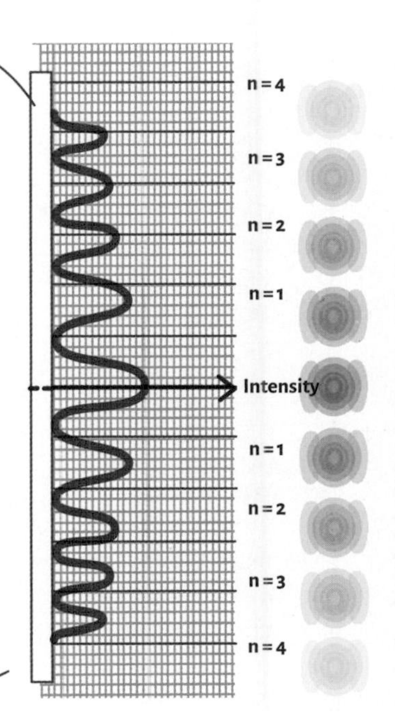

OPTICS

REFLECTION

INCIDENT RAY

A light beam that strikes a reflective surface.

ANGLE OF INCIDENCE

The angle that an incident ray makes with the normal.

REFLECTED RAY

A light beam reflected back from a surface.

ANGLE OF REFLECTION

The angle that a reflected ray makes with the normal.

LAWS OF REFLECTION

1. The incident ray, the reflected ray, and the normal ray are all in the same plane.
2. The angle of incidence equals the angle of reflection.

INTERFERENCE

THIN-FILM INTERFERENCE

Constructive interference caused by light beams reflecting from the outer and inner surface of a thin film, such as oil.

INTERFEROMETRY

The use of wave interference to measure distance and collect data.

MICHELSON INTERFEROMETER

Device that splits a light beam along two arms to create interference when they reflect back.

TWO-SLIT INTERFERENCE

Created when two coherent light beams interfere, causing a pattern of light and dark bars.

THE BEHAVIOR OF LIGHT

POLARIZATION

Selection of one plane of oscillation in a light source to the exclusion of others.

REFRACTION

SNELL'S LAW

Used to find the angle of refraction made when a beam passes between two transparent media with different refractive indices.

$$n_1 \sin \theta_1 = n_2 \sin \theta_2$$

ANGLE OF REFRACTION

The angle that a refracted ray makes with the normal at the point where two media meet.

REFRACTIVE INDEX

Measures the optical density, *n*, of transparent media. Air is n = 1; diamond is n = 2.5.

TOTAL INTERNAL REFLECTION

Light rays traveling through glass that hit the boundary at more than the critical angle are reflected back into the glass.

CRITICAL ANGLE

Reached when a light ray traveling from a glass to air has an angle of refraction of 90° to the normal.

FIBER-OPTIC CABLE

Uses total internal reflection within glass tubes to transfer superfast broadband signals.

MIRRORS AND LENSES

REAL IMAGE

Can be projected directly onto a screen and appears upside down.

VIRTUAL IMAGE

Formed when light rays are redirected and the object appears to be elsewhere; the image remains upright.

SCATTERING

A light beam can be reflected by an object in all directions, and be absorbed and redirected, sometimes at different wavelengths.

MIRRORS

CONCAVE MIRRORS

Used to focus light; give an inverted image; used in large reflecting telescopes.

CONVEX MIRRORS

Provide a wide field of view and an upright image; used for wing mirrors on cars.

LENSES

CONCAVE LENSES

Disperse light waves.

CONVEX LENSES

Focus light waves to a point.

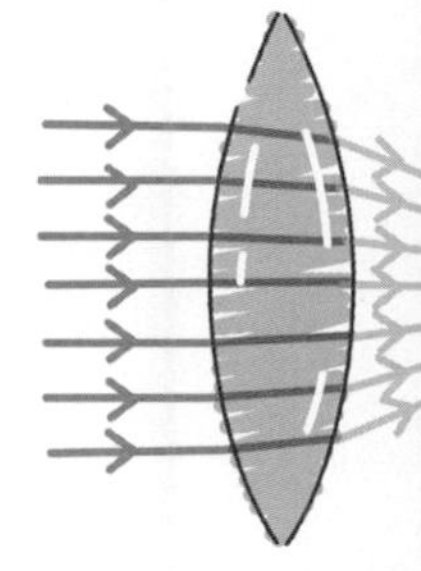

CHAPTER 10

THERMODYNAMICS

Thermodynamics is the branch of physics that studies the transfer of heat energy within a system via mechanical work, radiation, and conduction. All particles within a system contain kinetic energy by their movement as vibrations (in solids) and velocities (in liquids and gases). Solids have vibrating atoms that oscillate around a point and transfer energy to neighboring particles, whereas liquids and gases contain particles that are free to move. Energy is transferred around these systems as heat transfer.

TEMPERATURE

An arbitrary measure of the amount of energy that exists in a gas, liquid, or solid is known as its temperature. It measures the average amount of kinetic energy within the medium, defined as the movement of the particles. Particles can oscillate back and forth in a solid, such as a metal bar, or they can move from one place to another within a liquid, such as water or a gas.

Oscillations are cyclical in nature, and the variation is repeated over fixed time intervals, but the maximum magnitude (amplitude) of the variation may change over time as the system loses (or gains) energy. A simple harmonic oscillator will experience a restoring force directed back to its equilibrium position, the magnitude of which directly depends on the size of that displacement.

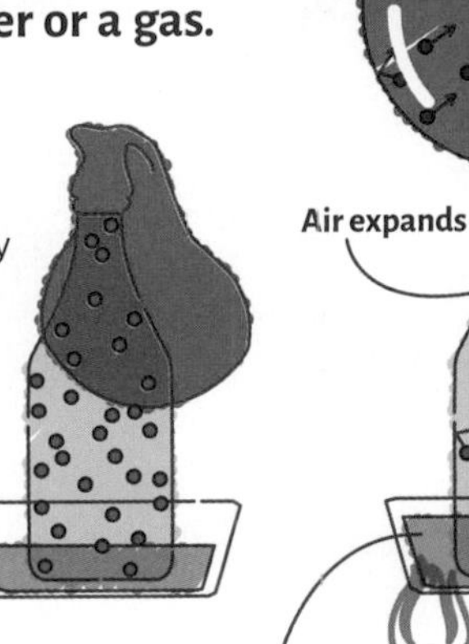

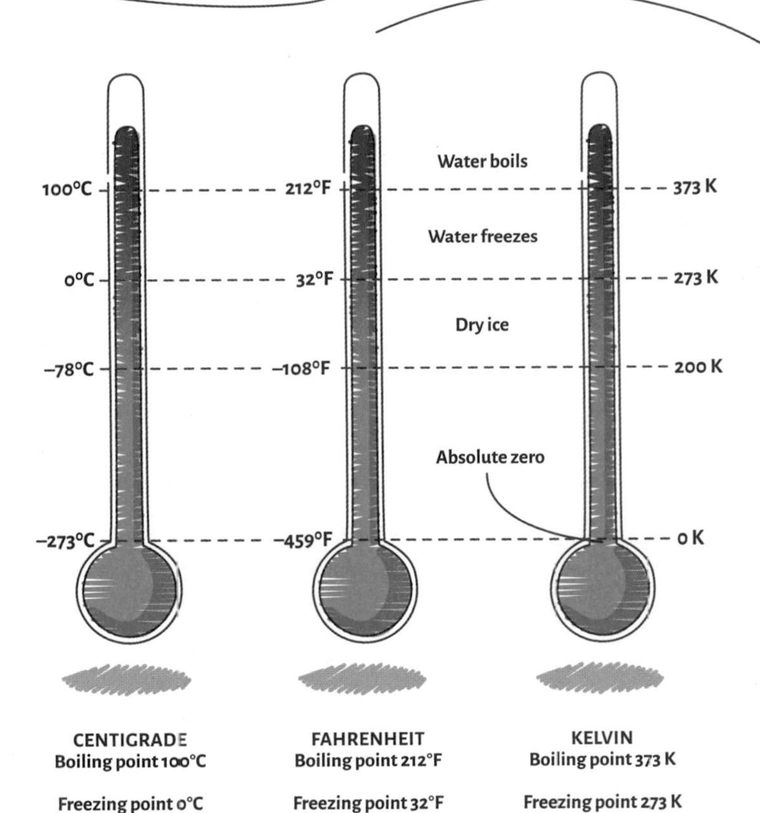

Temperature is usually measured in degrees **Fahrenheit** (F), degrees **Celsius** (C), or by physicists in degrees **Kelvin** (K).

Degrees Celsius and Fahrenheit are defined in relation to the freezing point and boiling point of water at one atmosphere, N/m^2 (defined as a **Pascal**). When water freezes, it turns from a liquid into a solid as the kinetic energy of the molecules decreases and the atoms are no longer free to move around.

Water boils when the kinetic energy becomes sufficient to break the intermolecular bonds and it becomes gaseous.

The Kelvin scale is based on **absolute zero**, the temperature at which a solid has no kinetic energy, so there is no vibration of the atoms within it.

THERMAL ENERGY TRANSFER

As matter is exposed to thermal energy (heat), it will respond in a number of ways: Its temperature will increase as a response to an increase in kinetic energy; it may also change state from a solid to a liquid, a liquid to a solid, a solid to a gas, or a gas to a plasma. Physically, its volume and pressure may also change.

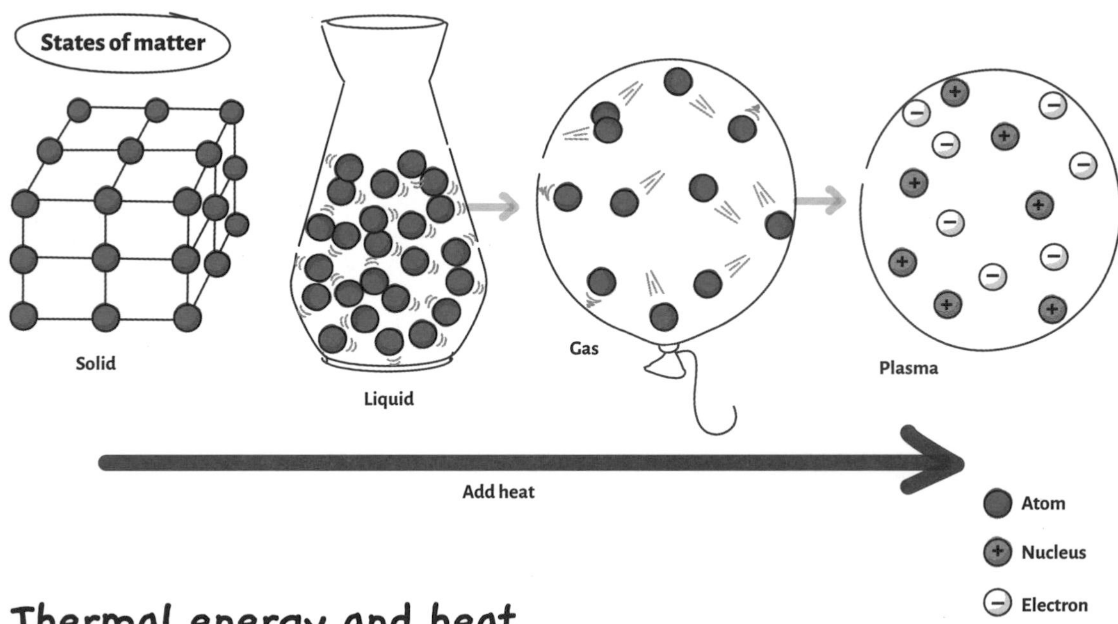

Thermal energy and heat

The amount of energy (measured in joules) that exists in a system is called the **thermal energy**. It is a measure of the kinetic energy of the particles moving in the system.

Heat defines the energy being transferred from a hot object to a colder one. Consider a tea kettle. It is filled with cold water, normally around 59°F/15°C, but to make good tea, you need water at boiling point. The kettle in this example is an electrical device that heats water by the heating of an element within it. As it comes into contact with the hot surface of the element, the water gains kinetic energy, increasing the movement of the molecules (this heat transfer is called **conduction**). The kinetic energy increases as heat is transferred from the hot element to the water, and the water boils. This is a direct result of two separate bodies (the water and the element) becoming thermally in balance as heat from the element transfers its energy to the water. Energy is also wasted as it escapes into the surroundings.

Making tea

Electrical energy

Heat energy

Sound energy

Thermal expansion

Liquids and solids do not expand by much as heat is transferred to them, so their temperature rises with little change in volume. For example, providing heat to an iron bar will cause a rise in the kinetic energy of atoms and the bar will get very hot but its volume will barely increase.

As gases absorb heat and the kinetic energy of the particles increases, they experience much greater movement, as they are not constrained like solids.

Expanding tire

F

Warming

If the volume being heated is free to expand, such as in an expanding car engine piston, the volume increases. Liquid fuel is lit by a spark plug, which causes hot gas to expand rapidly, pushing the piston outward.

If the volume of the gas is fixed, such as in a car tire, the pressure increases. This can happen when a car skids when braking, causing friction to heat the tires.

Piston

Exhaust gas

Expanding gas

Spark to light gas

Rotation

At the start of a squash game, the ball is easily squashed, as there is only a small gas pressure inside. It doesn't bounce much. As it is hit, the kinetic energy from the impact is transferred to the gas molecules inside. They move faster, heat up, and expand to exert greater force inside the ball. The surface becomes more solid and less compressible, so the ball bounces much higher.

Bouncing squash ball

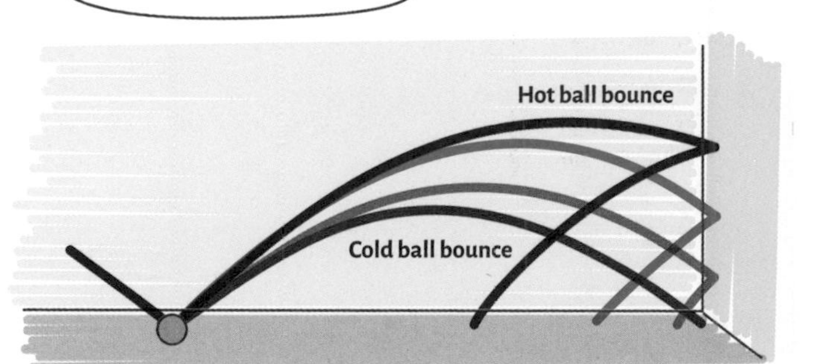

Heating and cooling

Heat can transfer into a body, raising its temperature, or flow out of a body, cooling it. The direction of heat flow will depend on differences in temperature, and its effects are seen in many places, from the simple heating of food to the complex systems that drive our weather.

Objects get hot and cool down depending on how much energy they contain: the **internal energy** of the system. As heat flows into an object, its temperature increases—by how much it increases will depend on the material.

Water, for example, is very good at storing heat and will absorb large amounts of energy while rising only slightly in temperature. This is why the sea takes a long time to warm up in spring and why water is used as a coolant in most engines.

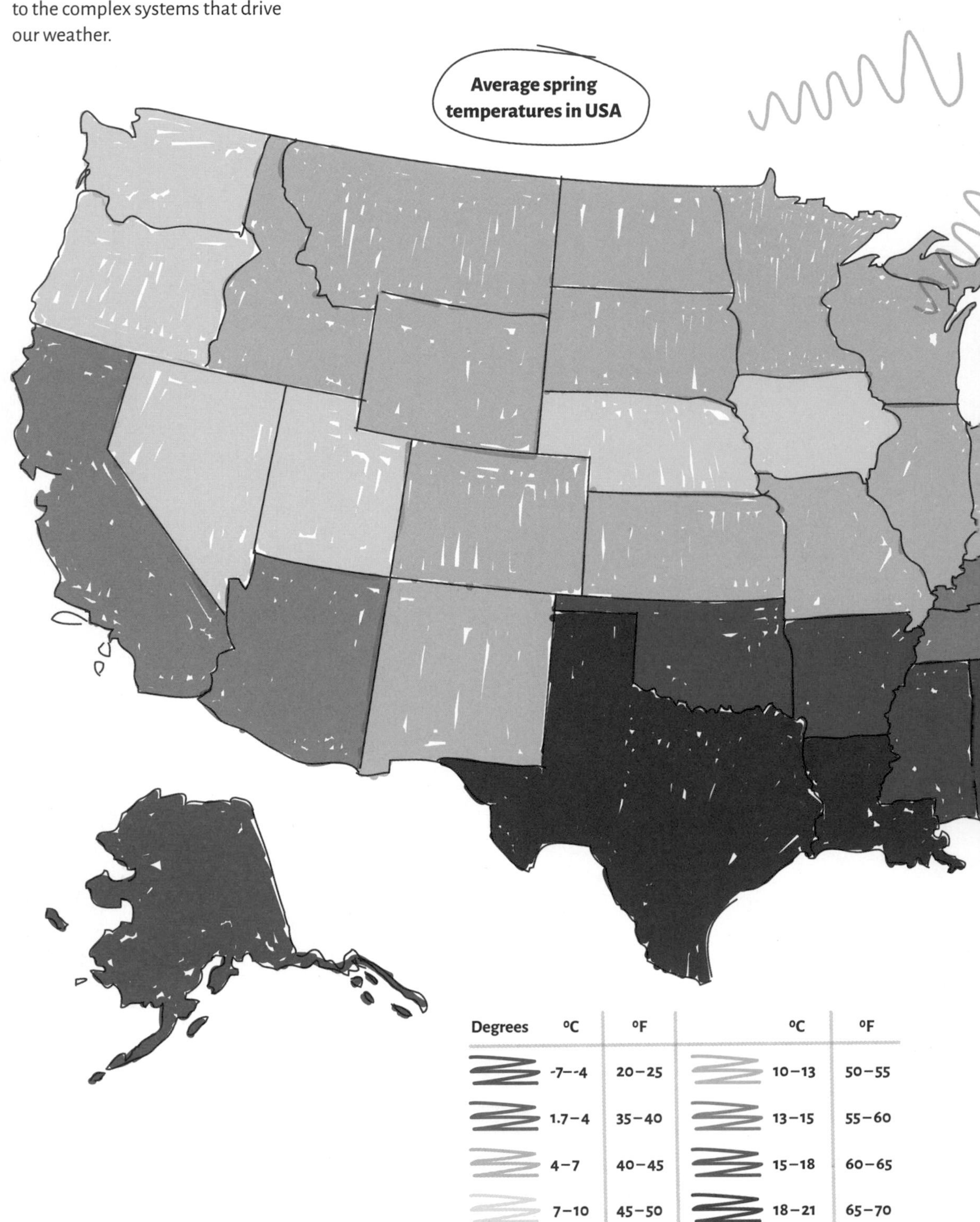

Temperature gradient

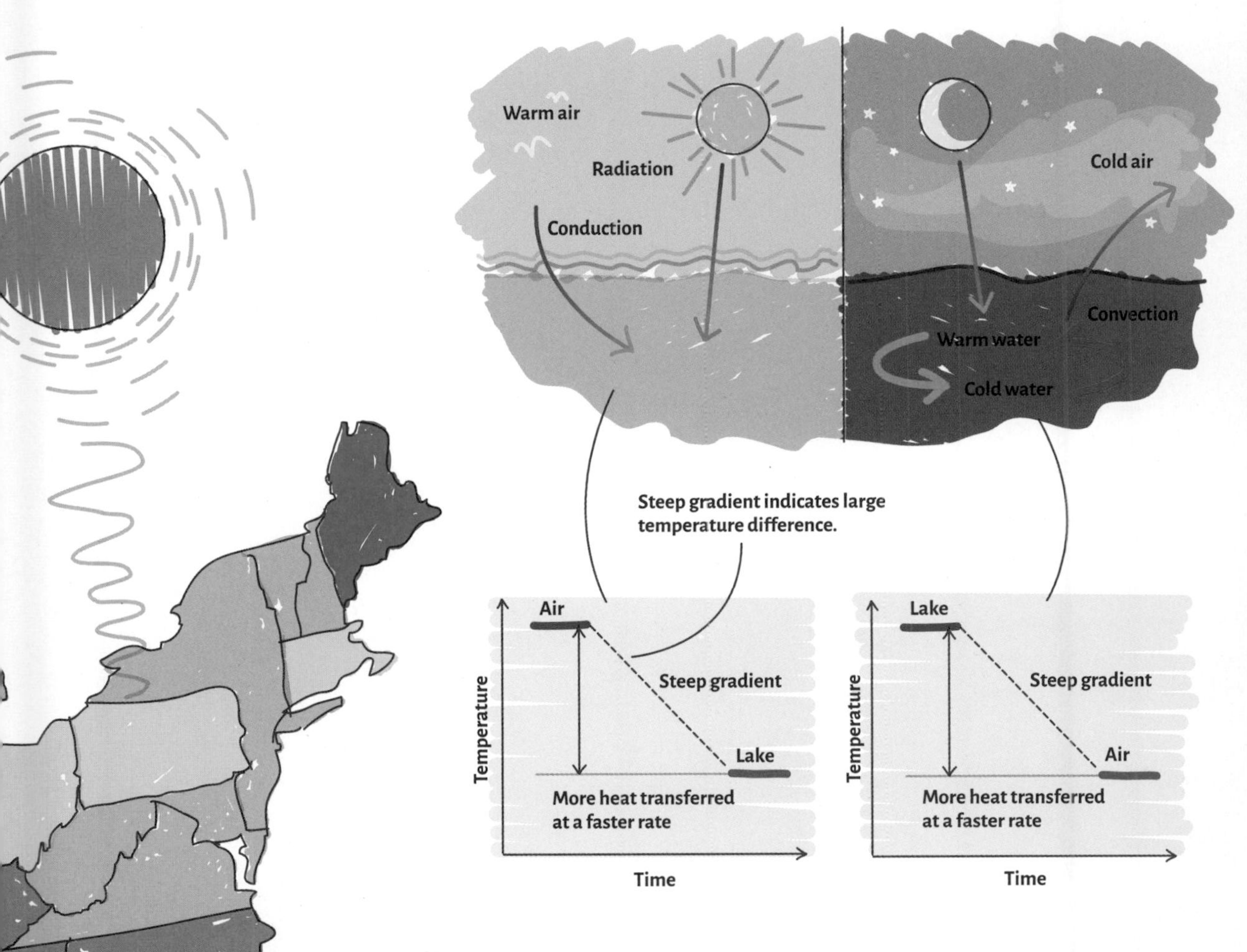

When the cool water in a lake is exposed to regular sunlight in the summer, it absorbs heat, via **radiation** from the sun, and kinetic energy from the warmer surrounding air is transferred to the water's surface via **conduction**, raising the water particles' kinetic energy and temperature. The water layers of different temperatures mix via a process called **convection**.

As the sun goes down, the outside air temperature decreases, dropping to below that of the lake water. The difference in temperatures between the air and water is called the **temperature gradient**. Heat flows from the lake into the night air, cooling down the water. Heat will always flow from a hotter region to a cooler one, until both regions become equal in temperature.

Solar energy is absorbed during the day by large landmasses, with higher levels of solar energy arriving closer to the equator and lower levels at higher latitudes toward the poles.

The map compares spring temperatures across the states of America. It is based on statewide averages from temperatures recorded throughout March, April, and May. They range from a high of 21.1°C (69.9°F) in Florida to a low of -4.1°C (24.7°F) in Alaska. Excluding Hawaii and Alaska, the entire United States spring season averages 11.1°C (52.0°F).

LAWS OF THERMODYNAMICS

There are four laws of thermodynamics. These govern the flow of heat in and out of systems and how the energy transfer changes the dynamics of the system. Let's take a look at the first two laws.

The first law of thermodynamics

The first law of thermodynamics deals with the flow of heat in and out of a system, the internal energy of the system, and any work done by the expansion of the system. Essentially, it is another statement of the law of conservation of energy.

Consider a gas-filled chamber that comprises an airtight cylinder fixed at one end, and a piston free to move at the other. The gas inside the cylinder has thermal energy and exerts a pressure on the cylinder walls and piston. The gas occupies a set volume for a specific temperature, so the system is static.

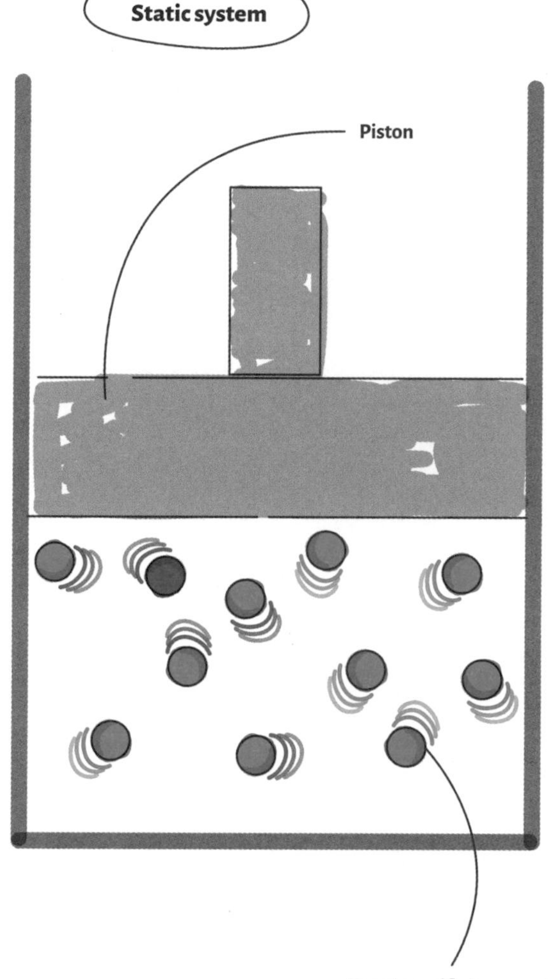

THE FIRST LAW
Heat is a form of energy, and so thermodynamic processes must follow the principle of conservation of energy.

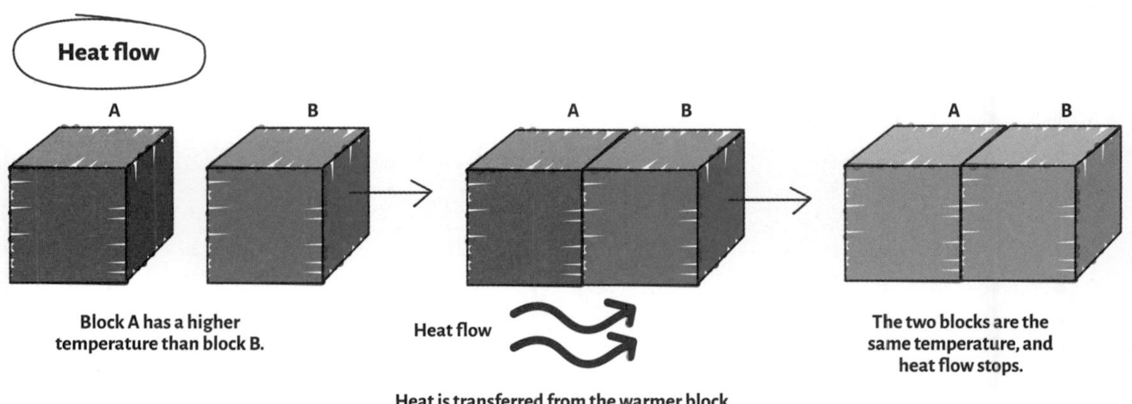

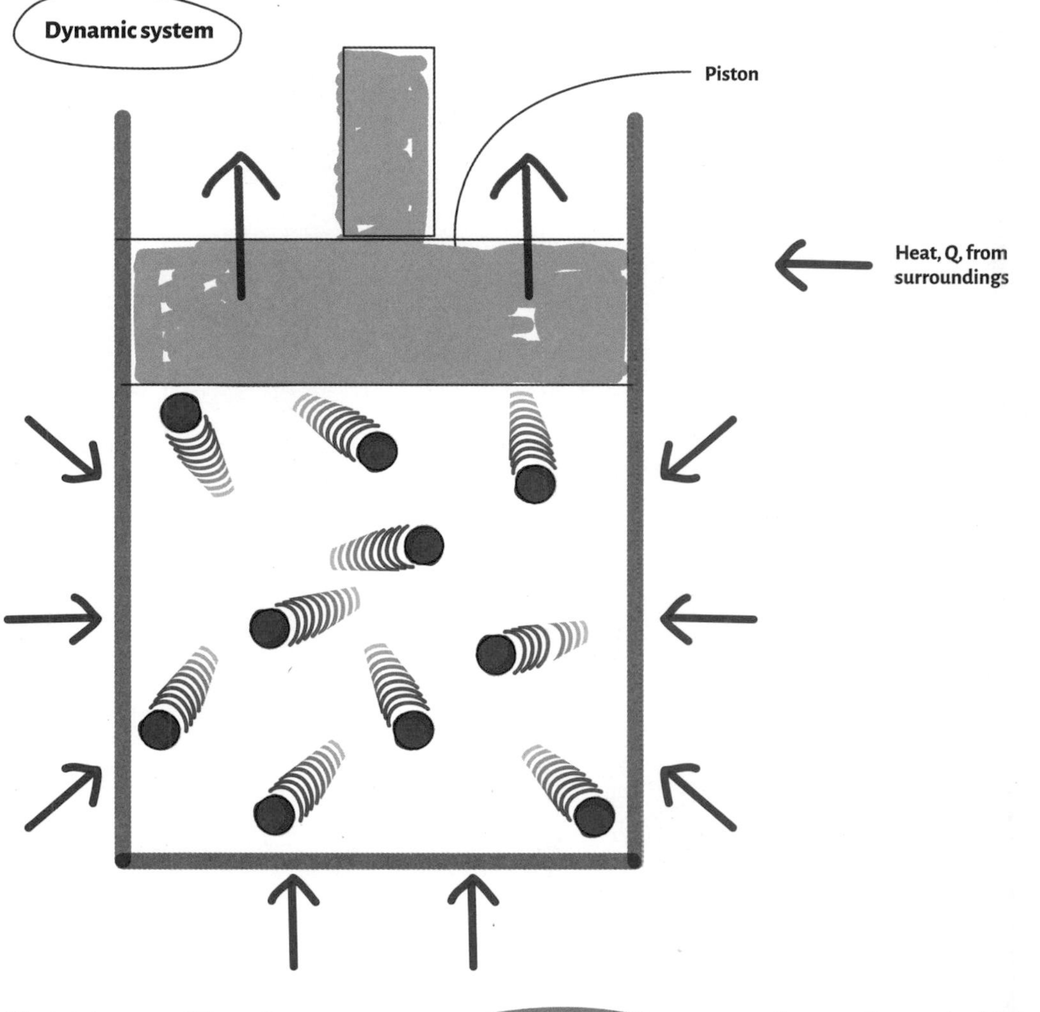

The total energy of the entire system cannot change, due to the fundamental law of energy conservation; therefore, the change in internal energy (ΔU) can be expressed by:

$$\Delta U = Q - W$$

Expressed in words, this formula states that any flow of heat into the system (Q) is converted into both the internal energy (U) heating the gas and mechanical work done (W) in moving the piston.

Work and heat transfer in gases

The piston within the previous system can move in both directions: Just as heat can flow into the chamber by raising the outside temperature, moving the piston outward due to the increase in gas pressure within the cylinder, the reverse can also be achieved. If mechanical work is done on the piston, compressing the gas in the chamber, work is said to be done on the gas.

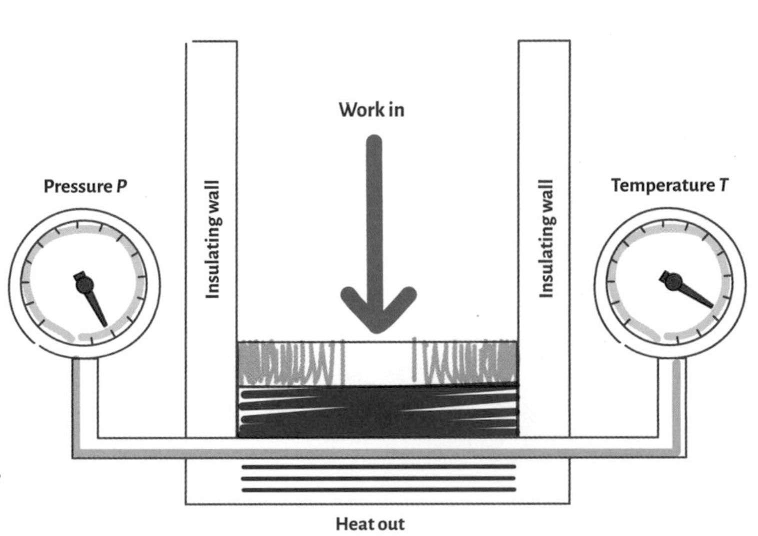

Again, this work is converted directly to raising the kinetic energy of the gas particles, thus raising the gas temperature. As the temperature of the gas in the cylinder rises, it becomes warmer than the outside temperature, creating a temperature gradient and allowing heat from the chamber to flow to the outside surroundings.

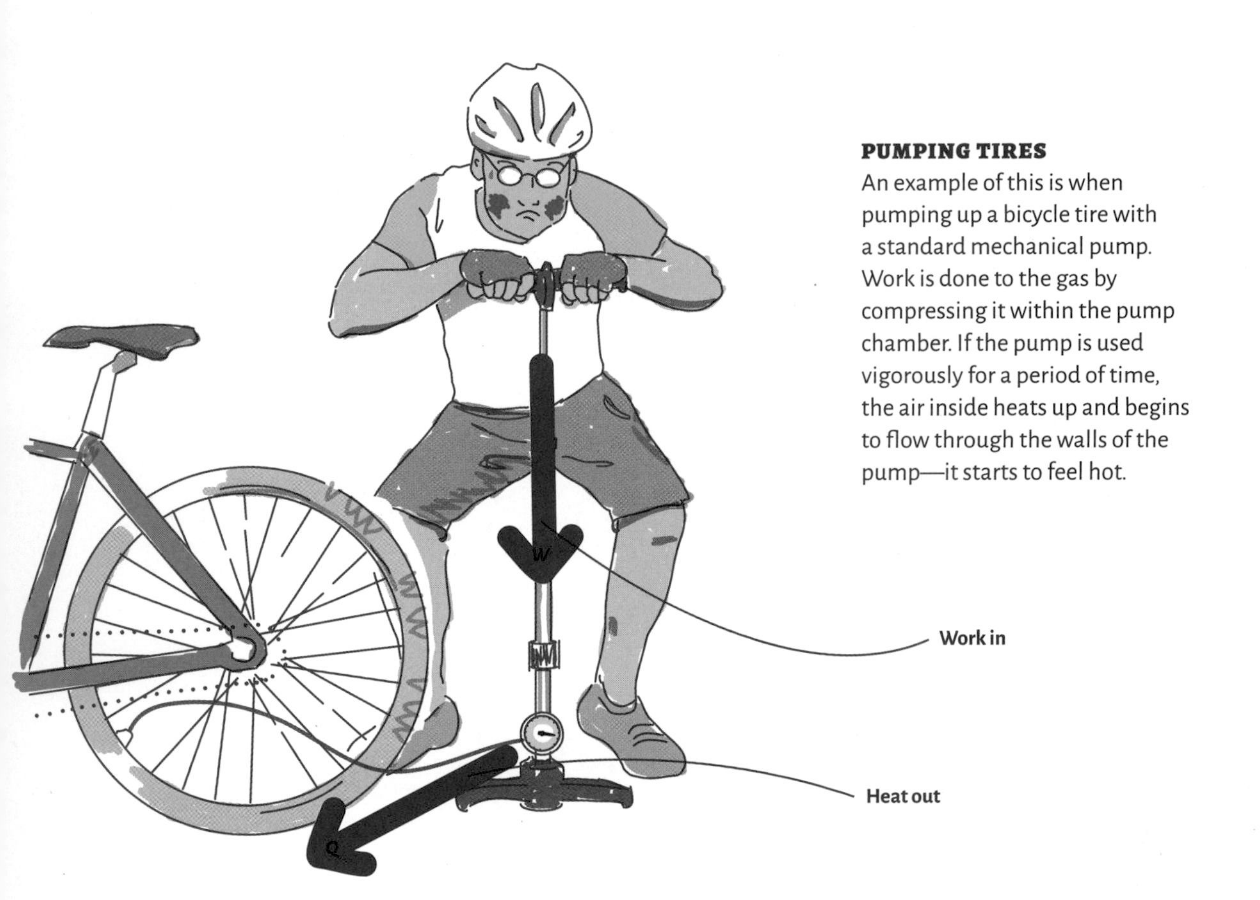

PUMPING TIRES

An example of this is when pumping up a bicycle tire with a standard mechanical pump. Work is done to the gas by compressing it within the pump chamber. If the pump is used vigorously for a period of time, the air inside heats up and begins to flow through the walls of the pump—it starts to feel hot.

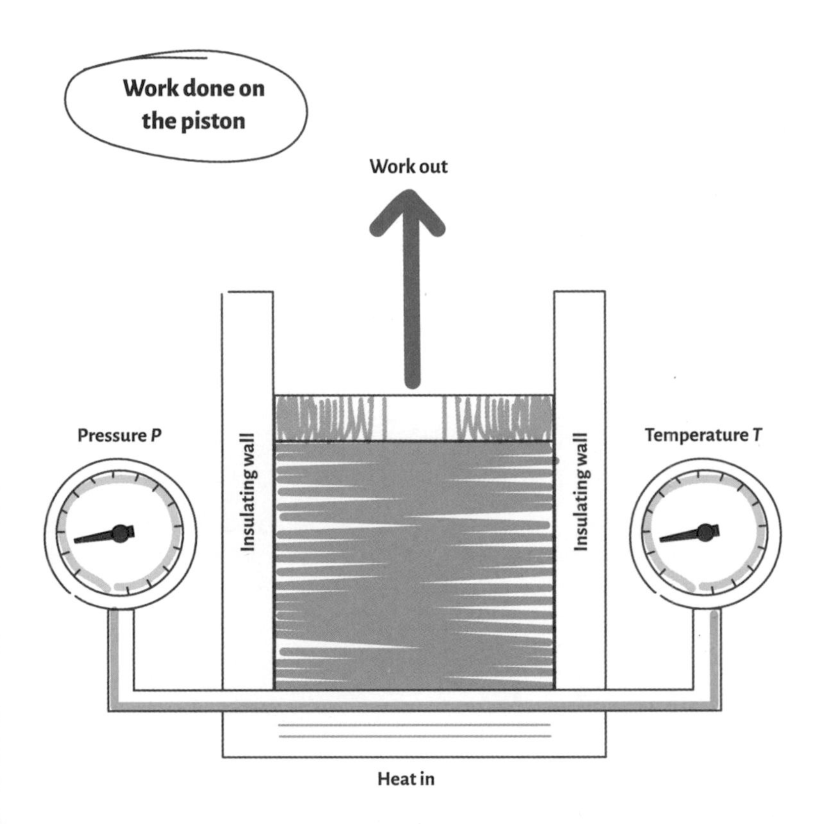

This transfers heat out of the system, so the flow of heat is said to be negative. This produces a revised statement of the first law:

$$\Delta U = W - Q$$

The change in internal energy (ΔU) of the gas is the difference between the work done (W) to the system and the heat leaving it (Q).

The hot gas inside the piston expands rapidly, pushing the piston outward and providing work out of the system as the gas cools and decreases in pressure while increasing in volume. This is how a piston in a car engine works.

Thermodynamics in biology

Thermodynamics can also apply to biological and plant-based systems. Energy is stored in humans when chemical energy in food is released. Work done by our muscles and the heat that generates is lost—the difference between food in and energy out is the change in internal energy (ΔU).

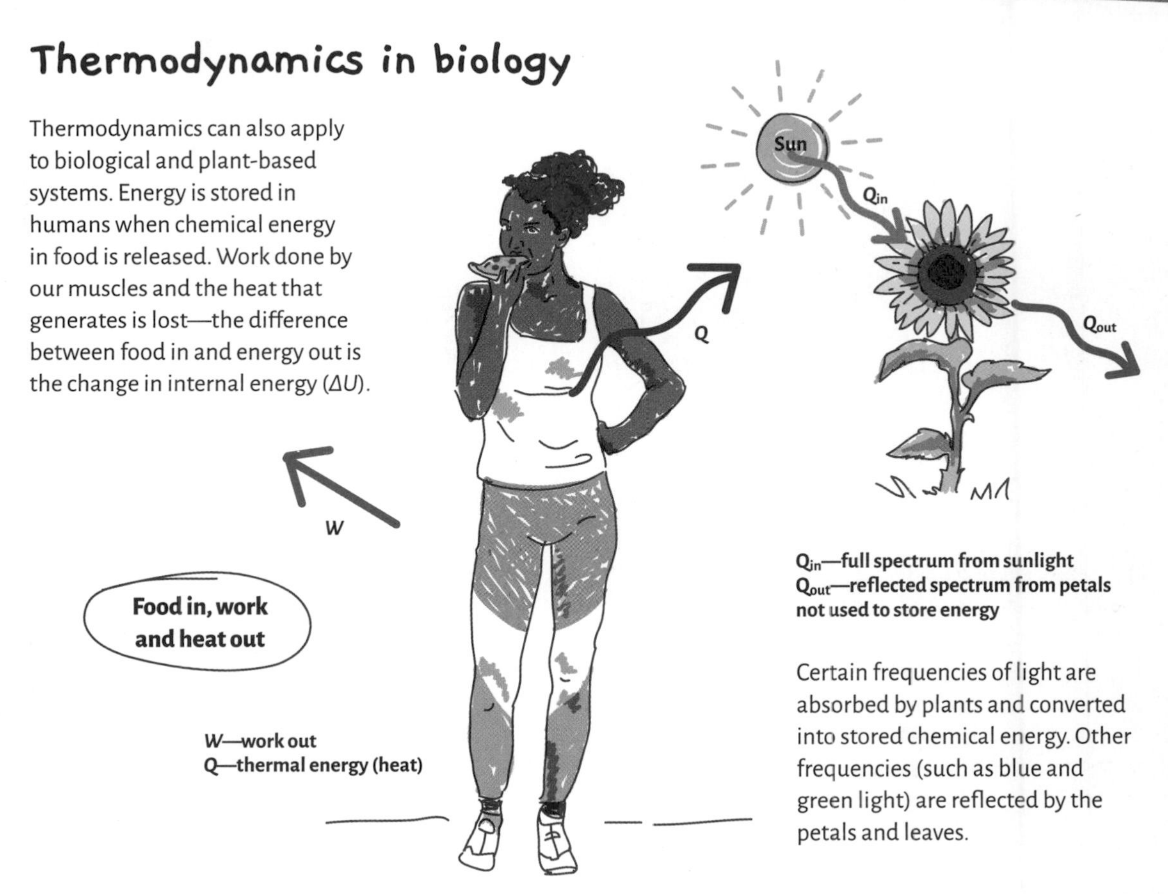

Certain frequencies of light are absorbed by plants and converted into stored chemical energy. Other frequencies (such as blue and green light) are reflected by the petals and leaves.

Entropy

The second law of thermodynamics dictates the nonreversibility of a process. Essentially, it states that if an isolated system undergoes a change due to heat flow within the system, that system becomes more disorganized when considered as a whole.

The measure of how organized a system is energetically is referred to as **entropy**. A highly organized state, such as a solid material, has a low entropy. As a solid is melted via heat transfer, the liquid and gas states have higher entropies due to the higher dispersion of particle movement.

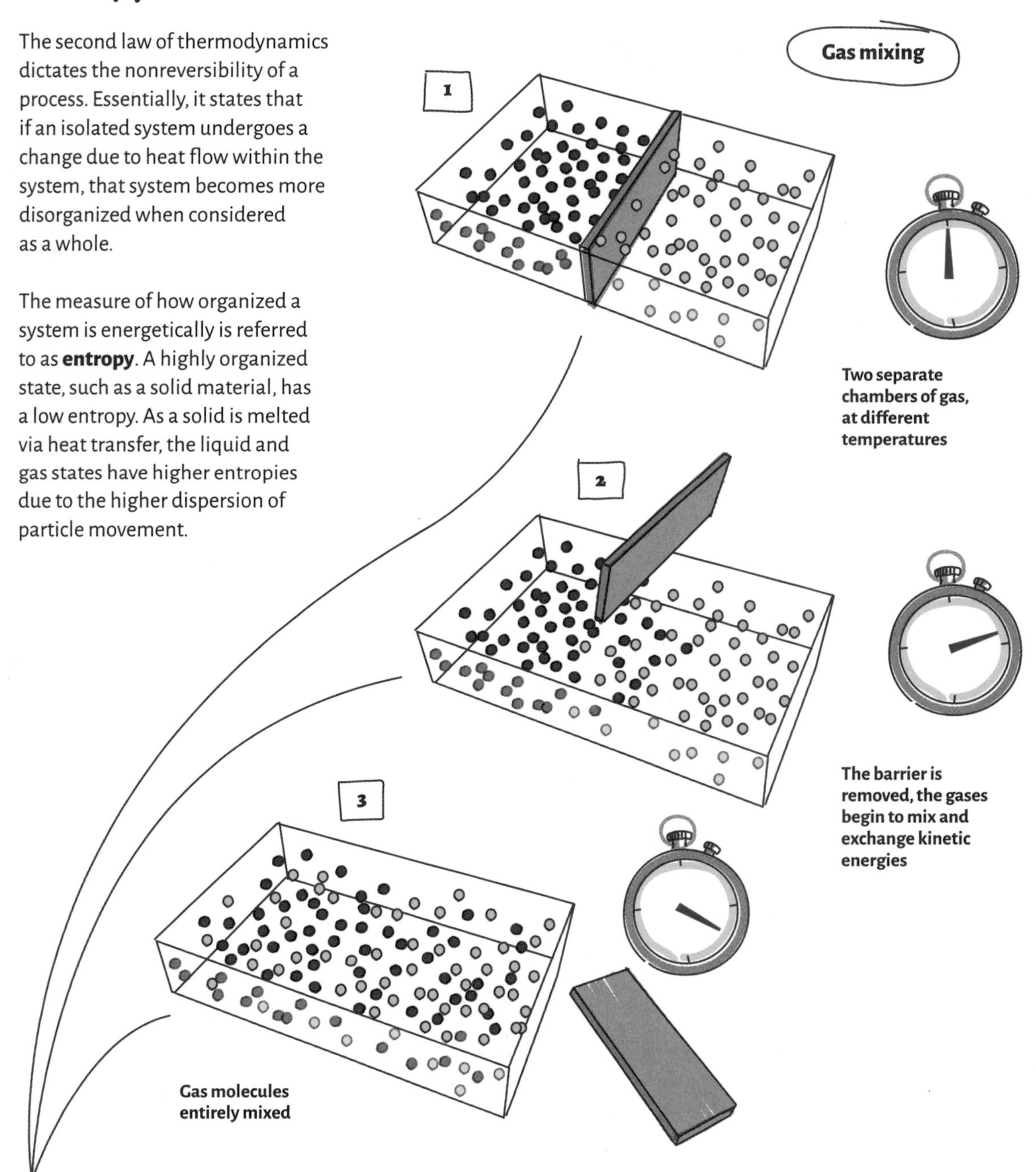

High entropy

In terms of gases, consider two separate chambers of gas, each at a different temperature. This system is highly organized and has a low associated entropy. When the gases mix, entropy increases as the system becomes more disorganized. When all the particles have mixed up, the result will be a warm gas with particles all moving at different speeds. This represents the **highest-entropy state**, which cannot be reversed back to its original state.

THE SECOND LAW
Processes involving the transfer or conversion of heat energy are irreversible.

Disordered states

To illustrate this, imagine a neatly stacked pile of bricks is accidentally tipped off a moving truck. It is highly unlikely that the bricks will land in a neat, ordered pile; it is far more likely that they will create a random, disordered pile. There is an infinite number of possible disordered states, each of which will look more or less like all the others.

All systems degenerate into a more chaotic higher-entropy state—a less organized system is statistically far more likely than an ordered one, as there are many more disordered states.

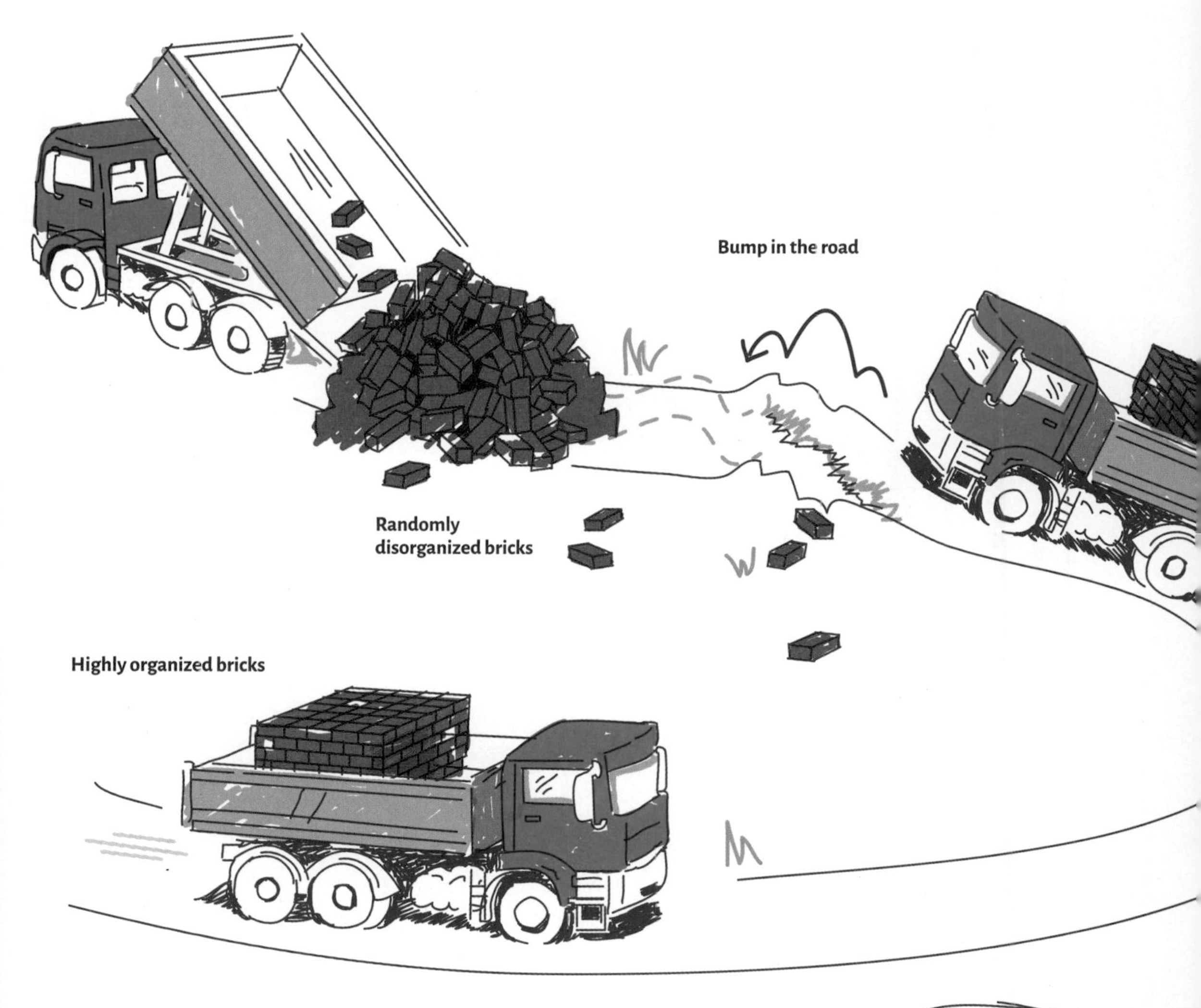

Low entropy

Solid materials, such as metal, have low entropy. The motion of their particles is very limited, and the individual energies of the particles are very similar. An iron horseshoe has to be heated to 2,300°F (1,260°C) before the particles agitate enough to induce change.

When the heat is removed, it reverts to much the same state as it was before, unlike mixed gases.

TEMPERATURE

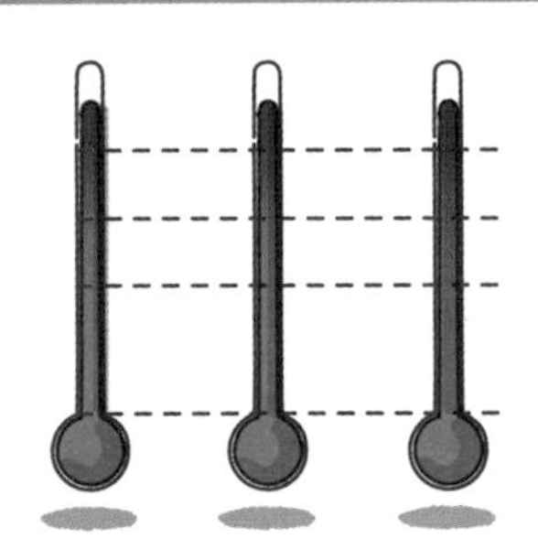

A measure of the average kinetic energy in a material.

WHAT IS TEMPERATURE?

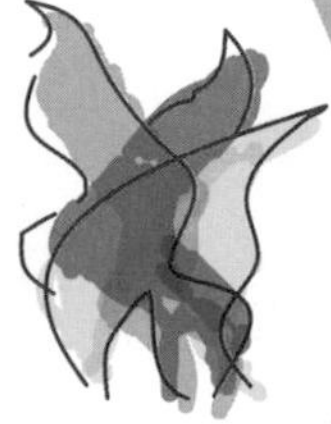

CELSIUS SCALE

Based on the freezing point (0°C) and boiling point (100°C) of water.

FAHRENHEIT SCALE

Based on the freezing point (32°F) and boiling point (212°F) of water.

KELVIN SCALE

Based on absolute zero (0K ≈ −273°C).

THERMODYNAMICS

LAWS OF THERMODYNAMICS

SECOND LAW

Any process of energy transfer within a closed system degenerates into a more chaotic state with higher entropy. This process is irreversible.

ENTROPY

The measure of how organized a system is energetically. A highly organized state, such as a solid material, has a low entropy; liquid and gas states have higher entropies.

FIRST LAW

Energy cannot be created or destroyed. As heat enters a system, it is turned into internal energy of the gas and work done by its expansion against a piston.

$$\Delta U = Q - W$$

THERMAL ENERGY TRANSFER

HEAT

The transfer of thermal energy from hot objects to cooler objects.

THERMAL ENERGY

The quantity of energy stored as the kinetic energy of particles in a material, measured in joules (J).

TEMPERATURE GRADIENT

Rate at which temperature changes, and in what direction, in a specific location.

Materials expand as they absorb energy. Most solids and liquids expand slightly; gases expand significantly.

THERMAL EXPANSION

Objects heat and cool as energy is transferred from a hot region to a cooler one.

HEATING AND COOLING

CONDUCTION

Heat transfer via direct contact with a heat source.

CONVECTION

Heat transfer via movement.

RADIATION

Heat transfer via electromagnetic waves.

HEAT FLOW

Heat always flows from a hotter region to a cooler one, until both regions become equal in temperature.

WORK AND HEAT TRANSFER IN GASES

Gases in an enclosed chamber can absorb heat by conduction or mechanical work done on them. This process can be reversed .

CHAPTER 11

FLUIDS

When people think about fluids, they tend to imagine liquids, such as water. However, a fluid is defined as a state of matter that is free to flow and change shape. This includes liquids but also gases. Fluids have various physical properties, such as their density, pressure, volume, and temperature, which are all interconnected. Fluids and our understanding of their dynamics have allowed us to cross oceans in ships and travel through the air in planes.

DENSITY AND PRESSURE

Fluids, particularly gases, can change depending on their temperature and the volume of the container in which they are enclosed. As their volume increases or decreases, providing the number of particles remains constant, the density of the gas will also change (mass per unit volume).

Consider a fixed mass of gas inside a flexible container. An example of this is a balloon that has been inflated and tied so no gas can escape. At room temperature, the balloon has a large volume because the air molecules have high kinetic energy due to their temperature and are moving around quickly. Each molecule collides with the sides of the balloon, producing a small outward force—these forces combine to hold the balloon in shape.

This is called **gas pressure**, *P* (measured in pascals or N/m^2), and is a function of temperature and volume: As the temperature of a gas increases within a fixed volume, its pressure will also increase.

Gas pressure (*P*), its volume (*V*), and its temperature (*T*) are connected by the relationship:

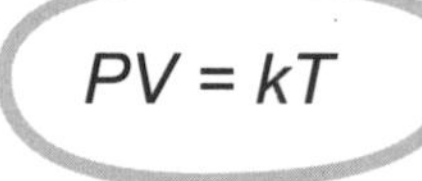

$$PV = kT$$

where *k* is a constant.

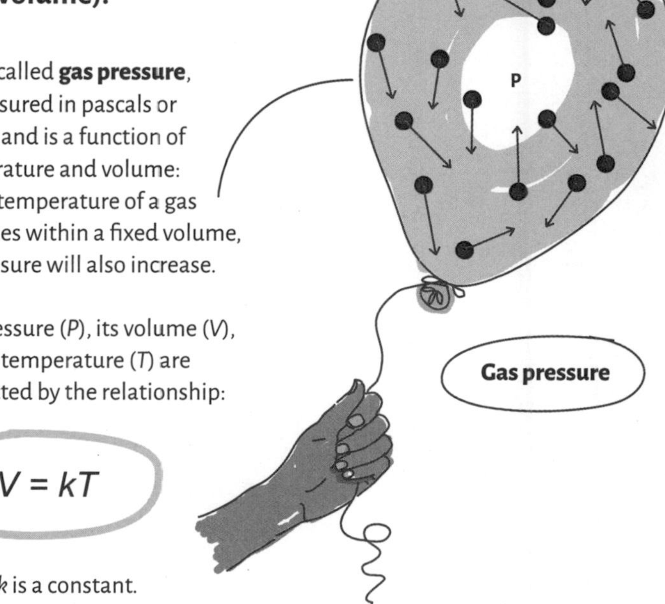

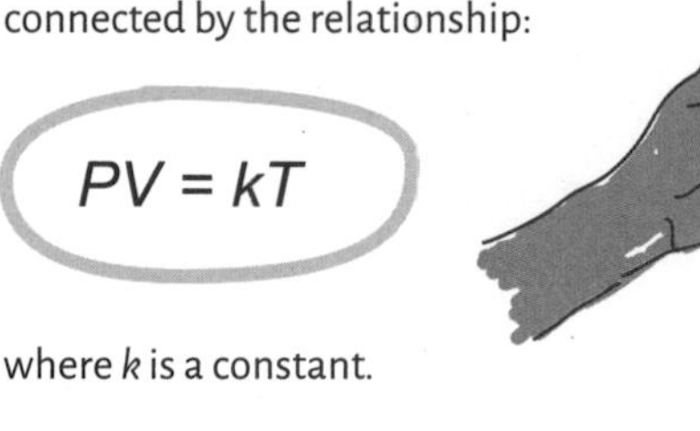

Balloon experiment

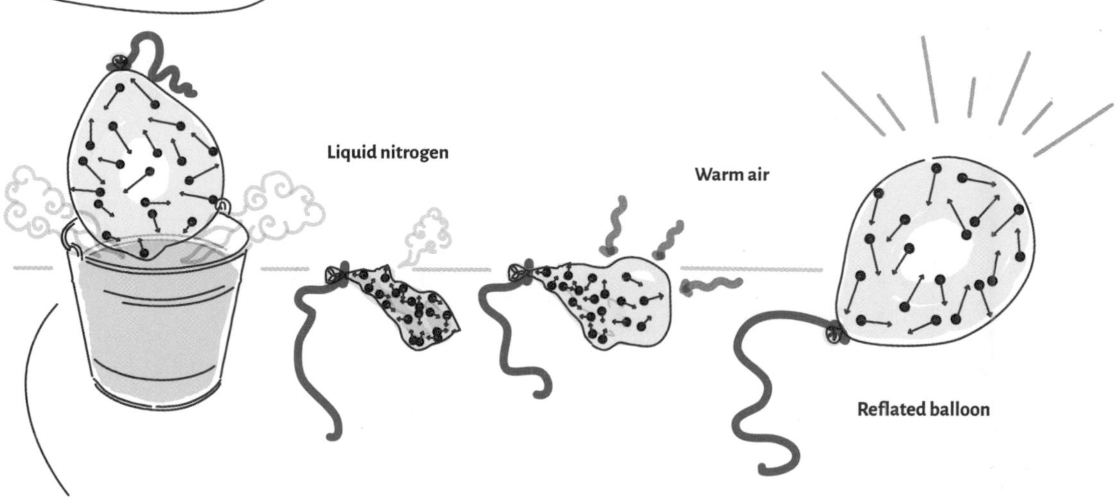

If the balloon is immersed in liquid nitrogen (at –320°F, –195.79°C, 77K), the air molecules suddenly cool to this temperature, and they lose almost all of their kinetic energy.

As the particles slow down, they become closer, so there are more in a small space. This is called density *ρ* (kg/m^3). The gas pressure exerted by the particles is much lower, so the balloon deflates.

When the balloon is removed from the liquid nitrogen, heat flows from the surrounding warm air into the cold balloon air, increasing the particles' kinetic energy and thus re-inflating the balloon.

PRESSURE DIFFERENCE, LIFT, AND BUOYANCY

The principle that a body of lower density than its surroundings will rise due to an upward force is known as buoyancy. Archimedes of Syracuse (287–217 B.C.E.) was a Greek mathematician, physicist, engineer, inventor, and astronomer who investigated this property and determined the link between the density of the medium surrounding a buoyant body and the buoyancy force experienced.

Buoyancy

When we think of buoyancy, we likely picture a boat on water foremost in our minds. Indeed, a floating buoy, used to indicate shallow regions in harbors, shares the same word root as "buoyancy."

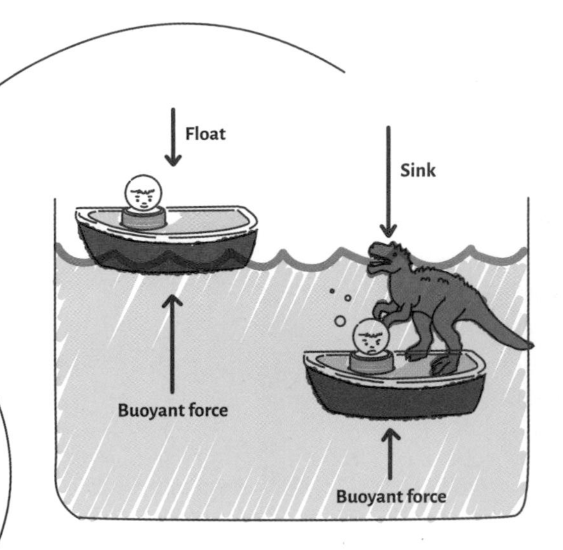

Helium balloons float because the density of helium is less than that of air. Similarly, a hot-air balloon rises because the air in the balloon is less dense than cool air around it.

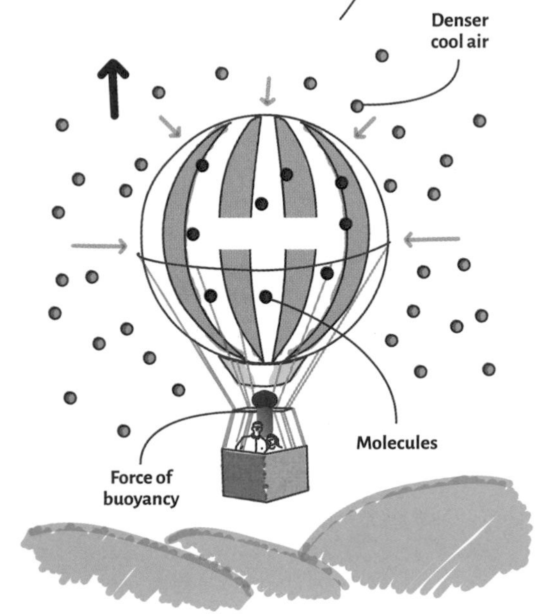

Imagine yourself diving down into the ocean. As you dive deeper, there is more water above you, so the weight of the water you experience also increases—this is called **water pressure**. Now imagine a column of water. There is a steady increase in water pressure as depth increases. This means that if an object is immersed in water, it will experience a larger water pressure, and therefore a greater force, at the bottom of the object than the top. This **buoyancy force** (or upthrust) gives rise to **Archimedes' principle**.

Archimedes' principle

There is only one way an object can float in a fluid, and that is when its weight due to gravity is balanced by an upward force.

An object placed in water will either sink or float, depending on its density relative to water.

Archimedes noticed that an object would float if it weighed less than the same volume of water—meaning it had a lower density. Objects such as wood and vessels filled with air floated, whereas stones and solid metal objects sank.

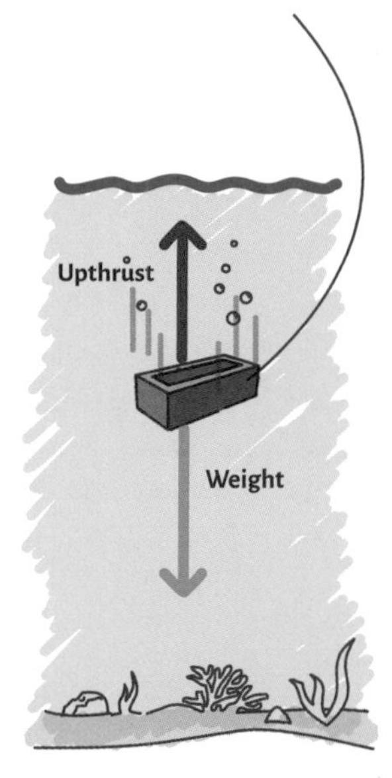

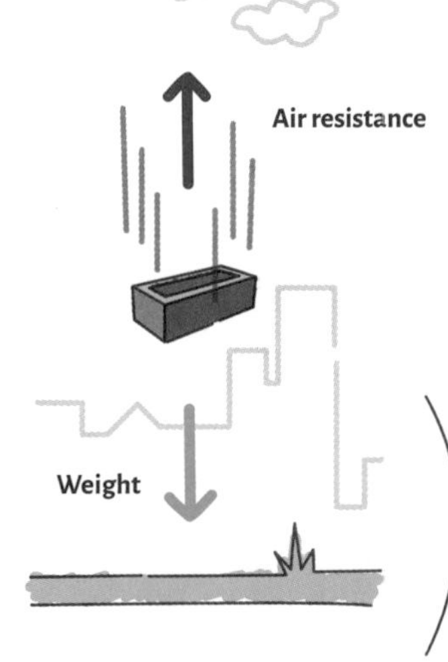

All objects, including bricks, experience an upward thrust in water, but only if this force is equal to the weight of the object will it float. A brick will sink far more slowly in water than in air.

ARCHIMEDES' PRINCIPLE
Upward buoyancy force = weight of displaced fluid.

EUREKA!
Archimedes discovered that the upward buoyancy force experienced by an object is exactly equal to the weight of the fluid displaced by the object.

Therefore, an object **less dense** than water will **weigh less** than the water it displaces, providing a buoyancy force greater than the object's weight. It will then rise until it floats on the surface, like a boat in the ocean. An object denser than water will sink like a brick.

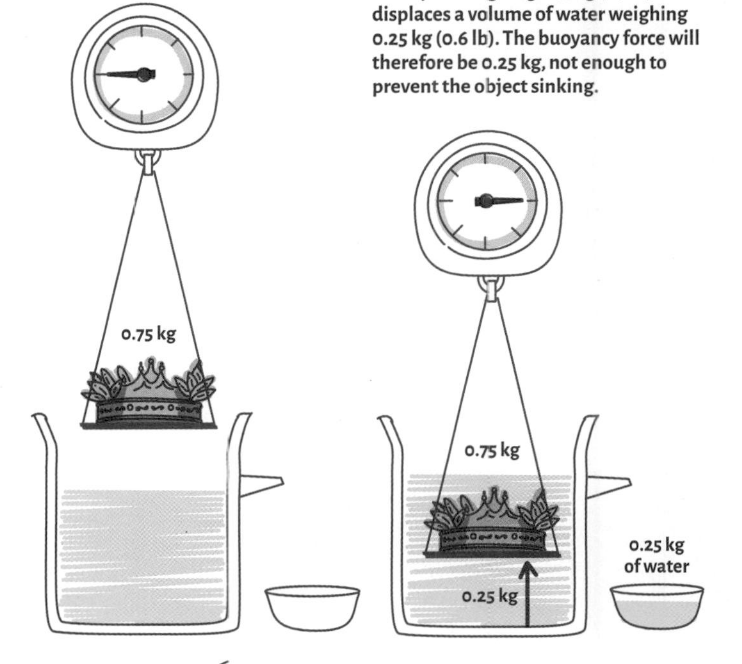

An object weighing 0.75kg (1.7 lb) displaces a volume of water weighing 0.25 kg (0.6 lb). The buoyancy force will therefore be 0.25 kg, not enough to prevent the object sinking.

FLUID FLOW AND BERNOULLI'S PRINCIPLE

Fluids can flow around objects due to their physical ability to change shape. If a rigid object passes through a fluid, the fluid responds by shaping itself around the object. There are many examples of this, such as the aerodynamics of a race car and the response of water around a propeller. Fluid dynamics is the branch of physics that explores the movement of fluids and the resultant forces as a response to this movement.

Fluid flow

Fluids such as gases and liquids flow according to forces that act on them. In a body of water, cold water will become denser down to approximately 4°C (39°F). (Water becomes less dense below this.) As the kinetic energy of the molecules decreases, the separation of each molecule also decreases. This will increase the density of the fluid, and the body of water will sink, due to Archimedes' principle.

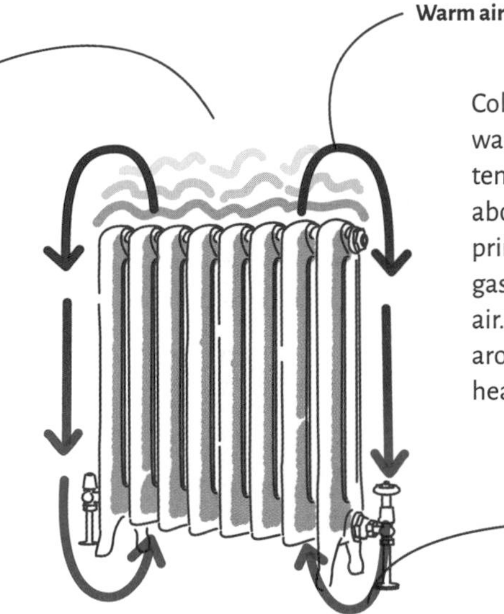

Cold fluids tend to sink below warmer fluids, causing layers of temperature, with warmer fluids above cooler fluids. The same principle governs the movement of gases: Warm air rises above cold air. Radiators warm the air directly around them, causing a flow of heat upward in a room.

This principle is the main driving force of our weather systems. As air warms, it rises. Cold air rushes in to replace the rising warm air, creating a **low-pressure system**, which we feel as wind as it passes over us. As the warm air rises, it cools in the atmosphere, and the moisture contained in the warm air condenses, producing rain.

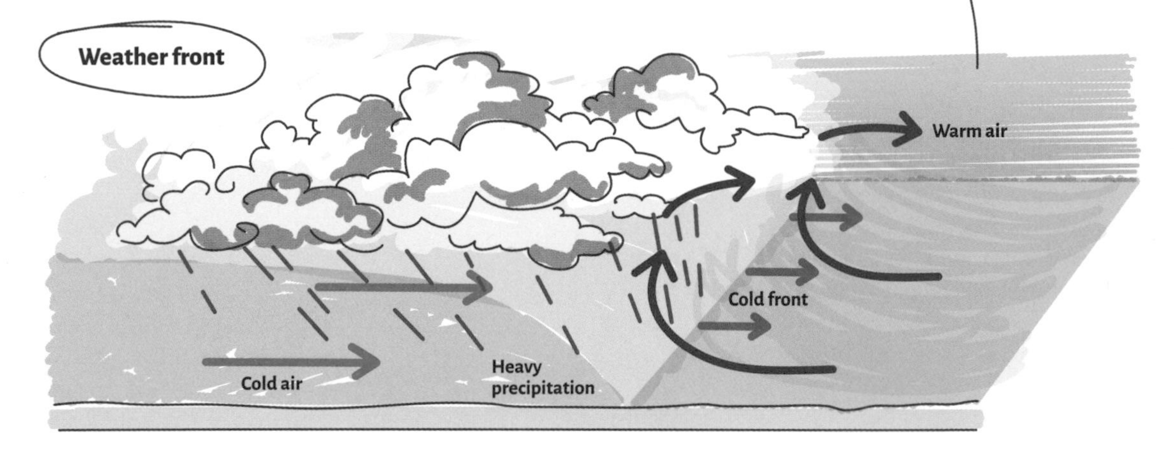

Bernoulli's principle

Named for Swiss physicist Daniel Bernoulli (1700–1782), **Bernoulli's principle** works in the same way as weather systems. Low pressure is created by the faster movement of air over the wing of a plane. The plane's wing shape is designed to direct the flow of air around it in such a way that the air flow over the top of the wing is faster than that underneath the wing.

The **difference in velocity** of the flow of fluid produces a **difference in pressure** between the top of the wing and the bottom of the wing. It is the relative velocity between the fluid flow and the aircraft wing that causes lift. An aircraft traveling through a headwind (an oncoming air flow) will experience a greater lift.

Pressure is defined as the force (*F*) per square meter of the area (*A*) to which it is applied:

$$P = \frac{F}{A}$$

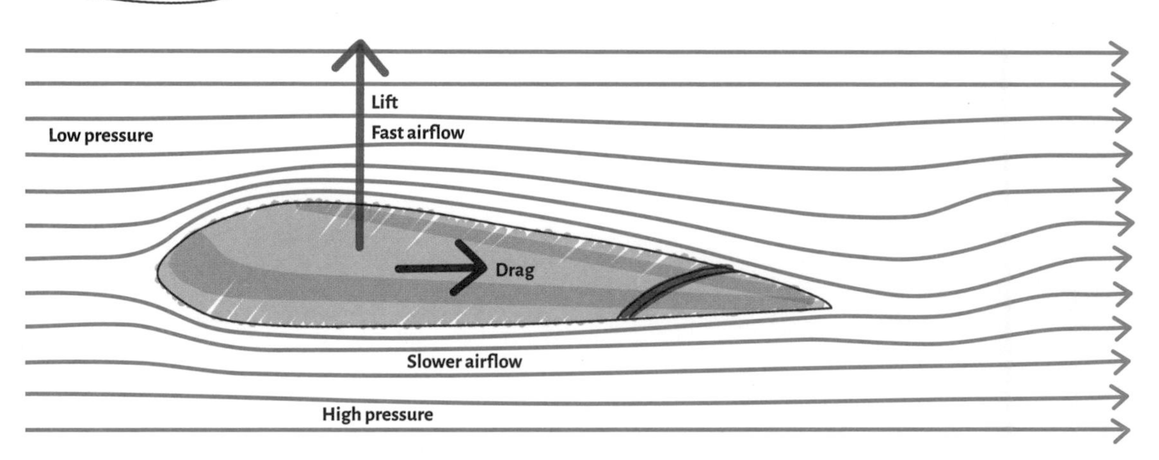

DRAG FORCE

Drag, D, or air resistance, is a frictional force on a moving object as it passes through a fluid. It is proportional to the size of the body (the bigger the body, the larger the drag force) and to the velocity of the object squared. A race car is designed to move through air as efficiently as possible, offering the least resistance and lowest profile, so that the fluid flows over it smoothly.

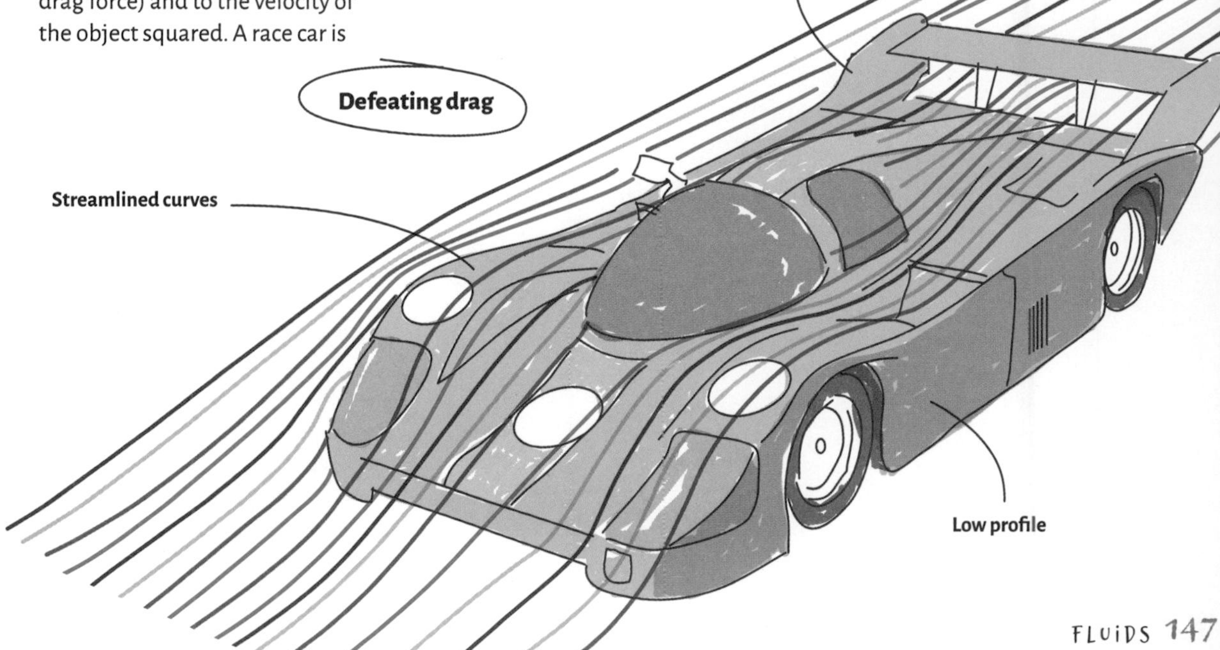

DENSITY

The measure of how close the particles in a gas or liquid are to one another. The more tightly packed the particles, the larger the mass of a fixed volume of fluid (kg/m³).

The force provided by moving gas particles colliding against the walls of a container, measured in N/m². A higher temperature creates faster particles, and more particles in a fixed volume increase the number of collisions; both increase pressure.

GAS PRESSURE

DENSITY AND PRESSURE

FLUIDS

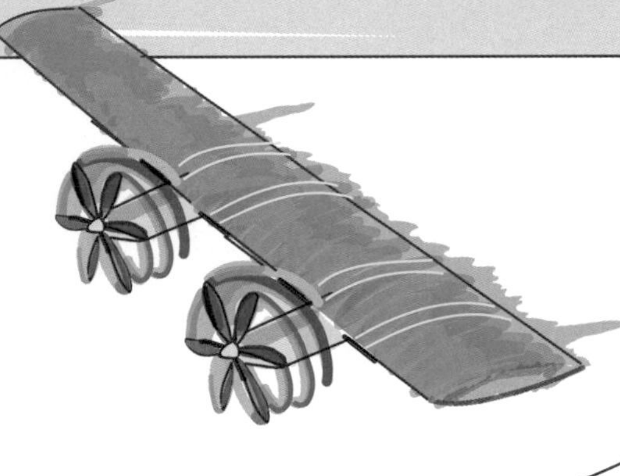

A difference in flow velocity around a fixed body produces a difference in pressures between the surfaces.

BERNOULLI'S PRINCIPLE

LIFT

Caused by the imbalance of forces as flow velocity changes. The body will move in the direction from high pressure to low pressure. This is how an aircraft wing produces lift.

FLUID DYNAMICS

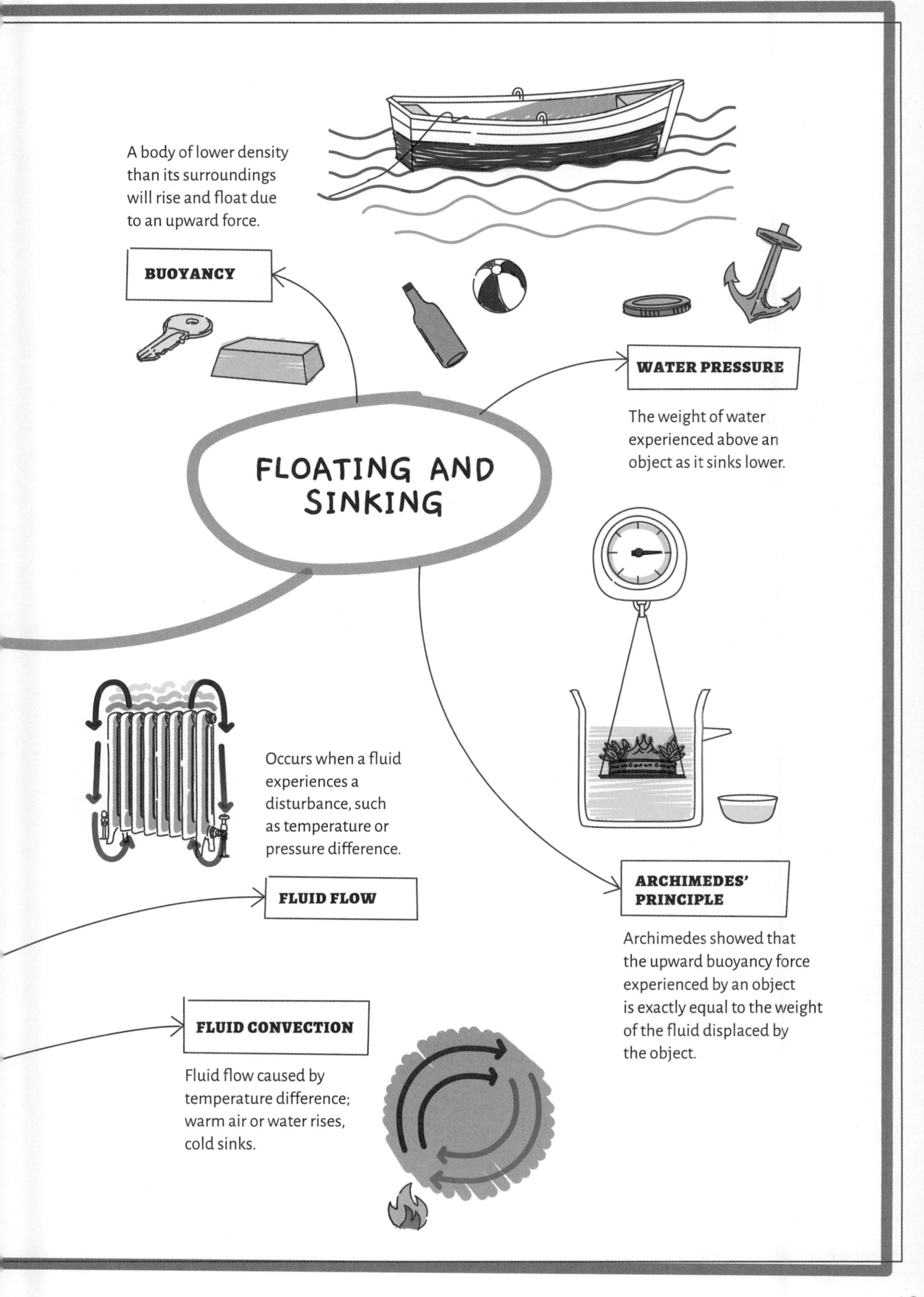
FLOATING AND SINKING
A body of lower density than its surroundings will rise and float due to an upward force.
BUOYANCY
WATER PRESSURE
The weight of water experienced above an object as it sinks lower.
Occurs when a fluid experiences a disturbance, such as temperature or pressure difference.
FLUID FLOW
ARCHIMEDES' PRINCIPLE
Archimedes showed that the upward buoyancy force experienced by an object is exactly equal to the weight of the fluid displaced by the object.
FLUID CONVECTION
Fluid flow caused by temperature difference; warm air or water rises, cold sinks.

CHAPTER 12

MODERN PHYSICS

Physics is the study of fundamental physical laws that govern the behavior of the universe. On a macroscopic scale, forces accelerate objects, and fluids flow from one place to another. However, there are areas of physics that confound our understanding, such as the behavior of particles at quantum (atomic and subatomic) level and the kinematics of particles traveling close to the speed of light. Newton formulated his laws of motion based on the movement of objects in the everyday sense, but these laws break down and need refinement when objects become very small or very fast.

SPECIAL RELATIVITY

Within all inertial (nonaccelerating) frames of reference (or moving regions in space), the laws of physics remain invariant (they do not change), and light speed is independent of an observer within the universe. This premise is known as special relativity.

Michelson and Morley used the Michelson **interferometer** (see page 124) to determine whether space was an empty vacuum or was in fact filled with an **ether**, a medium within which **electromagnetic waves** could propagate. The experiment was set to measure Earth's movement relative to the hypothetical medium of space and determine light's movement relative to this motion; rather like measuring wind speed relative to a moving vehicle. No variation was detected, suggesting that no such ether existed.

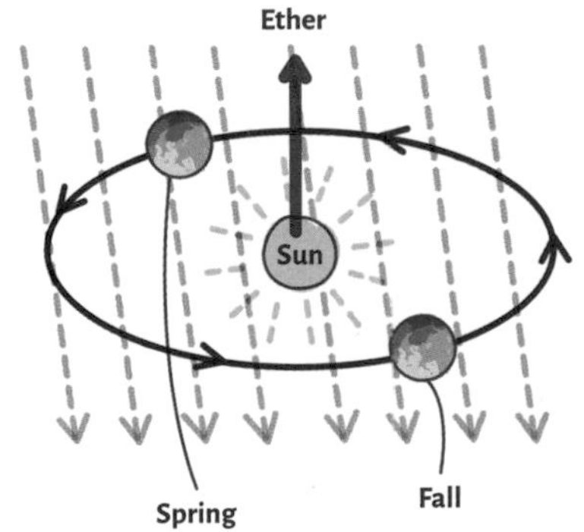

This null result was not a failure, instead it raised more questions: If there was no ether, and light speed measured relative to Earth's velocity didn't change, then what was different about the nature of light?

Albert Einstein (1879–1955) used this result to postulate that light speed was an absolute maximum within the universe and, regardless of an observer's motion relative to a light beam, the **relative velocity** measured could never exceed the **speed of light** ($\approx 3 \times 10^8$ m/s). In fact, the closer to light speed the observer was traveling, the slower that time within his frame of reference elapsed relative to the observer. This is the basic premise behind the fact that the relative speed of a particle outside of the observer's frame seems to slow.

Any two observers in different moving observatories will see each other with the same relative velocity. Regardless of their relative velocities toward each other, their combined speed of approach can never exceed light speed.

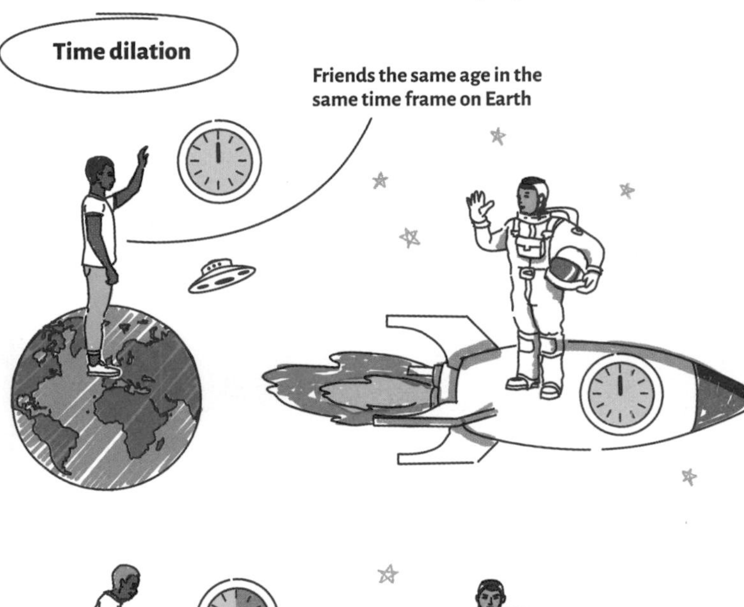

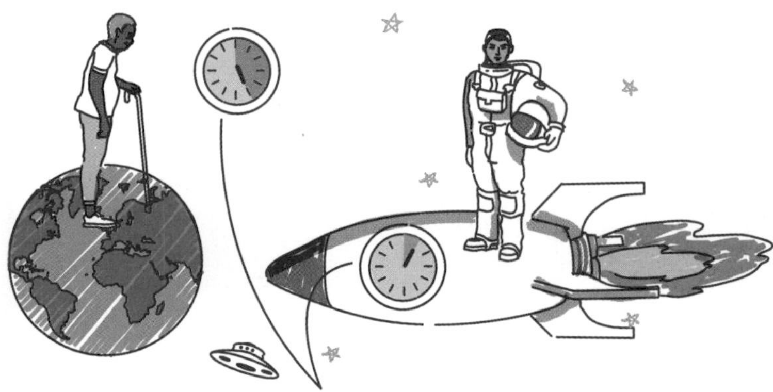

Time flows relatively faster on Earth than in a spacecraft traveling near to light speed.

GENERAL RELATIVITY

Mass and energy have an effect on the geometry of space and time, or space-time, causing local curvature and affecting the passage of light and time. This is known as general relativity.

For centuries, Newton's laws of gravity were the accepted treatment of the effect gravity has upon a mass. All observations confirmed his ideas that a body with mass within a gravitational field will accelerate at a rate dependent on the strength of the gravitational field. The important premise here is the existence of mass, which is accelerated by a gravitational field strength, g, based upon Newton's law of gravitation at a rate:

$$g = \frac{GM}{r^2}$$

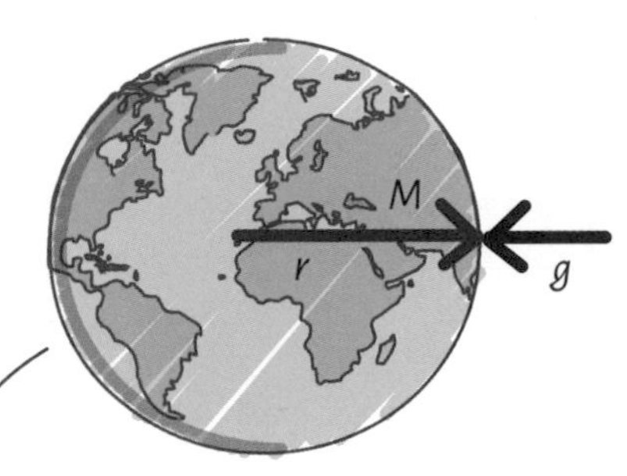

where M is the mass of the body causing the gravitational field, and r is the separation between the centers of mass of the falling body and the central mass.

The principle of equivalence

In 1907, Einstein developed a new idea, which he named the **principle of equivalence**. This states that there is no difference between an accelerating frame of reference in a gravitational field and one that is accelerated by an external force. This premise is a stepping stone to general relativity.

Consider a person in freefall in an elevator. They will feel weightless as both they and the elevator accelerate toward Earth at the same rate of 9.8 m/s^2, because acceleration is independent of mass. They would not know whether they were in an accelerating frame of reference within a gravitational field or stationary in space.

In fact, if they dropped a ball while the elevator was plummeting, it would also accelerate downward at the same rate and therefore remain stationary relative to the person.

Space-time curvature

Due to the visual nature of Einstein's mind, many of his ideas were born of **thought experiments** and later proved mathematically.

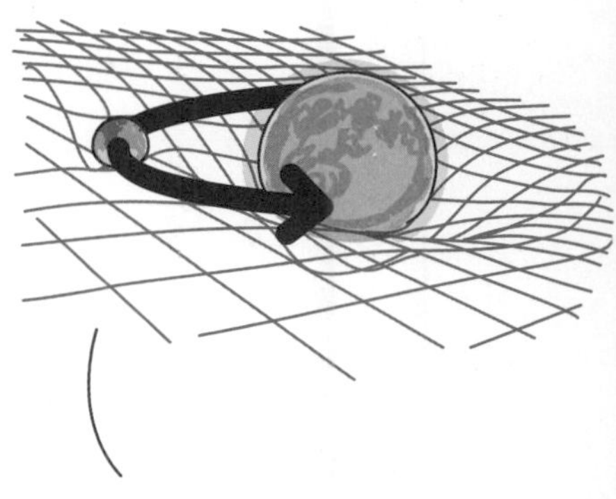

Consider the elevator scenario again, but in this instance, the system is no longer completely isolated, and two perfectly aligned round holes have been drilled into opposite sides of the elevator. As a light beam passes through the first hole and travels in a straight line toward the other hole, it will miss it because the elevator has moved up slightly. From the light beam's perspective, it has traveled in a perfectly straight line.

This simple thought experiment had profound implications when married to the principle of equivalence. If an accelerating frame is identical to the presence of a gravitational field, then gravity should affect the path of a light beam, which has no mass, contradicting Newton's law of gravitation. This prediction was unprecedented and controversial.

All masses distort the space around them—the larger the mass the bigger the effect. The sun creates a large distortion in space-time, which affects the motion of objects, such as Earth, around it.

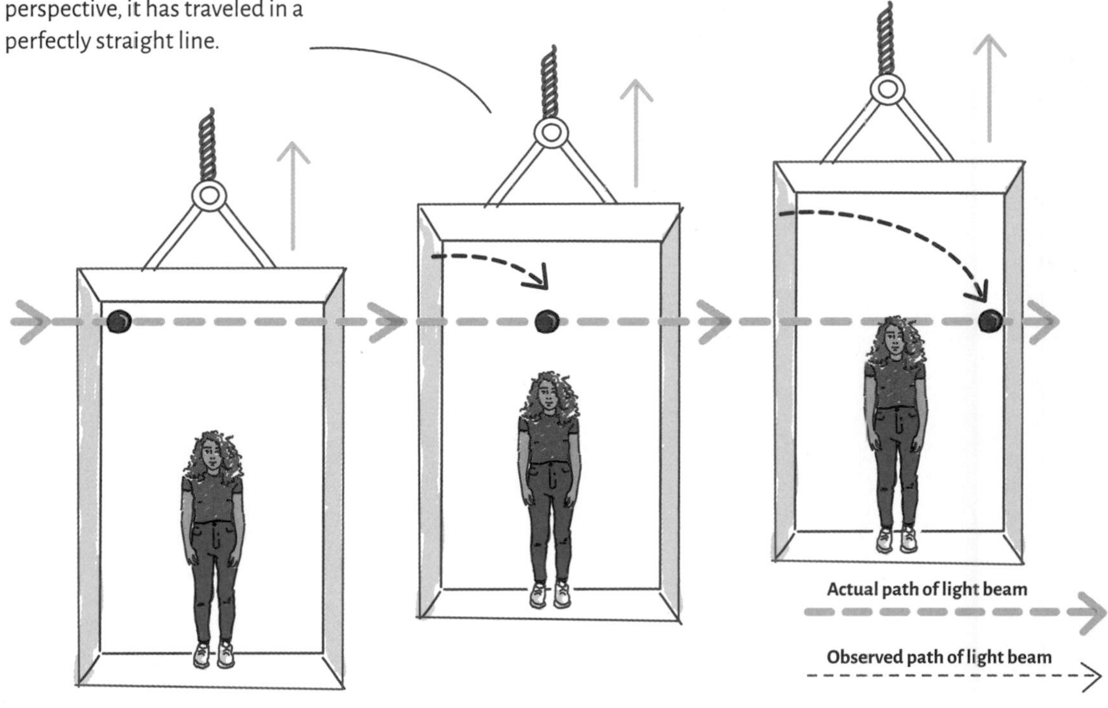

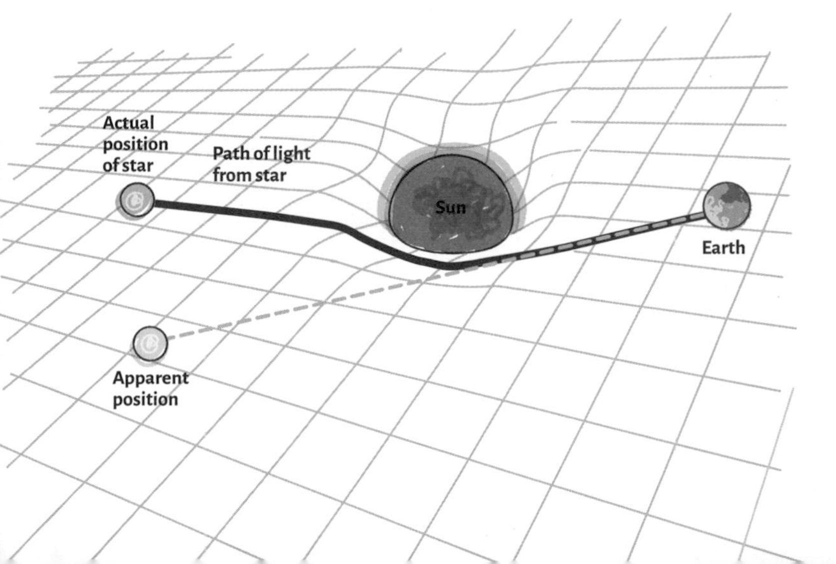

Einstein's theory predicted the redirection of light from the stars and galaxies caused by **gravitational lensing** (see page 182), the existence of **black holes** and **gravitational waves**, and the slowing of time in the presence of a gravitational field. Indeed, it has been confirmed by observations of all these phenomena and continues to accurately describe astronomical events that are altered by the presence of gravitational fields.

NUCLEAR PHYSICS

This is a relatively modern area of physics that has seen huge leaps forward in the last hundred years or so. Being so small, the nucleus cannot be observed directly, and it was the ingenuity of physicists in the early part of the twentieth century that developed our understanding of the nuclear realm.

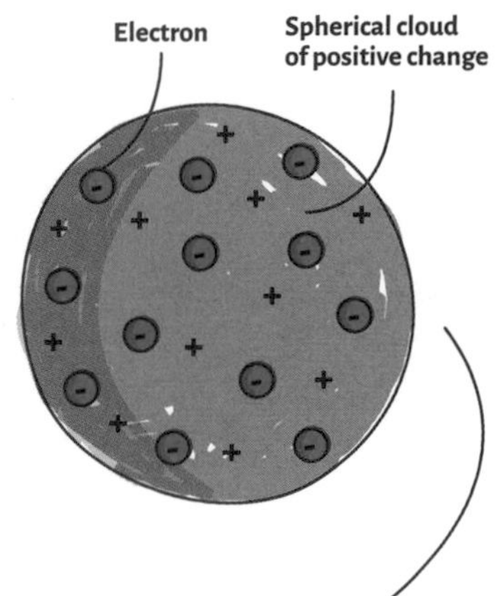

The Rutherford scattering experiment

Ernest Rutherford (1871–1937) was a New Zealand–born British physicist who transformed our understanding of the structure of the atom. In his famous **scattering experiment** of 1909, Rutherford fired alpha particles (helium nuclei) through a thin foil of gold and observed their paths. Based on the previous model—Sir John Thomson's "plum pudding"—the particles should have passed straight through with little deviation. However, observation revealed a very different and unexpected result: Most of the particles passed straight through, some were deflected through large angles, and a few rebounded completely.

ATOMIC PUDDING
Only five years earlier, in 1904, British physicist Sir Joseph John Thomson (1856–1940) had proposed the "**plum pudding model**." The existence of negatively charged electrons was known, and the overall neutral charge of an atom suggested that a positive charge of equal measure also existed. Thomson suggested that electrons were embedded within a volume of positive charge, like negative plums in positive pudding.

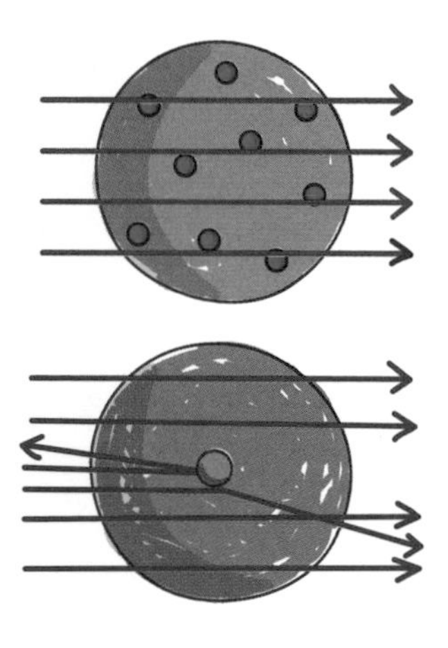

Rutherford likened this to firing a bullet through paper and the bullet rebounding! This result derived the conclusion that in fact an atom was mostly empty space with a very small concentrated positive charge in the center.

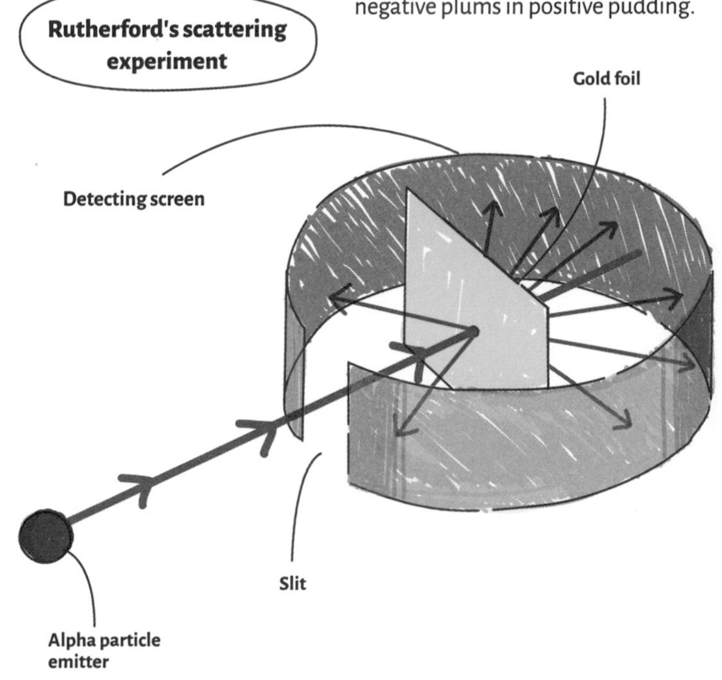

The atom and the nucleus

Rutherford's discovery completely revised the accepted model of the atom. It became clear that an atom consisted of a very small, very dense powerfully charged nucleus surrounded by orbiting electrons. The **Rutherford model** for the atom was proposed in 1911.

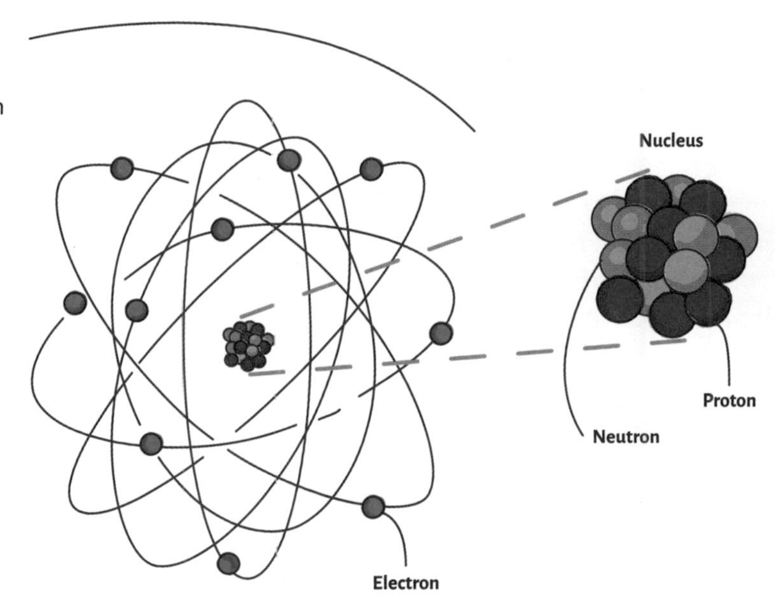

In 1913, Niels Bohr (1885–1962), a Danish physicist, revised the Rutherford model by including electron shells within which a specific maximum number of electrons could exist at very specific (**quantized**) energies. This revised model explained emission of radiation from atoms at very specific frequencies when their electrons dropped in energy from one shell to another.

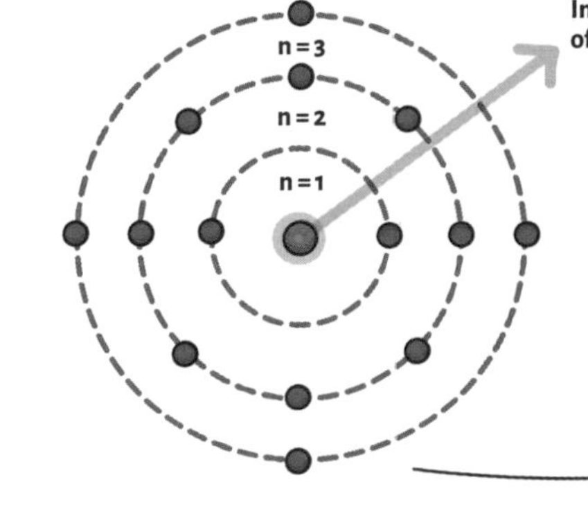

All elements are made up of atoms, and it is the number of protons with positive charge (called the **atomic number**) within the nucleus that defines the properties of the element.

Hydrogen isotopes

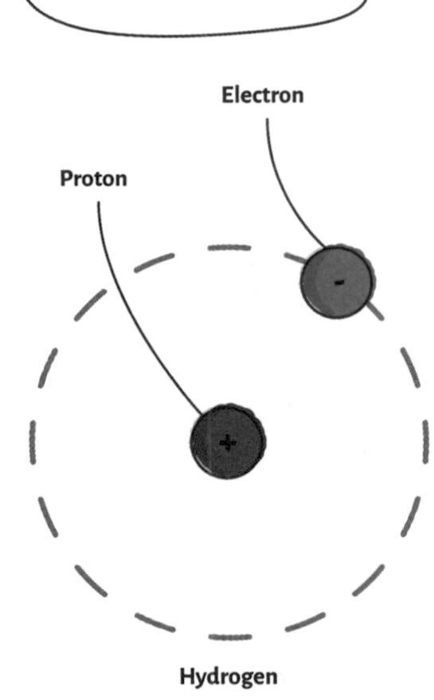

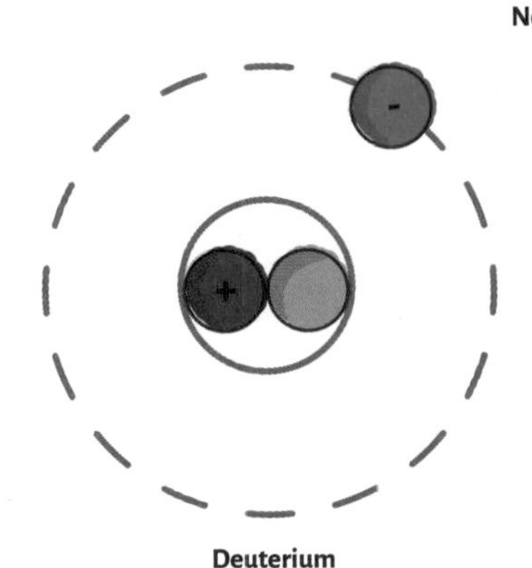

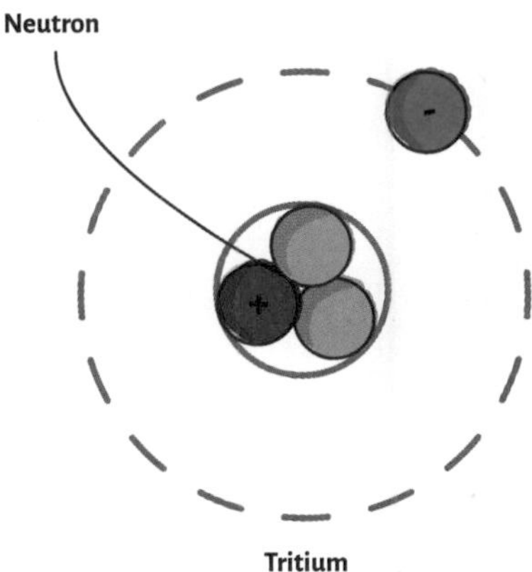

There is always the same number of electrons as protons to maintain overall charge neutrality. The number of neutrons can change for a fixed number of protons, which represents an **isotope** of the same element. Hydrogen, for example, has only one proton but three isotopes: atomic hydrogen, deuterium, and tritium. These isotopes form an essential link in the process of **nuclear fusion** within stars, turning hydrogen into helium and energy.

NUCLEAR REACTIONS

These events are defined as the interaction between two nuclei or a nucleus and another particle, such as a neutron, which then produces different nuclides as a result. Generally, a nuclear reaction will conserve the number of protons and neutrons (collectively, nucleons) in the reaction, as well as other quantities, such as charge and energy.

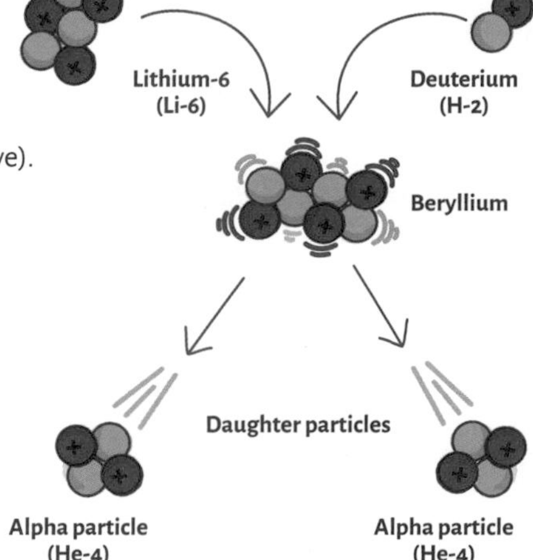

Nuclear decay

Nuclear decay occurs when a heavy isotope is unstable due to the strong electrostatic repulsion of the protons in its nucleus. The nucleus will break apart to create two (or more) **daughter elements** and a form of radiation: an **alpha particle** (helium nucleus), **beta particle** (electron), or **gamma radiation** (electromagnetic wave). Energy is released in the decay process as kinetic energy of the emitted particles and energy associated with the gamma radiation.

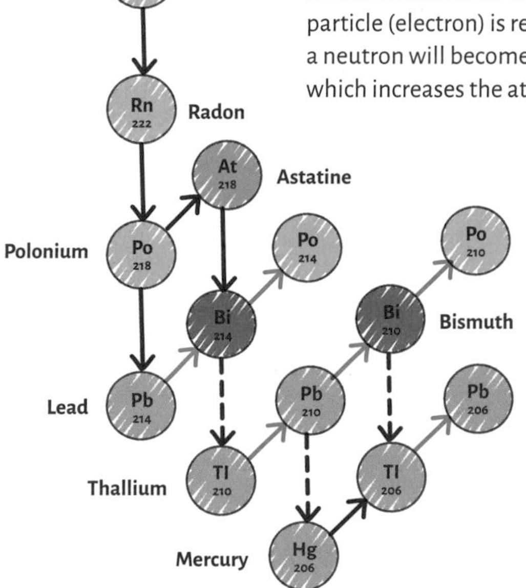

When an unstable nucleus releases an alpha particle, it loses two protons and two neutrons. Its chemical properties will change according to a decrease in its atomic number, and it will become a different element. If a beta particle (electron) is released, a neutron will become a proton, which increases the atomic number, again changing it into a different element. Gamma radiation is emitted from an unstable nucleus, which releases energy and stabilizes itself.

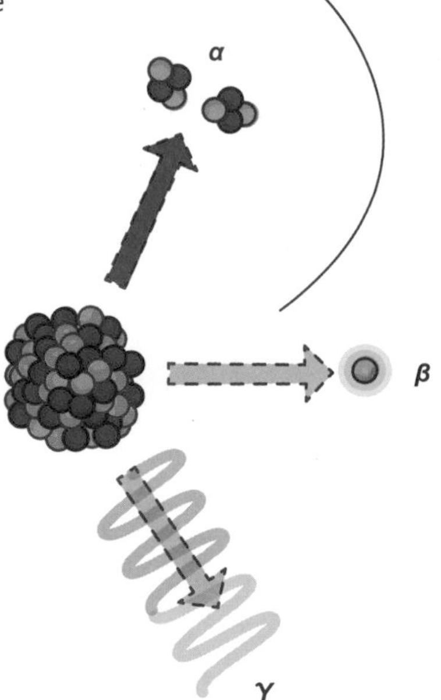

Common nuclear reactions

There are two types of nuclear reaction: **fission** and **fusion**. Fission occurs when an unstable element breaks apart into two or more fragments; it is used in **nuclear power stations**. Fusion occurs when two or more elements collide with sufficient kinetic energy to overcome the coulomb repulsion of protons and bind to form new elements. This process creates energy in stars.

Nuclear power stations predominantly use uranium-235, which is enriched with an extra neutron, making unstable, neutron-rich uranium-236. This unstable isotope of uranium breaks apart to form two radioactive fragments. As well as the two fragments, gamma radiation is released along with three **fissile neutrons** with a high kinetic energy. The result is a **chain reaction**, as each of the three fissile neutrons combines with three other uranium-235 nuclei.

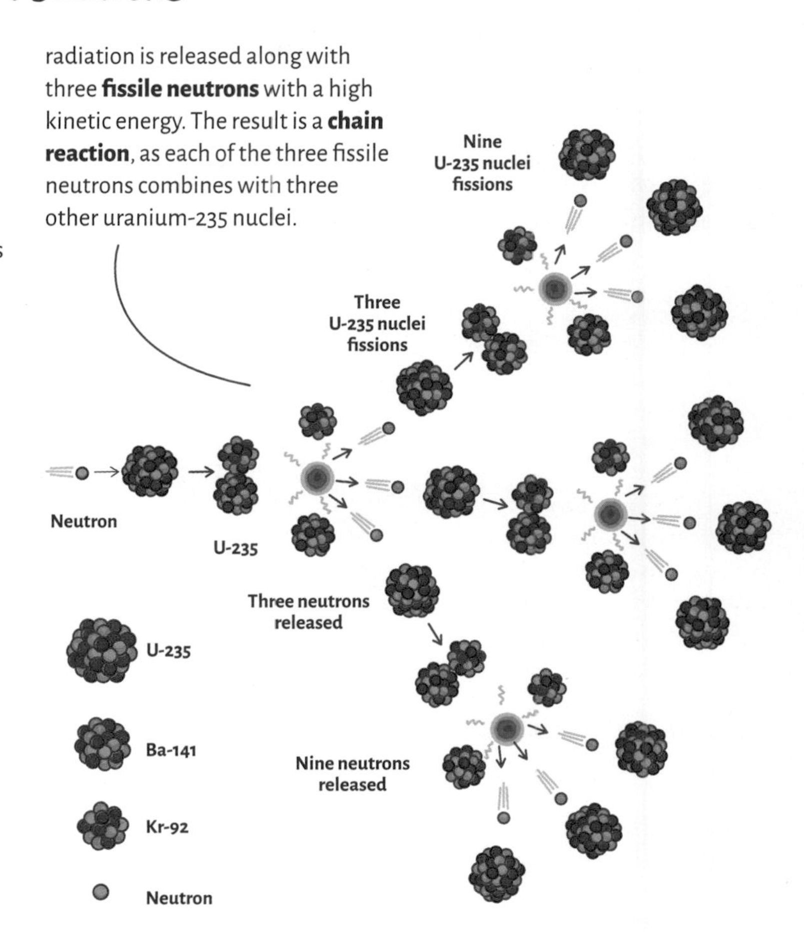

NUCLEAR FUSION IN STARS

Stars are fueled by nuclear fusion. They fuse hydrogen in their centers to produce deuterium, then helium-3 (He-3), a light isotope of helium, producing huge amounts of heat in the process. The temperature, densities, and pressures required to fuse elements in nuclear fusion are enormous and only exist at the centers of stars. Currently, scientists have been unable to beneficially harness the power of nuclear fusion on Earth.

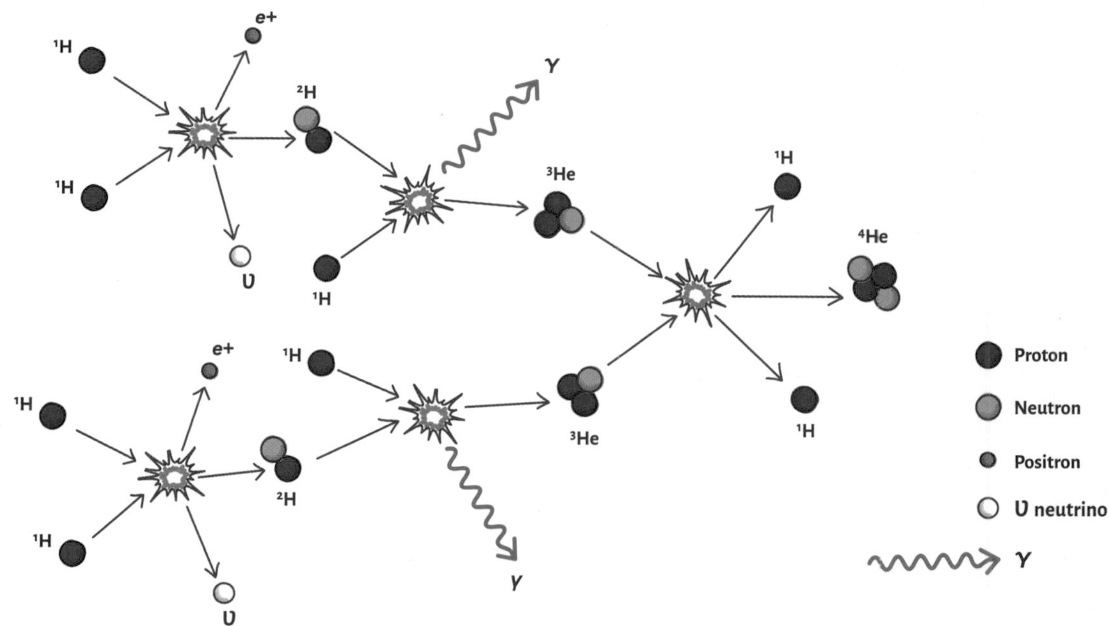

QUANTUM PHYSICS

The study of quantum physics is a culmination of a number of strange and nonintuitive phenomena that cannot be explained by classical physics.

It was Newton who first suggested that light was made up of particles he called corpuscles, which, he said, explained the nature of reflection. And it was Christiaan Huygens who, in 1678, proposed the wave nature of light. Curiously, they were both correct, as light exhibits both wavelike and particle-like behavior. This concept is referred to as **wave-particle duality**.

Wavelike

Wave and particle

Particle-like

It is now universally recognized that light is made up of tiny packets (**quanta**) of energy called **photons** and is not a continuous wave of energy. This idea began the journey of quantum physics.

Historical developments

Max Planck (1858–1947), a German physicist, speculated, in order to fit theory with observation, that the energy of a body when heated increases by fixed increments called quanta. He described this theory as "desperation," as it was so radical. In 1905, a young Einstein held onto this idea known as the **quantization of light**. Its implications went some way to explaining certain phenomena, including **spectral lines** at very specific wavelengths, such as the observed **hydrogen absorption lines**.

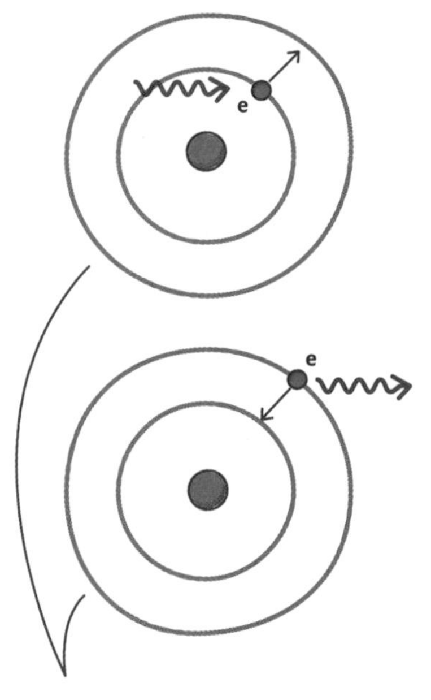

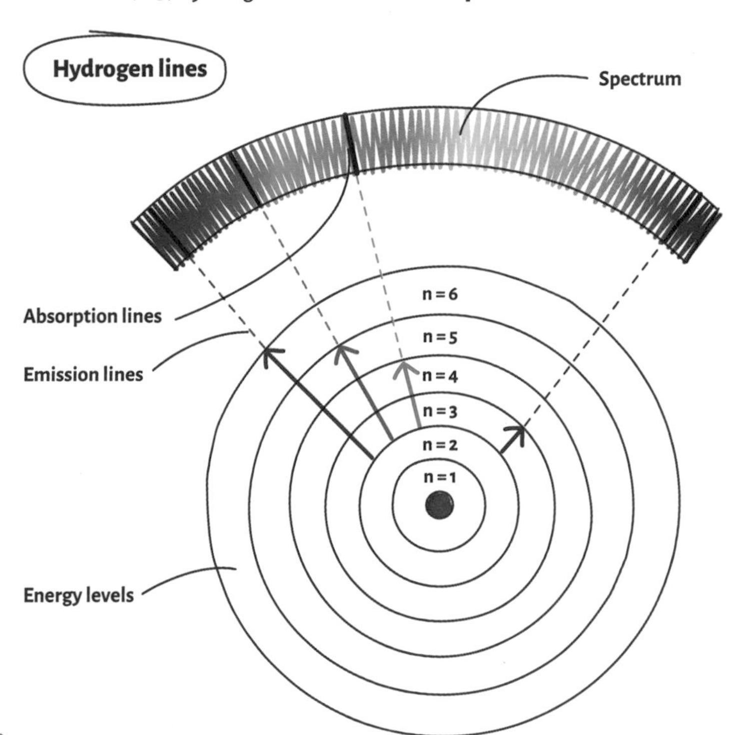

Niels Bohr helped resolve this problem in 1913 by stating that electrons in atoms exist only in specific energy shells and can only move between them if the exact quantity of energy is either absorbed or radiated, creating either dark absorption lines or emitting photons.

Hydrogen has only one electron but many energy levels. When the electron jumps levels, it releases a photon that appears as an emission line in its spectrum.

A collaboration of minds

In 1923, the French physicist Louis de Broglie (1892–1987) suggested that if waves possessed particle-like behavior as photons, then surely particles would display wavelike properties. In fact, he went further and stated that a particle has an associated wavelength based upon its momentum, *mv* (known as the **de Broglie wavelength**):

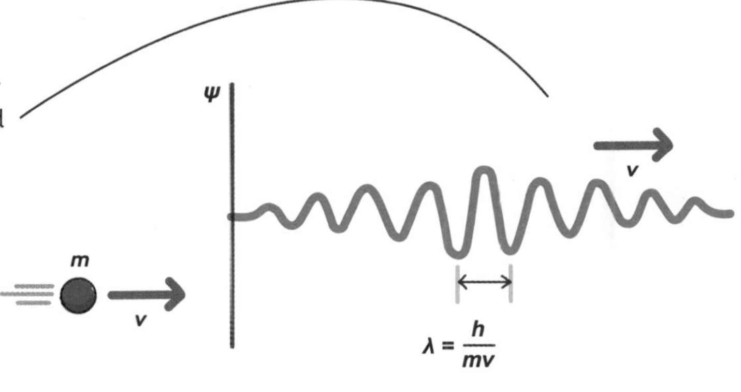

$$\lambda_{db} = \frac{h}{mv}$$

where *h* is Planck's constant.

If this were true, it should be possible to replicate wavelike properties, such as diffraction and interference. This was later confirmed by the diffraction of electrons based on the double-slit experiment of light (see page 125).

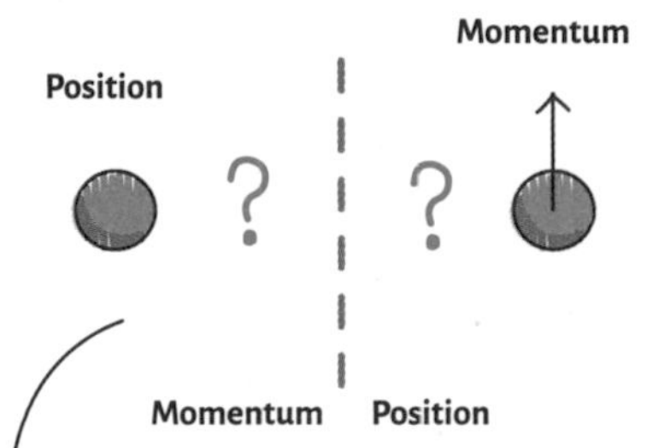

Werner Heisenberg (1901–1976) stipulated that it is impossible to know both a particle's position and momentum simultaneously in his **uncertainty principle**.

Wolfgang Pauli (1900–1958) postulated the **exclusion principle** that "two electrons cannot coexist in the same quantum mechanical state around a nucleus, thus fixing a maximum number of electrons able to occupy each concentric electron energy shell."

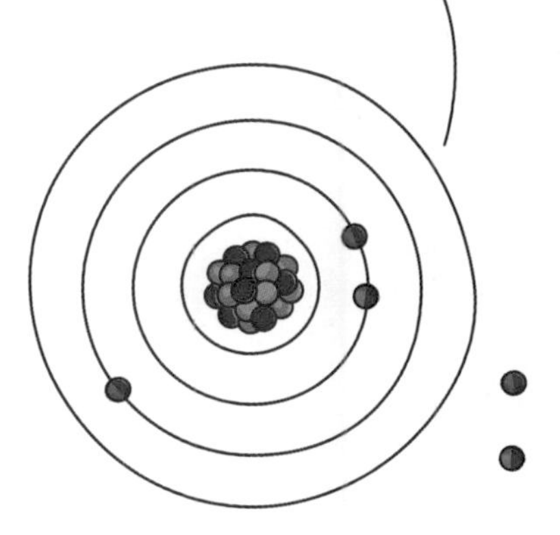

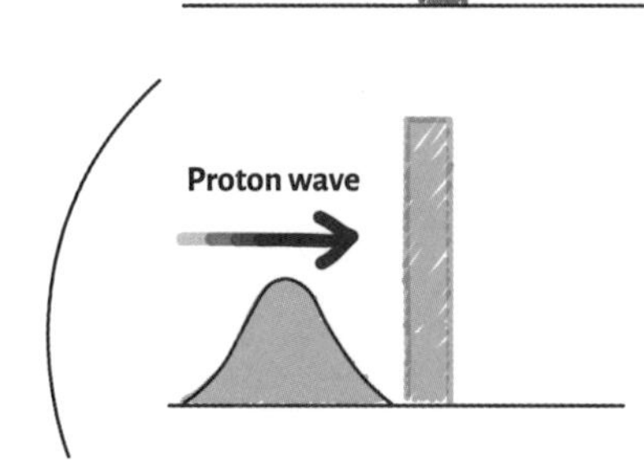

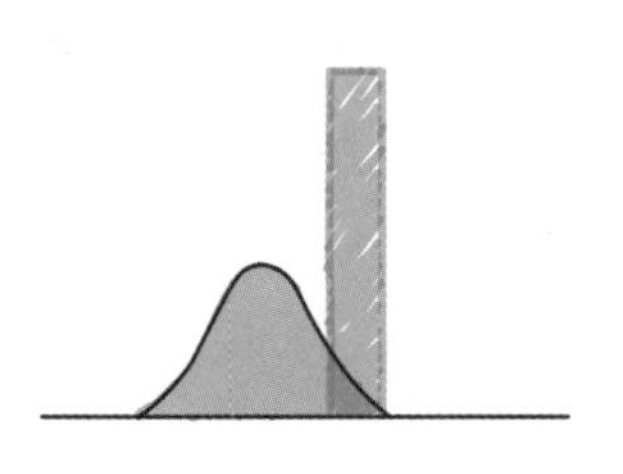

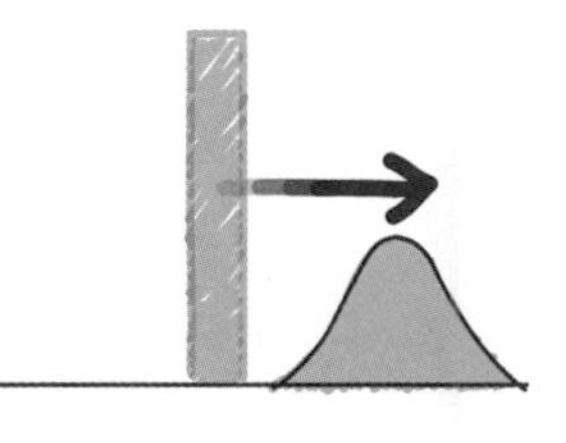

Erwin Schrödinger (1887–1961) invented **wave mechanics** and speculated the existence of a **probability wave function** to describe the likely position of a particle.

Phenomena that would be impossible under standard physics are made possible by the principle of **quantum mechanics**. The conditions for two protons to fuse in stars cannot be explained under Newtonian mechanics. It is the fuzzy wave nature of quantum mechanics that allows two protons to bind to form deuterium and so forth, via **quantum tunneling**.

THE STANDARD MODEL

So far, you have become familiar with particles within atoms, such as protons, neutrons, and electrons. The Standard Model breaks things down into smaller subatomic particles, classified as elementary particles: They are subdivided further into fermions and bosons. The Standard Model was proposed in 1970. All the elementary particles we know of were classified, and others were predicted.

Fermions and bosons

Fermions are the very building blocks of matter. They are further subdivided into **quarks** and **leptons**. There are many types of quarks: up, down, charm, strange, top, and bottom. Up and down quarks make protons and neutrons.

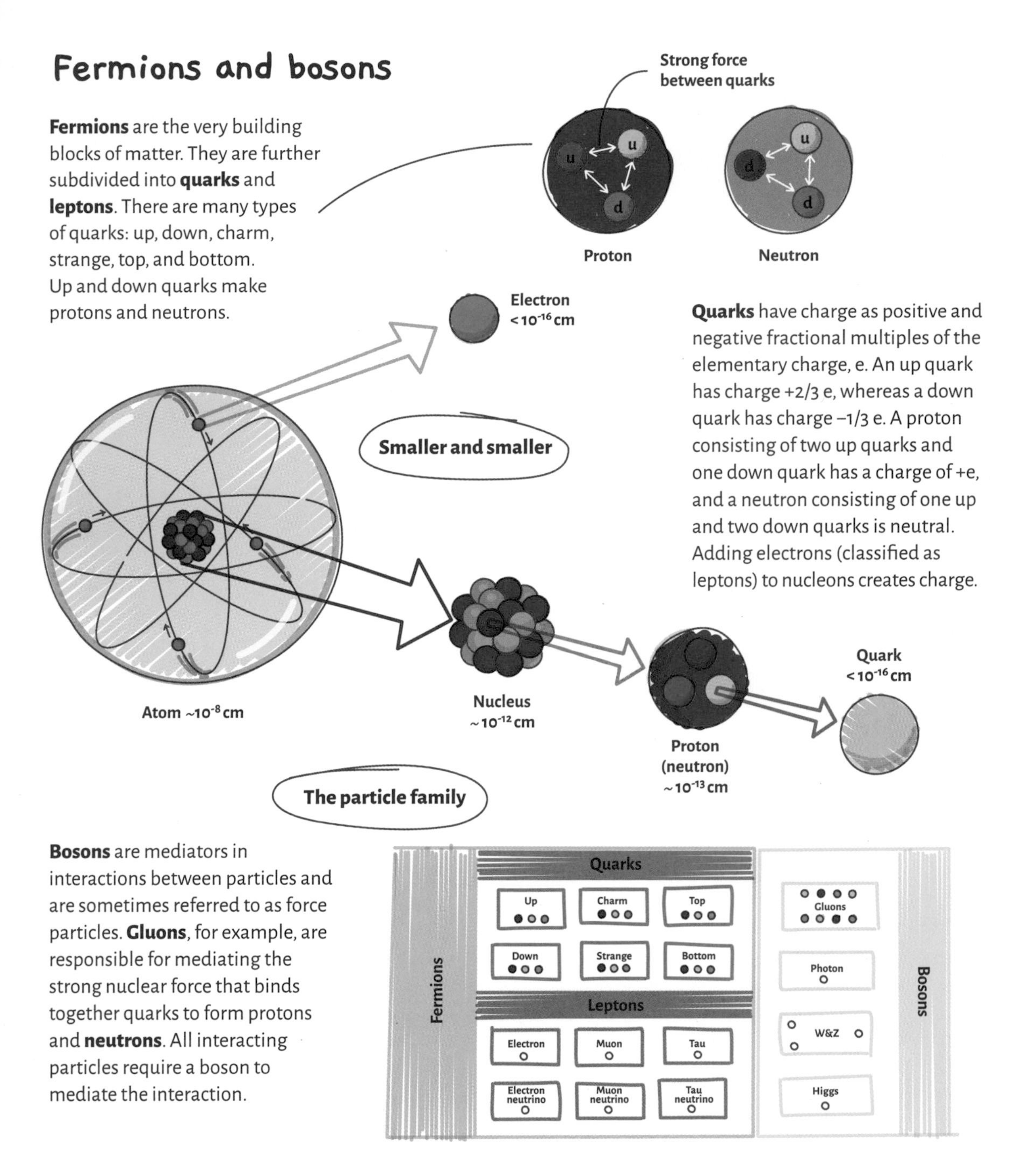

Quarks have charge as positive and negative fractional multiples of the elementary charge, e. An up quark has charge +2/3 e, whereas a down quark has charge −1/3 e. A proton consisting of two up quarks and one down quark has a charge of +e, and a neutron consisting of one up and two down quarks is neutral. Adding electrons (classified as leptons) to nucleons creates charge.

Bosons are mediators in interactions between particles and are sometimes referred to as force particles. **Gluons**, for example, are responsible for mediating the strong nuclear force that binds together quarks to form protons and **neutrons**. All interacting particles require a boson to mediate the interaction.

Subatomic particle properties

Fundamental particles have different quantum mechanical properties, such as **charge** (measured in fractions of e), **spin**, **color**, and **mass**.

This table summarizes the subatomic classification system and includes the year that the particle was predicted and when it was subsequently detected.

Family			Particle	Predicted/discovered		Charge (e)
Fermions	Quarks	u	Up quark	1964	1968	+2/3+
		d	Down quark	1964	1968	-1/3+
		c	Charm quark	1970	1974	+2/3+
		s	Strange quark	1964	1968	-1/3-
		t	Top quark	1973	1995	+2/3+
		b	Bottom quark	1973	1977	-1/3-
	Leptons	e	Electron	1874	1897	-1⅓-
		u	Muon	0	1936	-1-
		T	Tau	0	1975	-1-
		ve	Electron neutrino	1930	1956	-1-
		vμ	Muon neutrino	1940s	1962	-1-
		Vtau	Tau neutrino	1970s	2000	0
	?	p	Proton	1815	1917	0
		n	Neutron	1920	1932	0
Bosons	Vector	g	Gluon	1962	1978	+1+
		Y	Photon	0	1899	0
		w	W boson	1968	1983	±1±
		z	Z boson	1968	1983	0
	?	H	Higgs boson	1964	2012	0

SEMICONDUCTORS

Existing electrically between insulators, such as rubber, and conductors, such as copper, semiconductors are able to switch conductive properties on and off as needed. This electrical versatility makes them extremely useful.

Electrons in a conductor are free to move, so that when a voltage is applied across a conductor, a current flows. This is not the case with insulators, as there are no free electrons. **Semiconductors** can assume characteristics of both electrical conductors and insulators due to their atomic structure. A typical semiconductor material is **silicon**, and this is used widely in electronics.

Types of semiconductor

A silicon atom has 14 protons and 14 neutrons, with 14 electrons orbiting in their electron shells.

In the outer shell of a silicon atom, there are only four electrons where there could be eight (known as **valence electrons**). Neighboring silicon atoms share electrons in what is known as **valence bonding** and bond to form a crystal-like structure.

If an impurity with three valance electrons is added, such as boron, there will be an unoccupied shell hole. Neighboring electrons will periodically fill the hole, but it will keep moving from one place to another. This is called a **p-type semiconductor**. When a voltage is applied across the semiconductor, the random movement of the electron hole becomes organized in one direction, creating a measurable current.

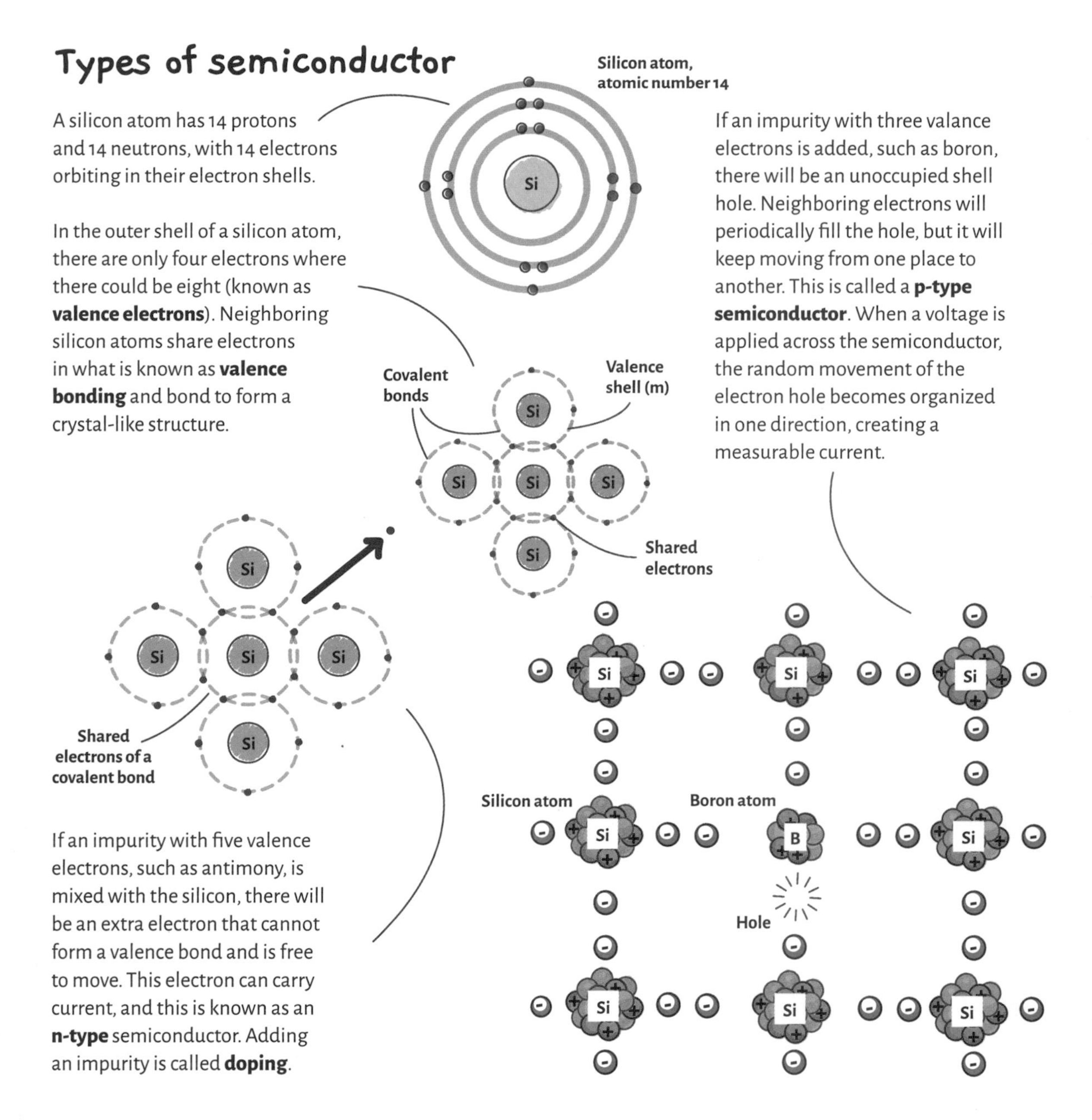

If an impurity with five valence electrons, such as antimony, is mixed with the silicon, there will be an extra electron that cannot form a valence bond and is free to move. This electron can carry current, and this is known as an **n-type** semiconductor. Adding an impurity is called **doping**.

Semiconductor uses

A semiconductor's ability to conduct electric current depends on several factors: the type of semiconductor, what it is doped with and the level of doping, and also the temperature of the semiconducting material.

Unlike metals, which increase in resistance as they get hotter, a semiconductor's resistance decreases rapidly.

These properties enable semiconductors to be used in different ways in a huge variety of electric appliances, such as cell phones, computer hardware, memory cards, and so on.

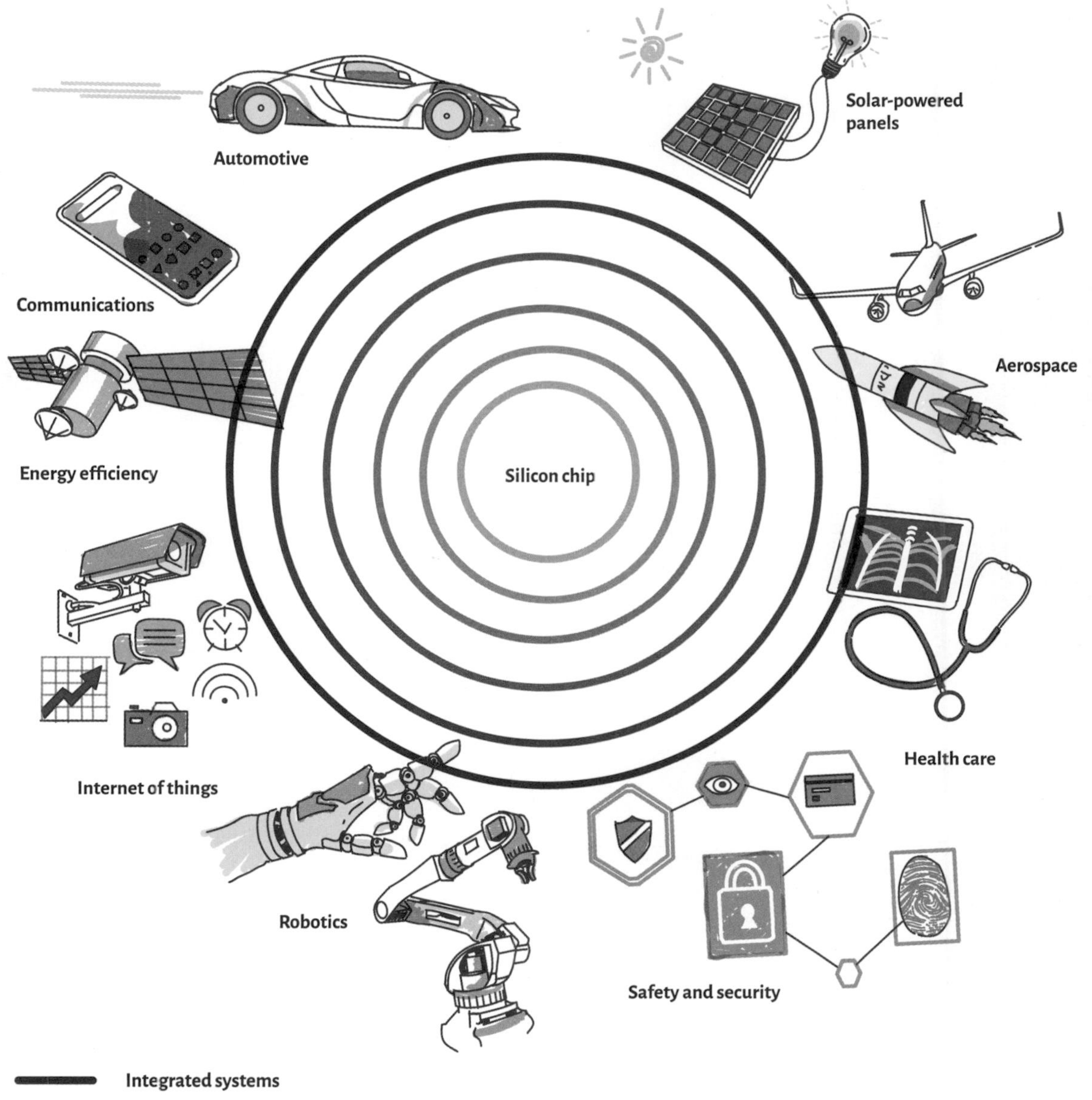

Silicon chip semiconductors are at the heart of modern technology. The concentric circles rippling out from the core indicate the layers of complexity that semiconductors have made possible and how they have graduated from powering modest integrated circuit boards to controlling highly sophisticated integrated systems. Almost everything you depend on in modern life owes its existence to silicon chips.

MODERN PHYSICS

SPECIAL RELATIVITY

EINSTEIN'S THEORY

The relative velocity between two separate observers can never exceed the speed of light.

TIME DILATION

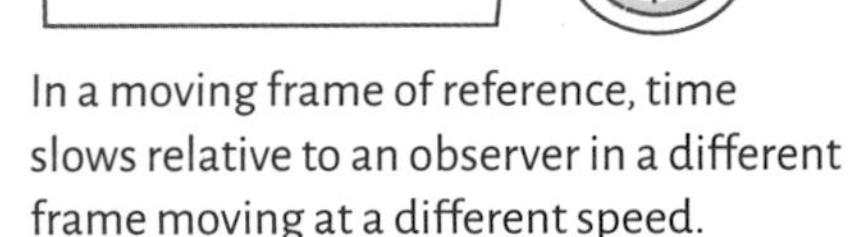

In a moving frame of reference, time slows relative to an observer in a different frame moving at a different speed.

GENERAL RELATIVITY

THE PRINCIPLE OF EQUIVALENCE

There is no difference between an accelerating frame of reference in a gravitational field and one that is accelerated by an external force.

SPACE-TIME CURVATURE

Gravity warps space-time.

QUANTUM PHYSICS

SEMICONDUCTORS

Switch between conducting and insulating states; used in all areas of electronics.

THE STANDARD MODEL

Classifies fundamental particles into fermions (building blocks) and bosons (force particles).

QUARKS

Subatomic particles possessing properties such as charge, spin, color, and mass.

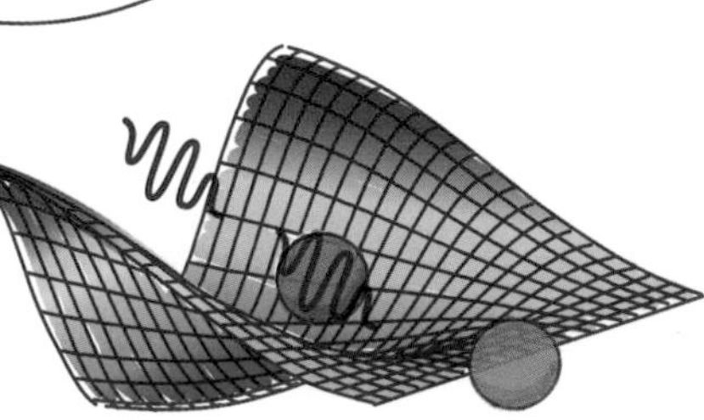

A COLLABORATION OF MINDS

Quantum physics was a series of discoveries and refinements.

MAX PLANCK

Quantum theory.

WOLFGANG PAULI

Exclusion principle.

ERWIN SCHRÖDINGER

Probability wave function.

WERNER HEISENBERG

Uncertainty principle.

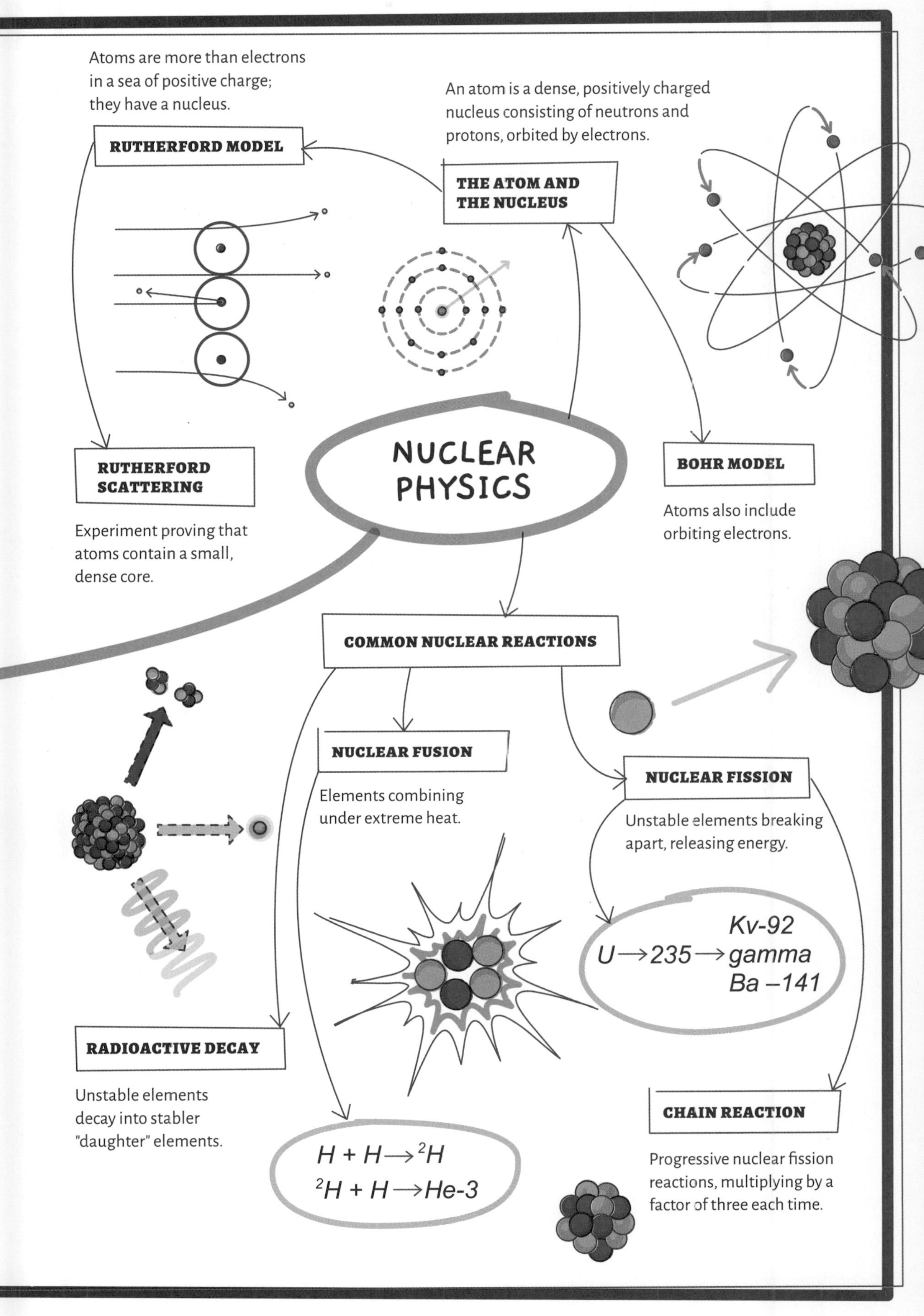

Atoms are more than electrons in a sea of positive charge; they have a nucleus.
RUTHERFORD MODEL
An atom is a dense, positively charged nucleus consisting of neutrons and protons, orbited by electrons.
THE ATOM AND THE NUCLEUS
RUTHERFORD SCATTERING
Experiment proving that atoms contain a small, dense core.
NUCLEAR PHYSICS
BOHR MODEL
Atoms also include orbiting electrons.
COMMON NUCLEAR REACTIONS
NUCLEAR FUSION
Elements combining under extreme heat.
NUCLEAR FISSION
Unstable elements breaking apart, releasing energy.
U⟶235⟶gamma
Kv-92
Ba –141
RADIOACTIVE DECAY
Unstable elements decay into stabler "daughter" elements.
H + H⟶²H
²H + H⟶He-3
CHAIN REACTION
Progressive nuclear fission reactions, multiplying by a factor of three each time.

CHAPTER 13

ASTROPHYSICS

Astrophysics is a relatively new and yet ancient area of physics. Astronomers have existed since the dawn of humanity, but they were limited by their equipment and could only make rudimentary observations. It was the advent of powerful new ground-based telescopes, such as the VLT (Very Large Telescope) in Chile with its 8.2 m (27 ft) diameter reflecting mirrors that enabled us to start to unlock and explore the wonders of the universe. When the HST (Hubble Space Telescope) was launched in 1990, with its unrivaled imaging capabilities, astronomy and astrophysics entered a new and exciting epoch.

STELLAR EVOLUTION

The closest star to us is the sun, defined as being one astronomical unit away (1AU). Every day, it radiates life-sustaining energy to Earth and has done so for billions of years with unfaltering consistency. Yet the sun is just an average star.

Disk

Protostar

Jet

The birth of a star

Nebulae consist of giant clouds of dust and gas, which are predominantly made up of hydrogen and helium, as well as traces of other elements. These heavier elements were formed billions of years ago in giant stellar explosions called **supernovas**.

Regions of gas within a nebula that are dense enough start to flow toward a center point due to the attraction caused by gravity. As huge regions of gas collapse toward a common center, the cloud's density increases with decreasing volume, and its rotation also increases as angular momentum is conserved. Eventually, the central sphere of gas becomes dense enough to ignite nuclear fusion, and a star is born!

The surrounding disk rotates around the new star, clumping together under gravity to form planets, which will orbit the star for the duration of its lifetime.

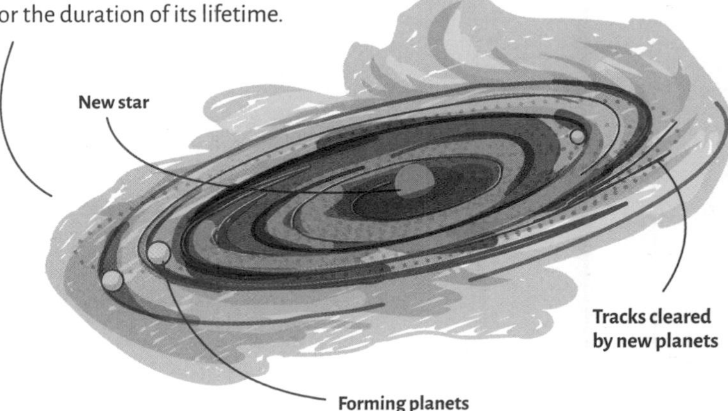

A star will predominantly burn hydrogen at its core by the process of nuclear fusion (see page 155), turning its hydrogen into helium. The rate at which this happens and the final fate of the new star is determined by its mass when it is born.

Horsehead nebula

The life and death of the sun

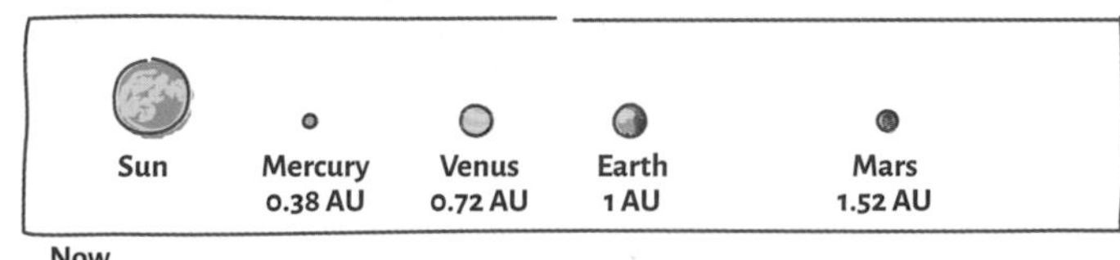

Now

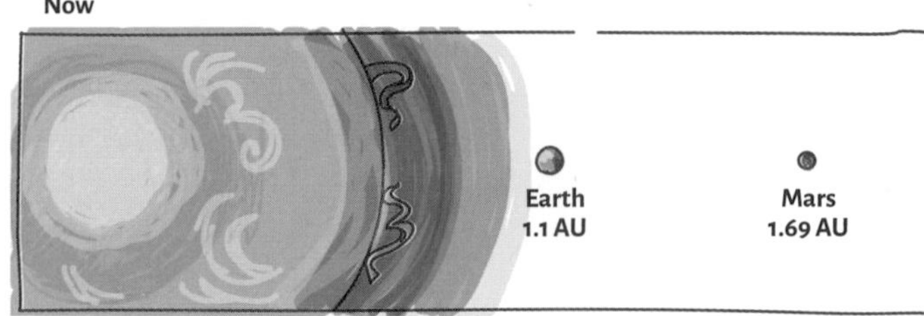

7.588 billion years in the future

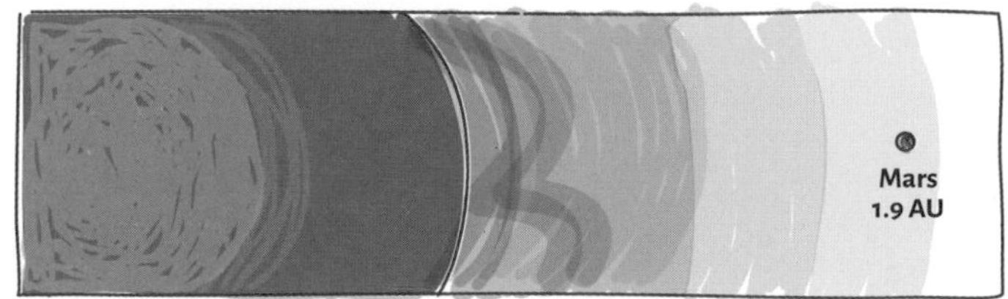

7.59 billion years in the future

Small stars such as our own sun tend to live a long and stable life, burning fuel at their centers at a moderate rate. This is ideal for sustaining life on orbiting planets, as the time frame for their stability is billions of years, allowing life to develop and flourish.

Our sun has been in its current state for more than 4.5 billion years and will continue to burn fuel at the same rate for about another 5–6 billion years. This stage of a star's life is known as the main sequence, and our sun is referred to as a **main sequence** star.

Our sun's surface temperature is currently about 5,800 K. The hydrogen fuel at its center will start to deplete in approximately 5–6 billion years. The central region will then become unstable and will start to collapse. The outer layers of the sun will begin to expand and cool, becoming redder in color—this is the **red giant** phase.

The expansion of the outer layers continues while the central hot region starts to gently radiate gas into space. As the outer layers become completely detached, the central hot region is exposed—this is known as a **white dwarf.** A white dwarf is about the size of Earth and is extremely hot (approximately 25,000K). White dwarfs do not generate their own energy but are simply the cooling remnants of the star's core.

As the outer layers recede from the white dwarf, the star enters the final phase of its life cycle, known as a **planetary nebula**—a hot glowing gas ring surrounding the white dwarf at its center.

Engraved Hourglass nebula (MyCn 18)

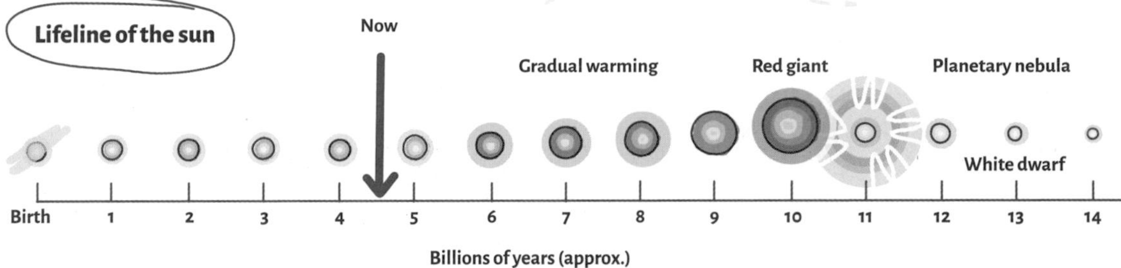

The life and death of massive stars

The life cycle of stars exceeding 10 solar masses is very different and far more spectacular. More massive stars burn their fuel at a much faster rate and have a surface temperature of 10,000–50,000 K. Regulus is a nearby **blue giant** star approximately 80 light-years from Earth, with a surface temperature of nearly 13,000 K.

Blue giant stars usually remain stable for between 100 million and 1 billion years, before their fuel starts to run out. At this point, they become very unstable and will terminate in a violent death: a **supernova explosion**.

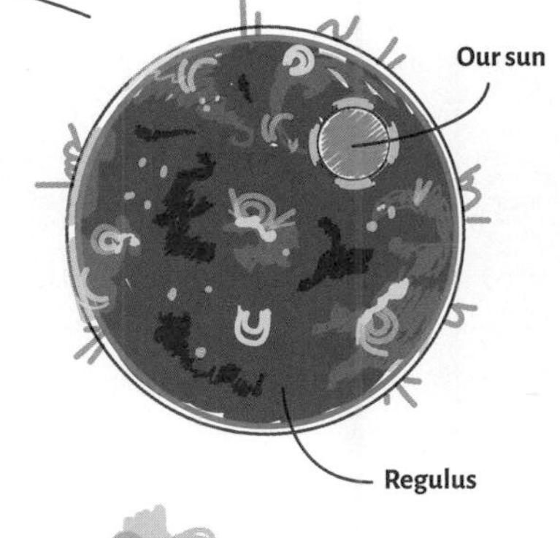

The exhausted core suddenly collapses, which causes a very large surge in temperature. The result is an explosion, which creates massive shock waves radiating out through the star's outer layers, causing very high-density ripples. These high-density regions provide the mechanisms to fuse elements heavier than hydrogen and helium, which forge all the other elements that exist in the universe.
A supernova explosion lights the sky with the power of an entire galaxy for weeks or even months.

The leftover remnant may be a **neutron star**, with the mass of approximately 1.4 solar masses compressed into a sphere about 20 km (12 miles) across.

If the mass of the star is even higher, the remaining core will not be able to stop its own collapse, and the result will be a **black hole**, with a central region of such high density that all light within its grasp is unable to escape.

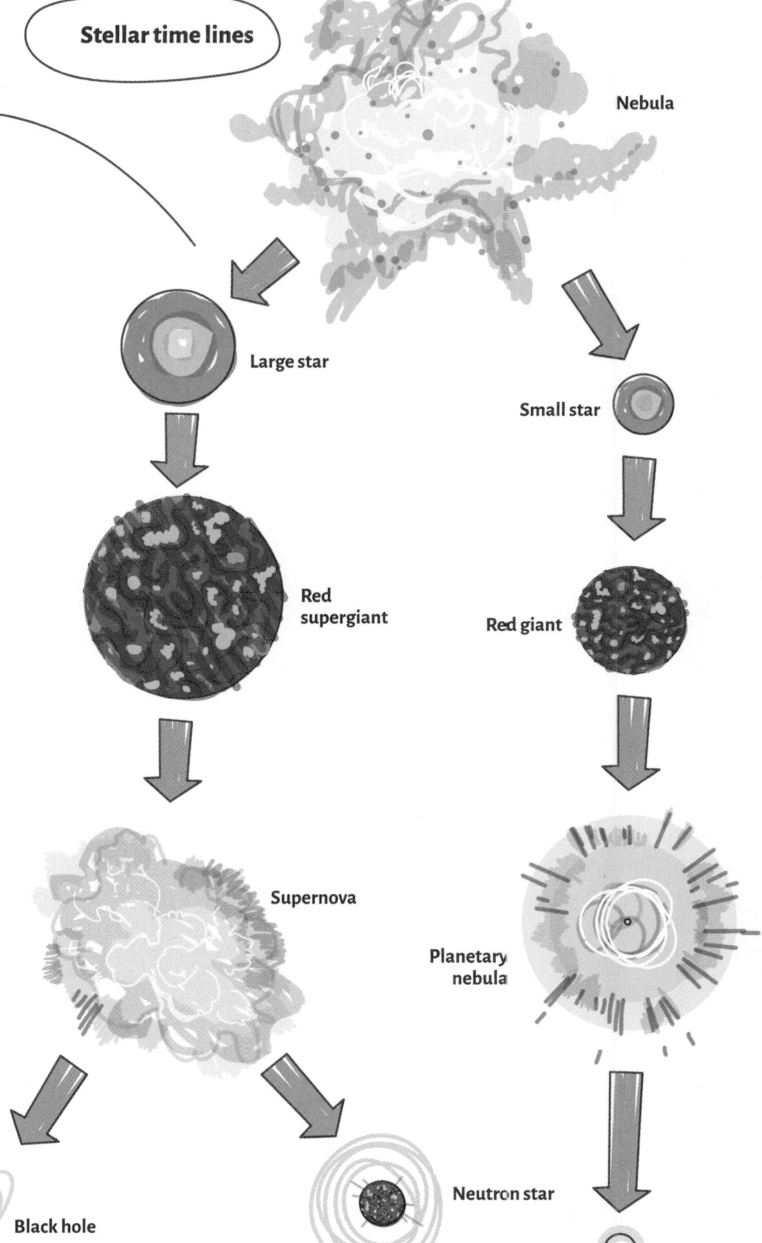

THE HERTZSPRUNG-RUSSELL DIAGRAM

On the Hertzsprung-Russell diagram, the temperature of a star is represented by the *x*-axis, and the luminosity (power output) of the star by the *y*-axis. The scale of the axes is logarithmic (nonlinear) due to the vast range of temperatures and luminosities exhibited by the range of stars present in the galaxy.

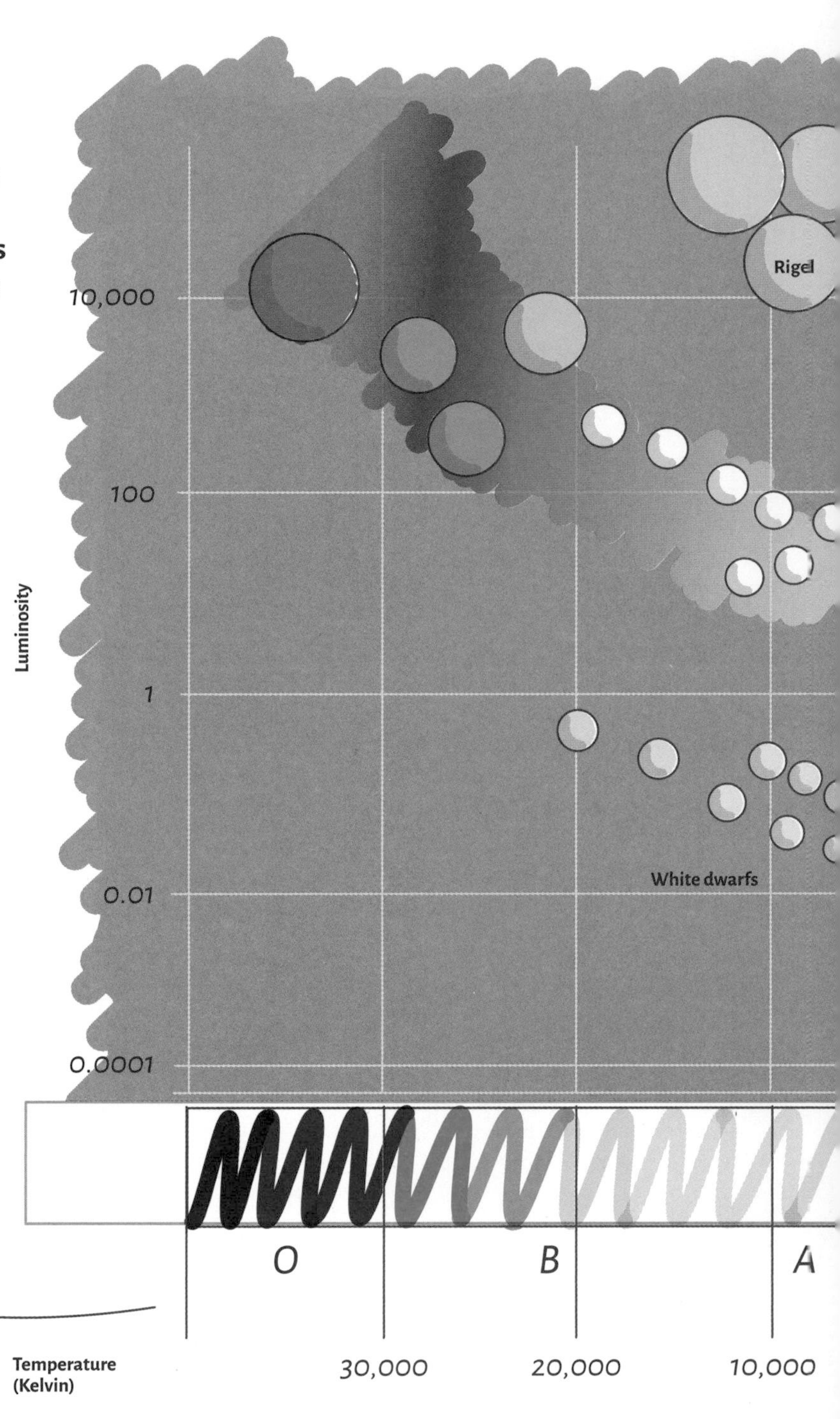

HEAT AND LIGHT

This system is somewhat difficult to visualize, and so stellar type is plotted against stellar luminosity, which groups stars together and shows their relative brightness and temperature. This is called the **Hertzsprung-Russell (HR) diagram**. It was created independently around 1910 by Ejnar Hertzsprung and Henry Norris Russell and represented a major step toward our understanding of stellar evolution.

The scale is unusual in that the horizontal axis shows decreasing temperature from left to right, whereas the vertical axis increases as multiples of 10 solar luminosities. More complex versions show diagonal lines sloping downward from left to right—these group stars according to their radii, shown as multiples of solar radii.

The classification of stars depends on their size, color, temperature, and **luminosity**. A star's luminosity is a measure of its brightness or power and is given in watts, *W* (joules/s). The sun has a luminosity of 4×10^{26} W, whereas the blue supergiant Rigel has a luminosity of approximately 120,000 times that of our sun.

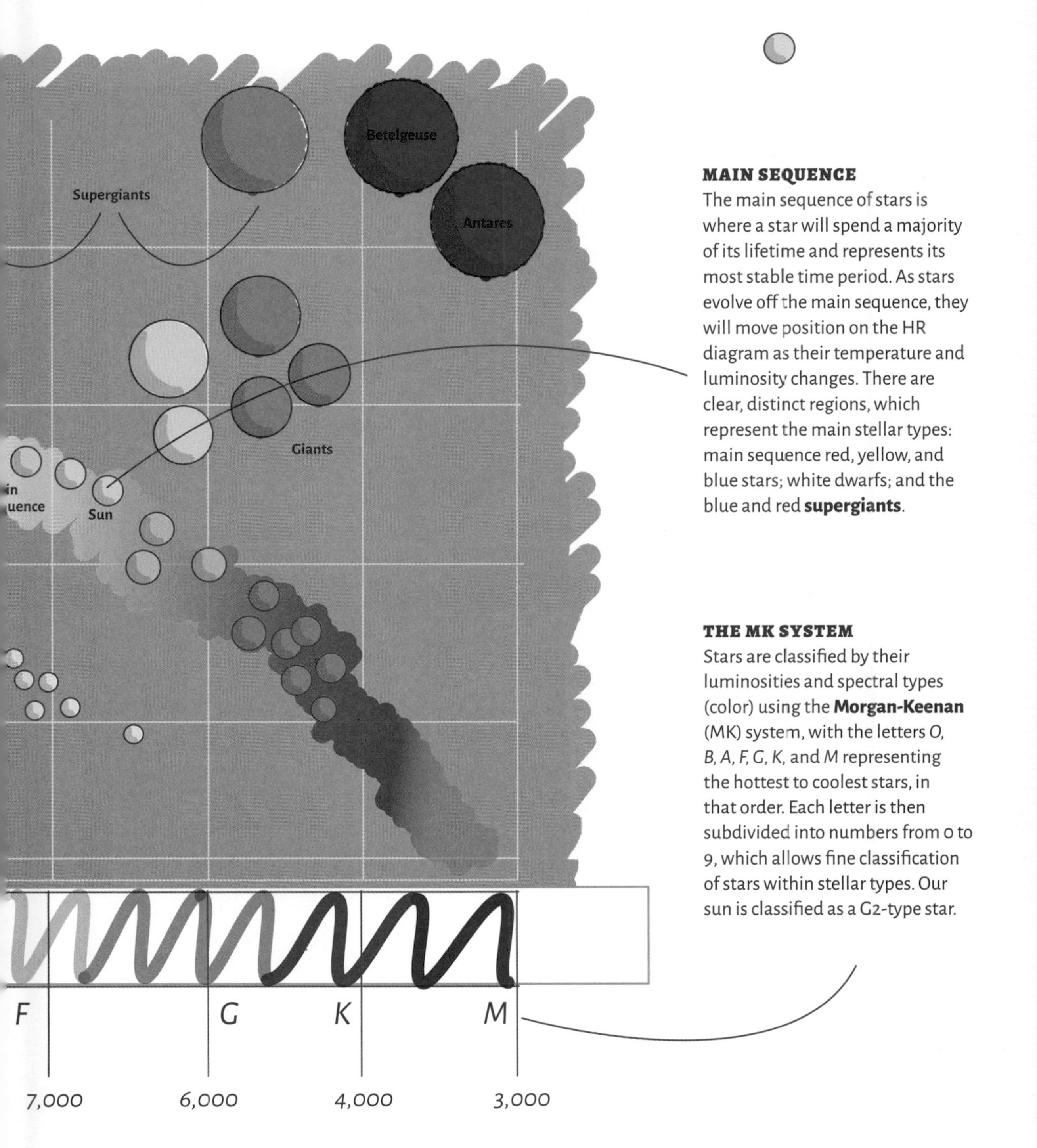

MAIN SEQUENCE
The main sequence of stars is where a star will spend a majority of its lifetime and represents its most stable time period. As stars evolve off the main sequence, they will move position on the HR diagram as their temperature and luminosity changes. There are clear, distinct regions, which represent the main stellar types: main sequence red, yellow, and blue stars; white dwarfs; and the blue and red **supergiants**.

THE MK SYSTEM
Stars are classified by their luminosities and spectral types (color) using the **Morgan-Keenan** (MK) system, with the letters *O, B, A, F, G, K,* and *M* representing the hottest to coolest stars, in that order. Each letter is then subdivided into numbers from 0 to 9, which allows fine classification of stars within stellar types. Our sun is classified as a G2-type star.

THE DYNAMICS OF GALAXIES

Our galaxy is called the Milky Way. It is a fairly large spiral galaxy consisting of billions of stars and is approximately 100,000 light-years across and 1,000 light-years thick at its center. Our solar system is in one of the Milky Way's spiral arms, at approximately 26,000 light-years from its center.

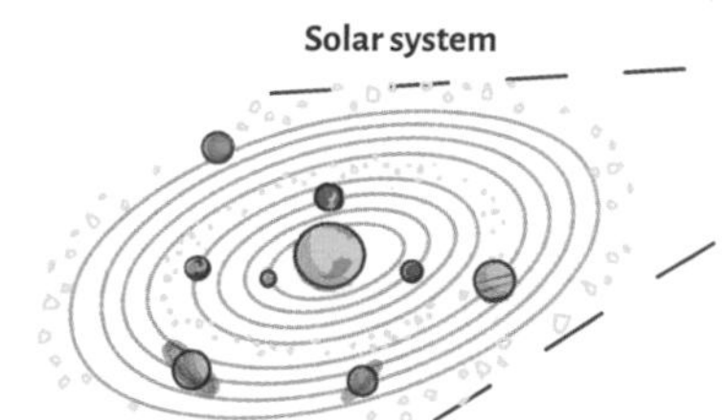

Galaxies

The universe consists of billions of galaxies, each containing millions or billions of stars, depending on their size. Galaxies started to form about 2–2.5 billion years after the big bang. Gas had then cooled enough to start to collapse into regions of space, increasing in density and creating the first clusters of stars.

As the regions collapsed, their rotation speeds increased as angular momentum was conserved, and galaxies started to display their disklike spiral structure. Stars rotate around the centers of spiral galaxies at speeds in excess of 800,000 km/h (approximately 500,000 mph).

Very old distant galaxies are usually elliptical in nature and do not have a spiral structure. This is probably due to galactic collisions in the early universe causing galaxies to combine their rotational dynamics, creating **elliptical galaxies**.

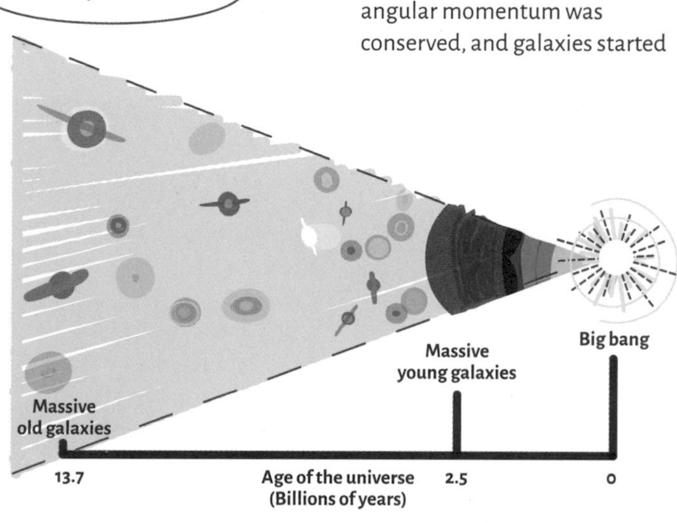

GALAXY SHAPES

It is thought that there is a central massive black hole in all galaxies. Indeed, this may be essential to their creation, their dynamics, and their evolution into different types. Galaxy shape is known as **morphology** and can be categorized into distinct classifications.

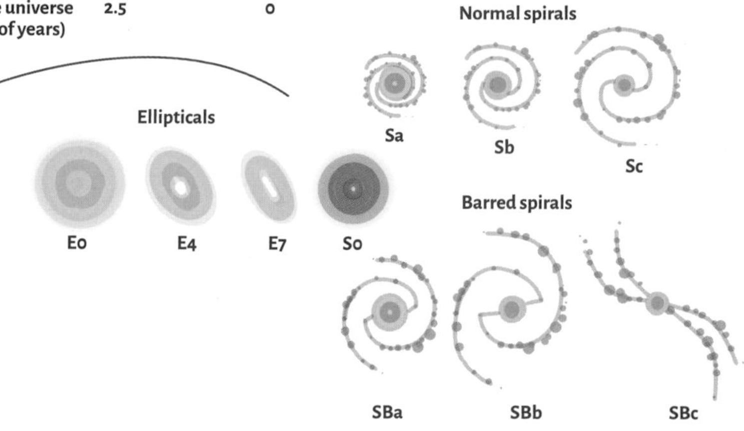

Rotational velocity

The rotation of galaxies is a legacy of the dynamics that created them. In the same way as an ice skater spins faster as she pulls her arms in toward her body (see page 58), a shrinking mass of rotating gas that will form a galaxy also spins faster.

The force due to gravity of all the mass inside the galaxy holds it together, while its **rotation speed** tries to tear it apart, in the same way as a child on a spinning roundabout has to hold on so as not to fly off. The sun is moving around our galactic center at approximately 240 km (150 miles) per second and is held in rotation by the gravitational pull of all the mass inside its circular path.

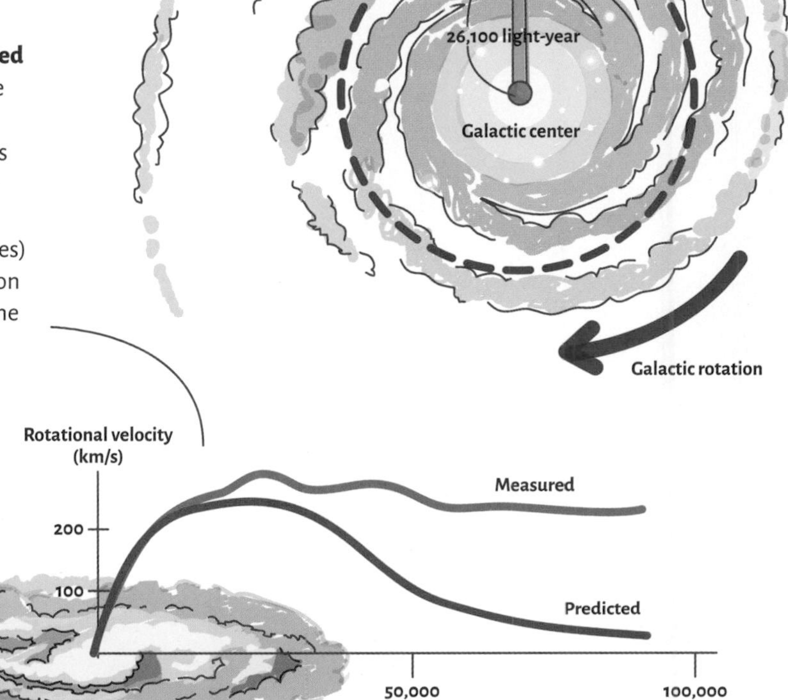

DARK MATTER

Astronomers can estimate the total mass within the galaxy by measuring the brightness of all the stars. Based on this observation, they then make an estimate of the number of stars there are in the galaxy. The number of stars in the Milky Way falls short by a long way, and the mass we see cannot account for the mass needed to keep the galaxy from flying apart. The predicted rotation velocities of stars along the galactic radius is far lower than the actual observed velocities, particularly at larger distances from the galactic center.

The question then arises, what is keeping the galaxy together? There are a number of proposed possibilities, the most popular of which is the existence of **dark matter** that we cannot see.

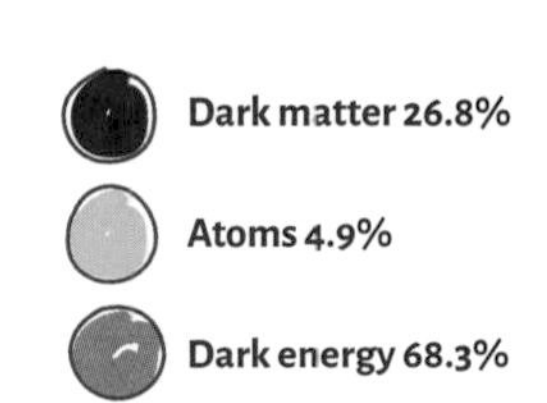

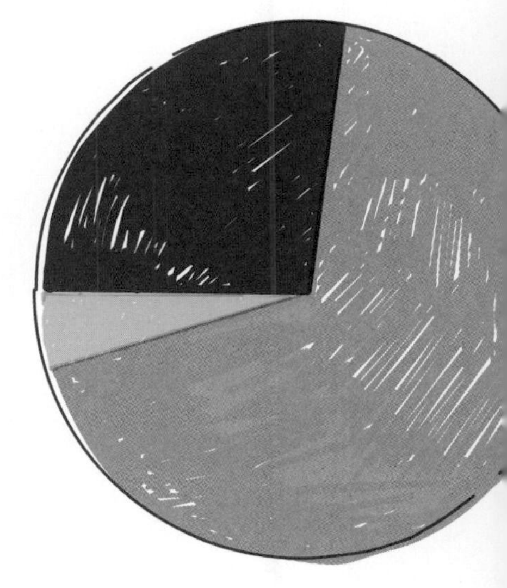

REDSHIFT AND RECESSION VELOCITY

Galaxies are moving at very high speeds as the universe expands. Light from the galaxies observed using large ground-based telescopes allows astronomers to measure their speeds relative to Earth. They do this by splitting their light up into all its wavelengths, through a process called spectroscopy.

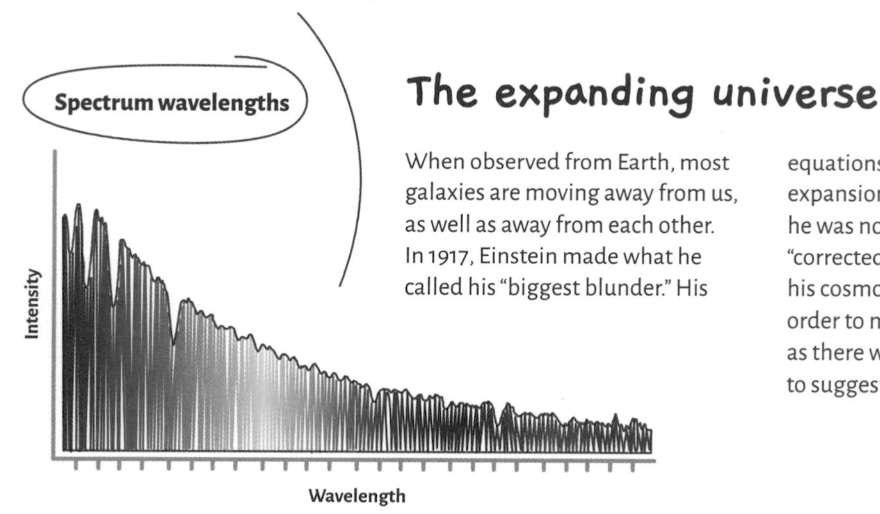

The expanding universe

When observed from Earth, most galaxies are moving away from us, as well as away from each other. In 1917, Einstein made what he called his "biggest blunder." His equations correctly predicted the expansion of the universe, a result he was not convinced by. He then "corrected" this mistake by creating his cosmological constant, Λ, in order to make the universe static, as there was no obvious evidence to suggest it was expanding.

Today, we can measure the expansion of the universe and the speeds at which galaxies are receding from us by measuring **redshift**. Redshift is similar to Doppler shift (see page 111) and has the effect of stretching light emitted by galaxies.

Expansion rate

Expansion

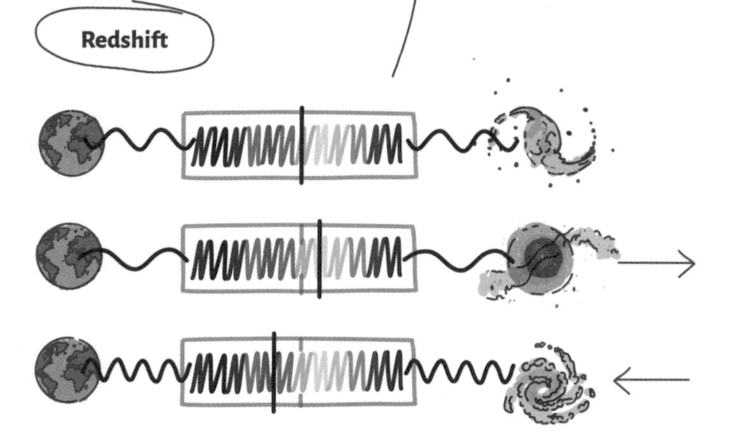

The term redshift refers to the effect of blue light emitted from the observed galaxy being stretched to a longer wavelength and appearing redder to the observer. A galaxy moving toward the observer appears blue due to its spectrum being blue-shifted. Only very localized galaxies and stars exhibit blue shift, as their local motion in the universe is toward Earth.

Recession velocity

Consider a number of galaxies receding or speeding away from us as the universe expands. Those galaxies that are farther away are experiencing more rapidly expanding space, which is observed as higher **recession velocity**—the rate at which the galaxies are moving away from Earth. The farther away they are, the faster they move.

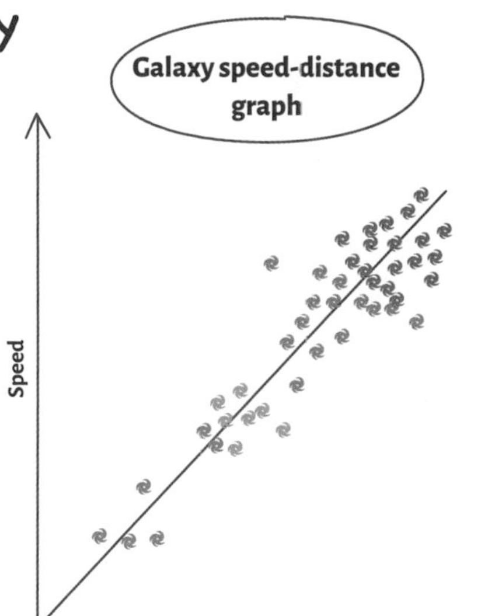

In a similar way to the ambulance moving away from us on page 111, the frequency of light observed by telescopes on Earth is of a lower frequency and a higher wavelength than that at which it was emitted from the galaxy. The shift in wavelength is proportional to the recession velocity, allowing astronomers to determine the galaxy's speed as it is moving away from us.

The spectrum of a galaxy has various known reference points, which occur as **emission** and **absorption lines**. These lines are caused by elements within the gas of the galaxy and can be easily identified. On Earth, the lines appear at exact known wavelengths, called the **laboratory reference frame**. When the light from a distant galaxy is observed, the lines appear redshifted to a different lower frequency and larger wavelength. The amount of shift measured corresponds to the recession velocity of the galaxy.

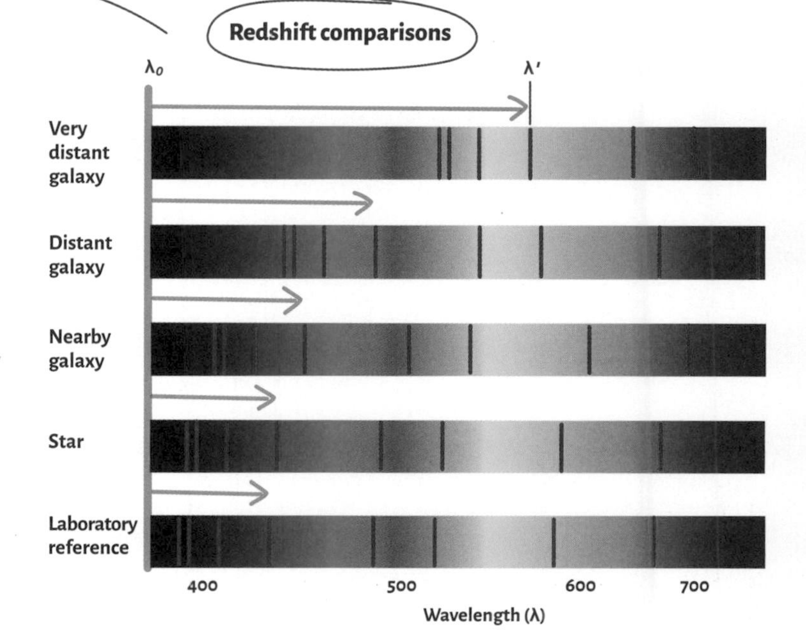

Redshift, z, is calculated by measuring the change in wavelength, $\Delta \lambda$, of the known emission or absorption line and dividing it by the emitted wavelength, λ, giving the relationship:

$$z = \frac{\Delta \lambda}{\lambda}$$

Redshift, z, is also equal to the ratio of the recession velocity and the speed of light, c (providing v is much less than c), giving the further relationship:

$$z = \frac{v}{c}$$

Using these two equations, we can determine the speed of a distant galaxy fairly accurately.

HUBBLE'S CONSTANT

Edwin Powell Hubble (1889–1953) was an American astronomer and a contemporary of Einstein. Hubble is universally regarded as the scientist who proved categorically that the universe is made up of billions of distant galaxies that are receding from us at high velocity. The Hubble Deep Field (HDF), an image taken of a tiny region of "dark sky" by the Hubble Space Telescope (HST), confirms this.

Hubble also discovered through careful observation that the recession speed of distant galaxies, v, is directly proportional to their distance, d. This is known as Hubble's law and can be stated by the following formula:

$$v = H_0 d$$

where H_0 is Hubble's constant.

1 Megaparsec = 3.1×10^{22} km or 1.9×10^{19} miles

The value of this has been hotly discussed over the decades since its emergence and has varied between 50 and 100 km/s/Mpc.

This is due to inconsistencies in observational data and a lack of reliable methods to determine the distance to galaxies. It is generally accepted to be 71 km/s/Mpc based on more accurate data today.

REDSHIFT AND RECESSION VELOCITY

Hubble's constant can be used to calculate the age of the universe. Rearranging his equation gives the following relationship:

$$\frac{d}{v} = \frac{1}{H_0}$$

Since distance over velocity gives the unit of time, converting the unit of Hubble's constant into standard units of distance yields what is referred to as the Hubble time, which is approximately 14 billion years.

Hubble's law

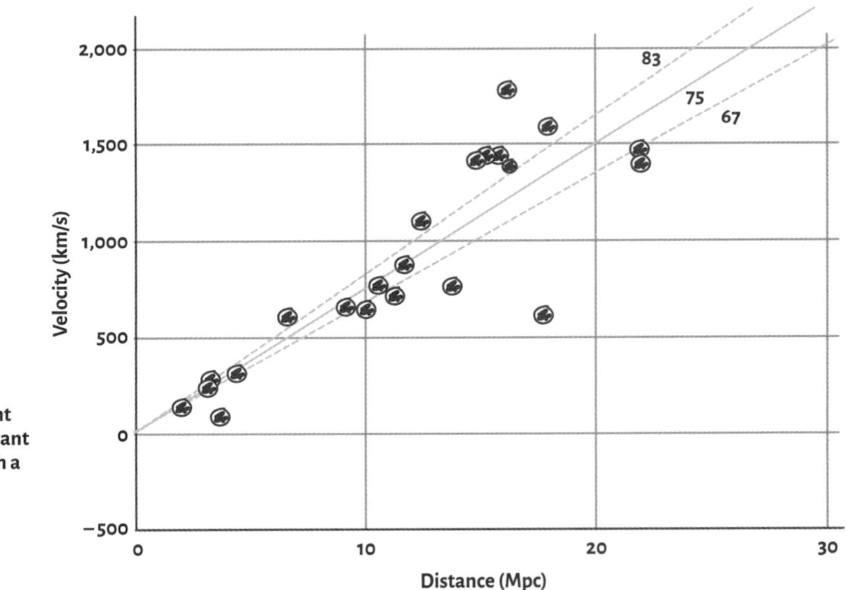

The unit of Hubble's constant states that for each Mpc distant a galaxy is, it is receding with a velocity of 71 km/s.

FINDING QUASARS

Quasar J043947.08+163415, the brightest **quasar** yet discovered in the early universe, was identified by combined high-resolution images from ground-based observatories and the Hubble telescope.

Quasars (quasi-stellar objects) are supermassive black holes, emitting up to a thousand times the energy of the entire Milky Way across the whole electromagnetic spectrum.

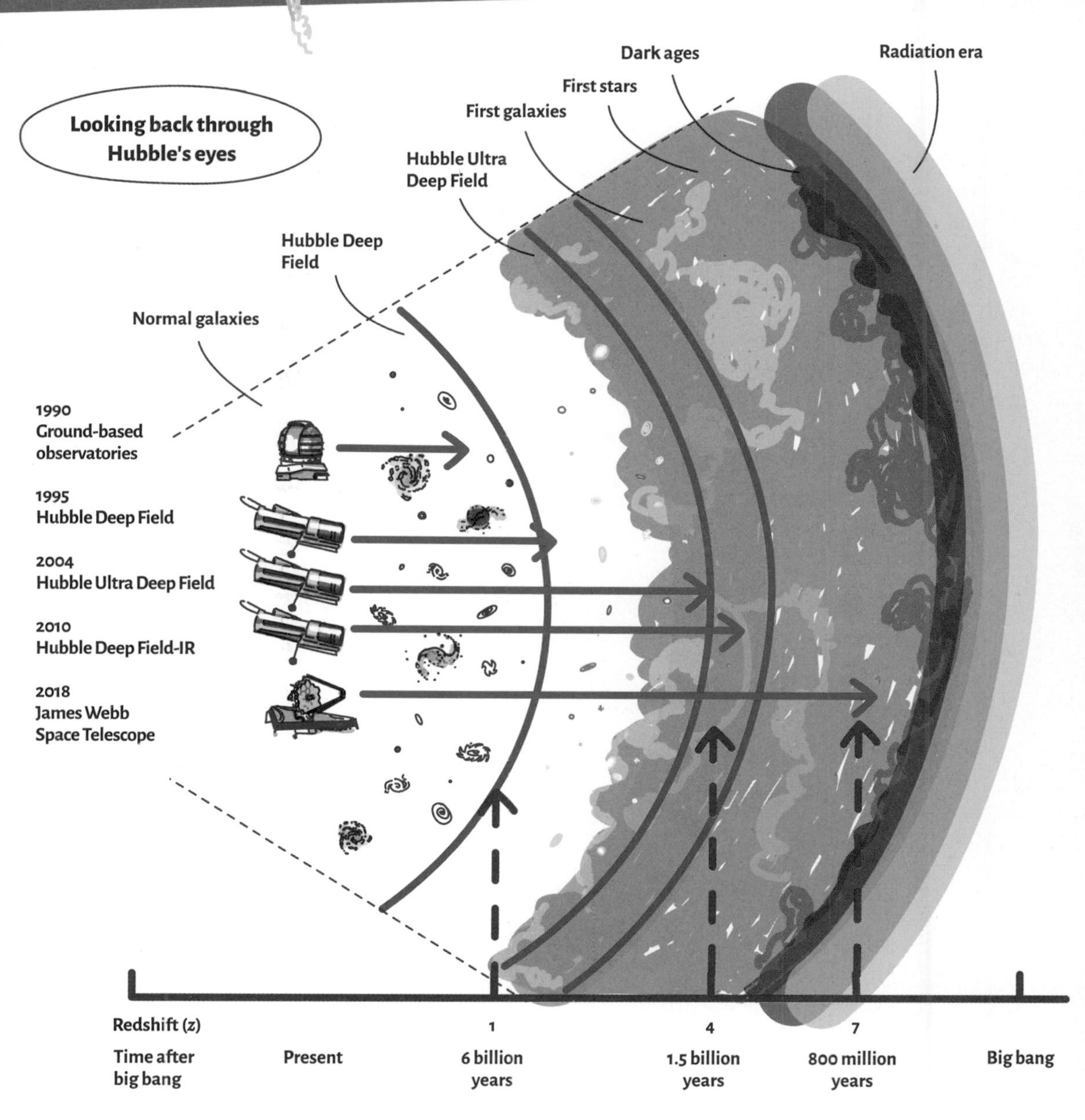

THE BEGINNING OF THE UNIVERSE

The universe had a beginning. Astrophysicists believe that at one point in time, the entire mass of all we see and all we cannot see was concentrated in an infinitely small and dense point called a singularity, and that this point of infinite mass exploded into being, creating the universe.

The big bang theory

The **big bang** is actually a misnomer—it had no size and made no noise. It was infinitely small, and there was no air to transmit any sound.

Astrophysicists believe that approximately 14 billion years ago, a singularity of infinite density exploded, and the universe came into being.

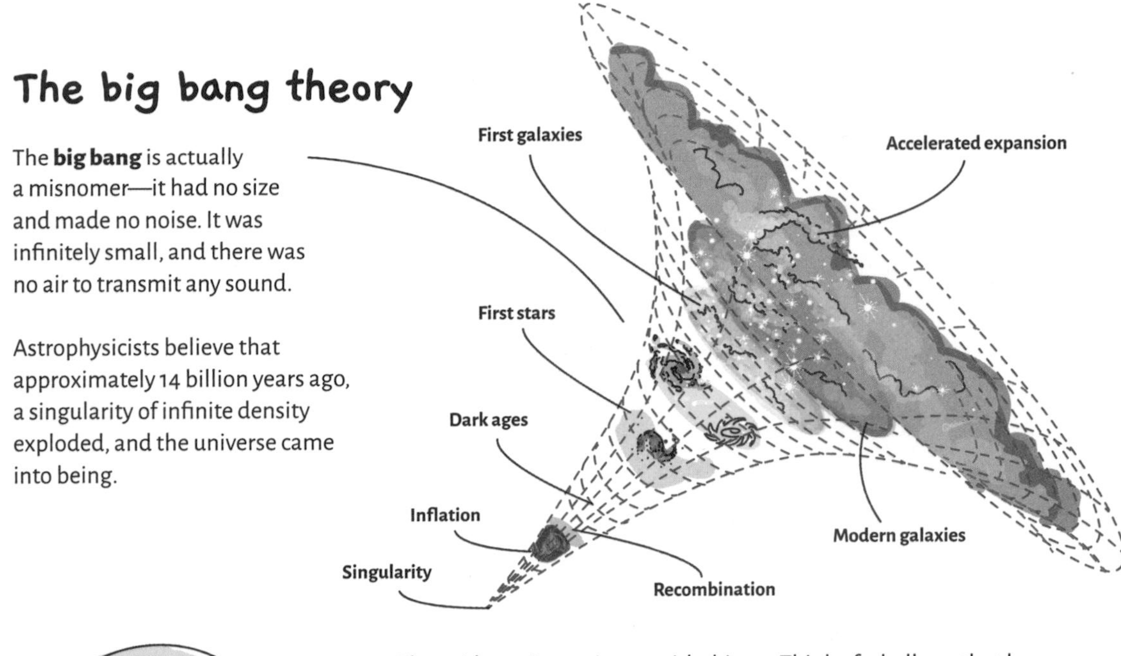

The evidence is consistent with this theory. All galaxies observed from Earth are moving away from us as their light is redshifted according to their recession velocities. The more distant a galaxy is, the faster it is receding.

Think of a balloon that has a number of dots drawn on its surface, representing galaxies in the universe. As the balloon is inflated, it expands, and the distance between dots increases. This is a model of the expanding universe.

Every dot or galaxy observed from every other dot or galaxy is moving away with a speed that is proportional to the distance of their separation. Observations of galaxies from Earth support this idea—the entire universe is expanding, and all galaxies are, on average, moving away from each other.

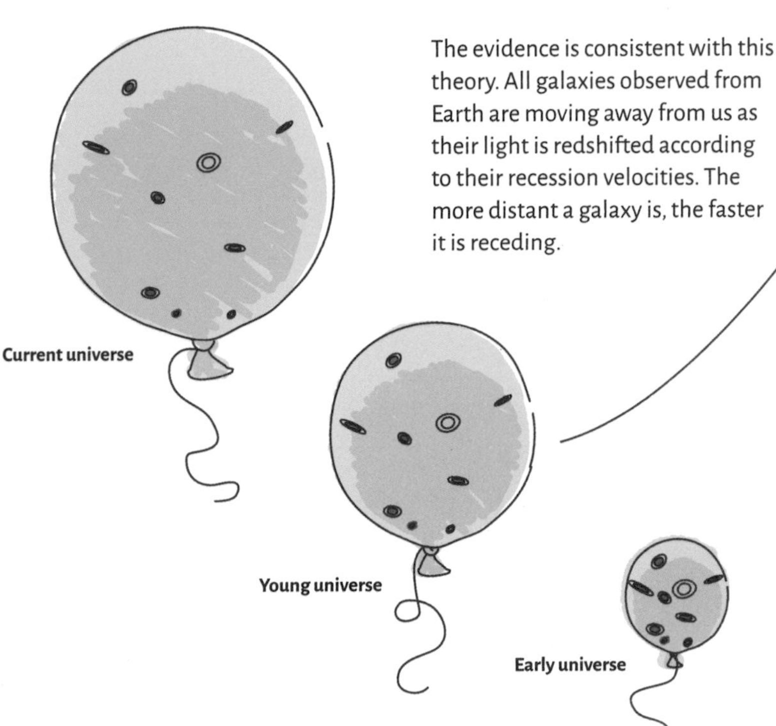

The cosmic microwave background

In 1964, American radio astronomers Arno Penzias (b. 1933) and Robert Wilson (b. 1936) accidentally discovered an echo of the big bang, for which they received the Nobel Prize in Physics in 1978. Scanning the sky for radio frequencies, the two astronomers recorded a consistent radio signal. This was anomalous, and they initially thought it to be the result of contamination by pigeon droppings. After careful cleaning of the equipment, Penzias and Wilson realized that the signal was authentic; they later confirmed that it was emitted from all regions of the sky and varied slightly in intensity.

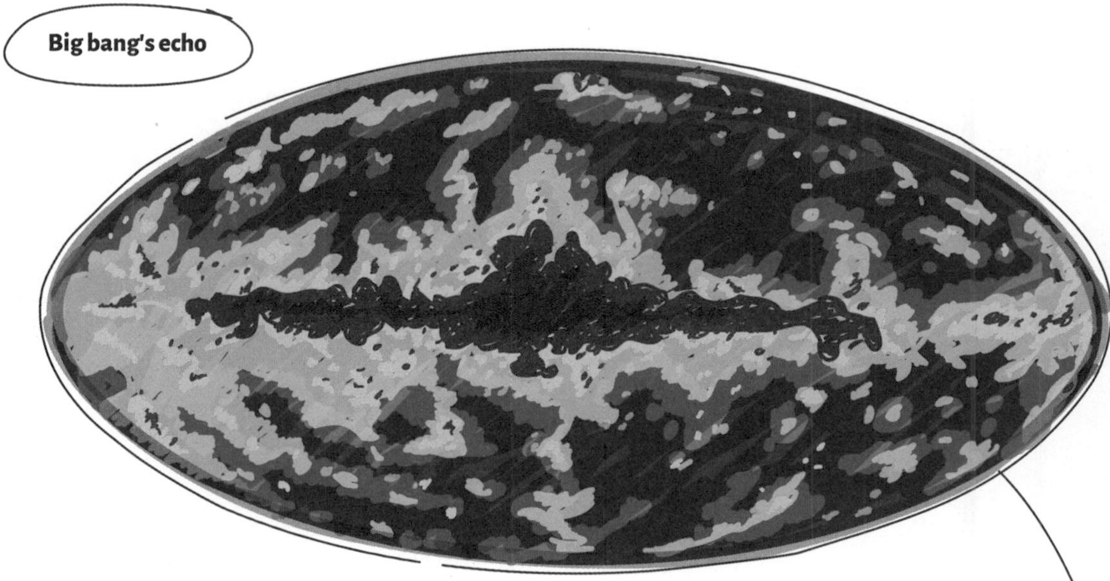

The microwave signal was consistent with an ambient temperature in space of approximately 3 K and was attributed to the afterglow of the big bang. It was christened the cosmic microwave background (CMB) based on its frequency in the electromagnetic spectrum (see pages 92 and 94).

The temperatures and energies immediately after the big bang were enormous. Radiation emitted during this epoch has been stretched as the universe has expanded, up until the present day. The wavelength of the microwave background is consistent with very high energy radiation from shortly after the big bang (≈400,000 years ago) redshifted over a period of expansion equivalent to the Hubble time (see page 176) of approximately 14 billion years.

Back to the beginning

Expanded universe

Galaxies

Hot ionized gas

Back to the big bang

THE END OF THE UNIVERSE

As is clear from observation, the universe is expanding, but what we are unsure of is whether or not this expansion is slowing down. Much like a ball thrown in the air, which is slowed down and returns to Earth due to the force of gravity, the universe may be slowing itself down due to its own gravitational force. The elusive and cryptically named "dark energy" will undoubtedly complicate things further.

Visible matter

Dark matter

Critical density

Astronomers are in debate over the average density of the universe. All that you can see as starlight does not make up all that there is in the universe.

As the universe is a dark canvas, any dark objects are almost impossible to see, and their presence can only be inferred from the effect they have.

There is a very specific value of the density of the universe, which is the tipping point for its ultimate fate. It is called the **critical density**, and will determine one of the following fates for the universe: an open and accelerating universe with infinite expansion, or a closed and slowing universe followed by a collapse into the big crunch.

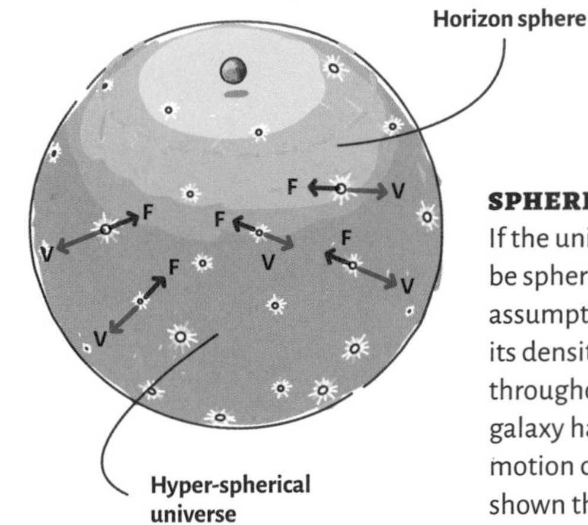

SPHERICAL UNIVERSE

If the universe is considered to be spherical (a highly flawed assumption due to relativity) and its density assumed to be constant throughout, and each expanding galaxy has force acting along its motion due to gravity, it can be shown that:

$$Pc = \frac{3H_0}{8\pi G} \approx 1.5 \times 10^{-26}\ kg/m^3$$

where H_0 is Hubble's constant.

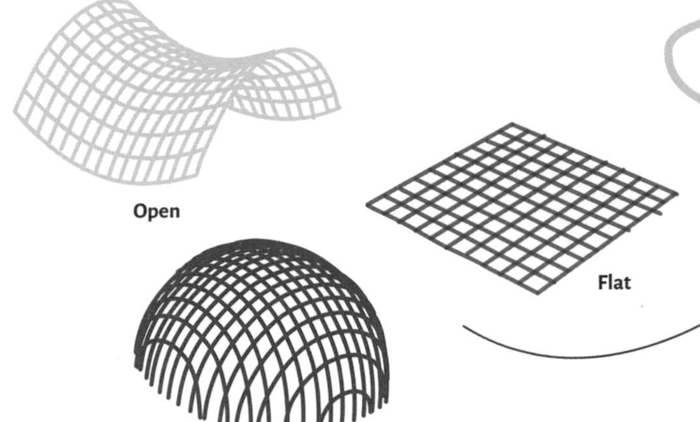

FLAT UNIVERSE

Geometrically, the universe can be viewed as a flat surface. Three states exist: open, flat, or closed, depending on the value of the density of the universe compared with the critical density.

repeating cycle.

The fates of the universe

This depends entirely on the value of the critical density.

If it is too small, there will not be enough gravitational force to slow down and reverse the universe's expansion, and galaxies will continue to expand forever. All the stars in every galaxy will eventually burn out, leaving a vast cold void of darkness. Philosophically, this seems to be at odds with the cyclical rebirth of nature and the grand design of the universe.

If it is too big, the expansion of the universe will slow, stop, and reverse, accelerating back toward another singularity.

The entire universe being crushed into a singularity may not sound favorable, but it may just be another beginning of the cycle.

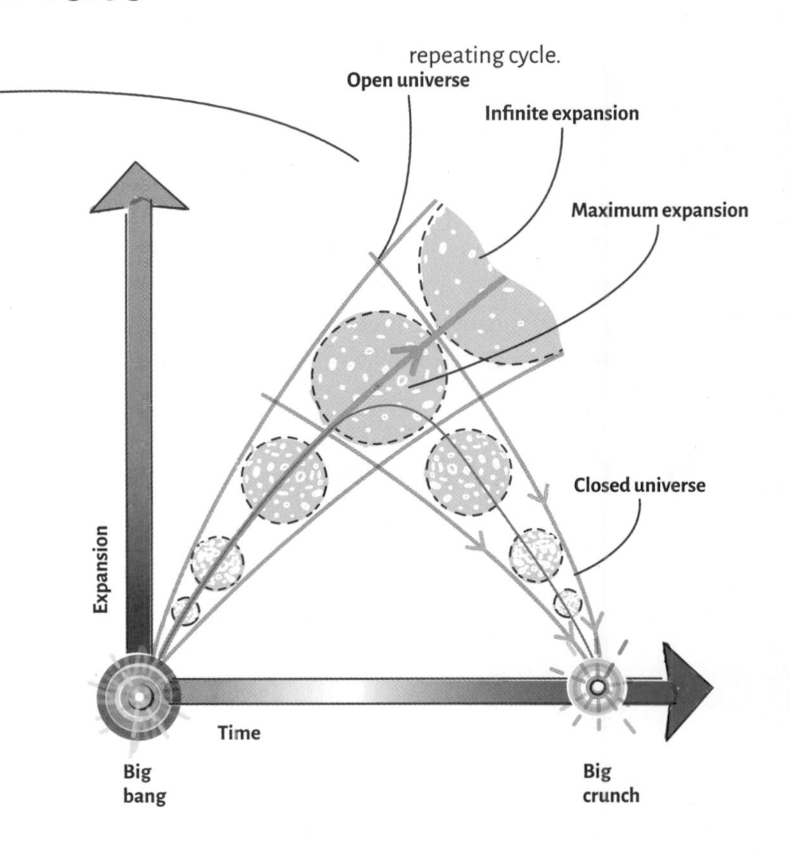

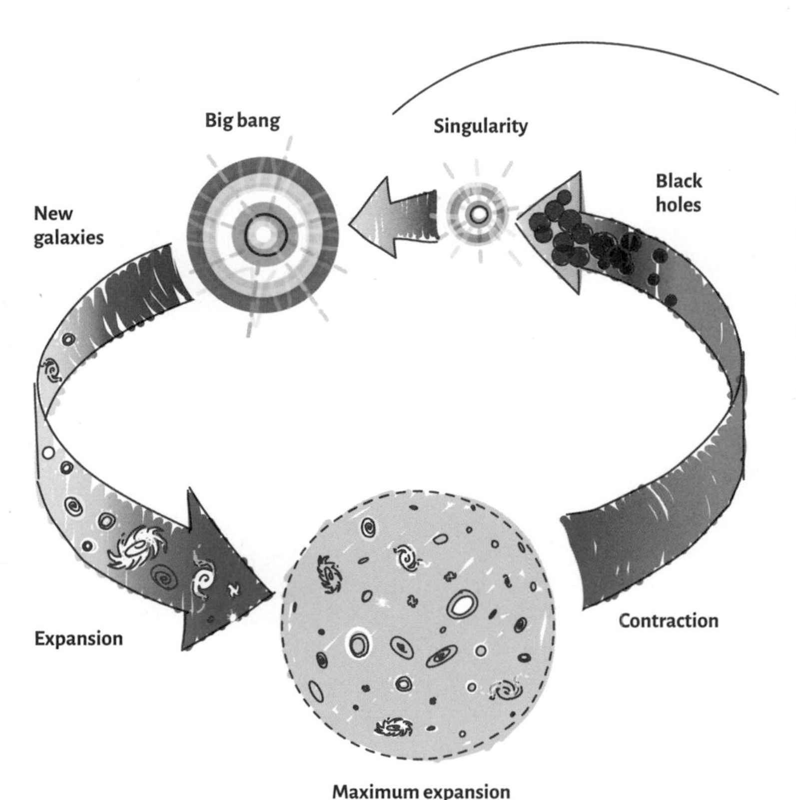

THE BIG CRUNCH

Once Einstein recovered from Hubble gaining credit for the discovery of the expanding universe, he set the value of his cosmological constant to zero, and in 1932 proposed, with his collaborator Willem de Sitter, that the universe would cycle through expansion and contraction, big bang to **big crunch**. This was known as the Einstein-Sitter model of the universe.

GRAVITATIONAL LENSING AND GRAVITATIONAL WAVES

General relativity connected space and time and led to the renaming of the three dimensions of space and the fourth dimension of time collectively as space-time. This reframing had profound effects on our understanding of the nature of the universe and how gravity, light, and time interact.

Gravitational lensing

Lensing is a property of waves that distorts their path and changes the resulting image of the object emitting the lensed light (see page 153). Gravity affects the path of light rays in space, changing their direction and lensing their light.

Under perfect conditions, the lensed object can be seen as a circle around a lensing galaxy. This is known as an **Einstein ring**.

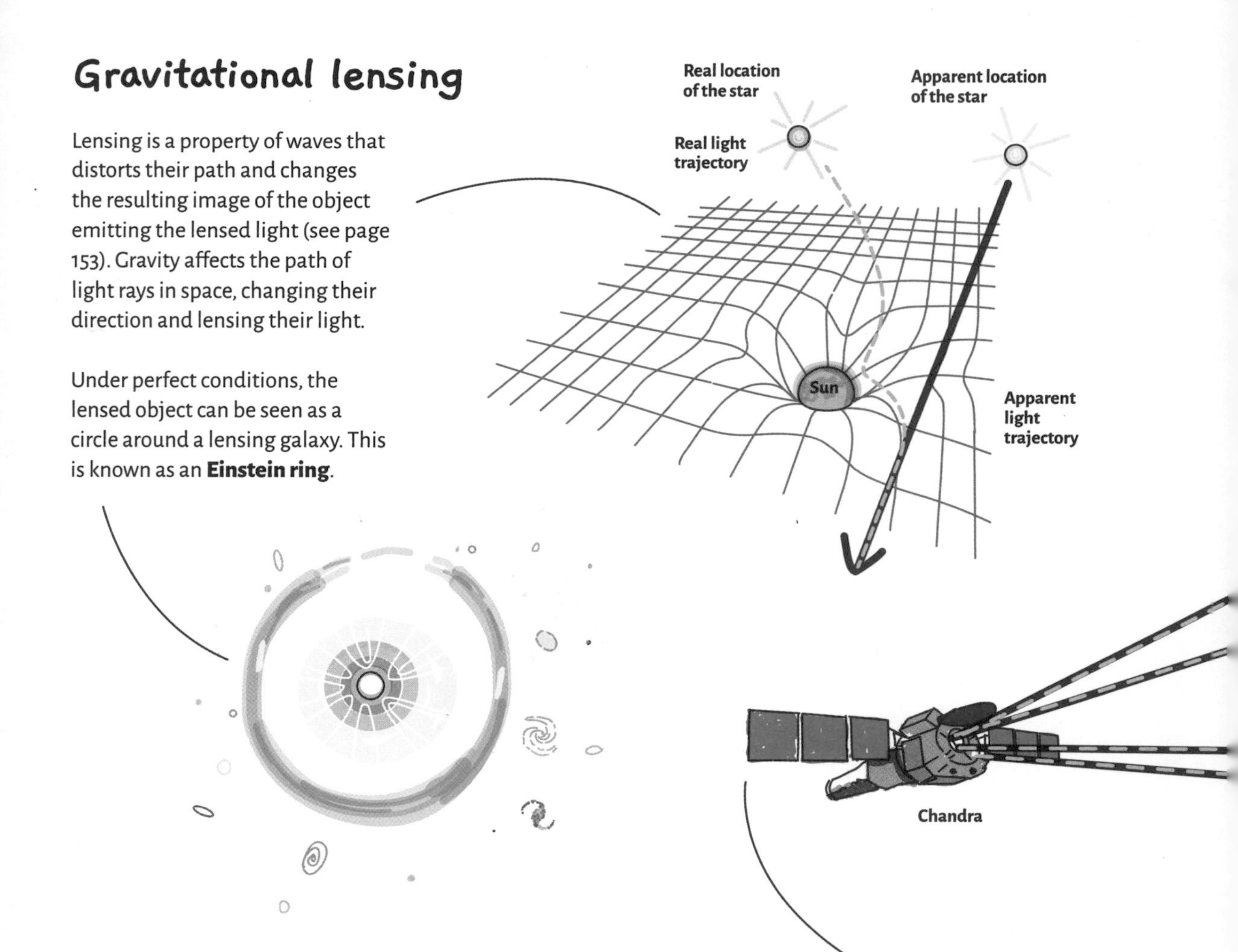

You can see this effect around our sun. A ray of light from a distant star passing close to the sun will be bent toward it and will change direction. Einstein predicted this effect, but it was difficult to prove due to the brightness of the sun. It was during a solar eclipse that the relative positions of three known stars close to the sun were recorded. (The only time stars are visible to us is in daylight). Their positions were then rerecorded at night, when their light was not passing close to the sun and their positions had changed.

Large galaxies also lens light. NASA's Chandra X-ray Observatory recorded such an event, where a known source was observed in multiple positions at once.

Gravitational waves

It was Einstein, once again, who postulated the existence of gravitational waves. These are ripples through space-time that radiate out from their source at light speed, according to Einstein's general theory of relativity. Gravitational waves occur when two very massive bodies (such as binary black holes) interact, losing energy and radiating it away as a disturbance in the fabric of space-time.

Despite their existence being predicted in 1915, gravitational waves were highly elusive and detectors were not sensitive enough to register any wave disturbances passing Earth, as these were diluted over the vast distances they traveled. They change the size and shape of objects as they pass, but the effect is so small, it is almost impossible to detect.

The LIGO project at Caltech in Pasadena, California, uses a giant Michelson interferometer (see page 124) with arms 4 km/2.5-miles long. The size and sensitivity of the instrument has enabled many gravitational wave events to be recorded, the first of which was in 2015.

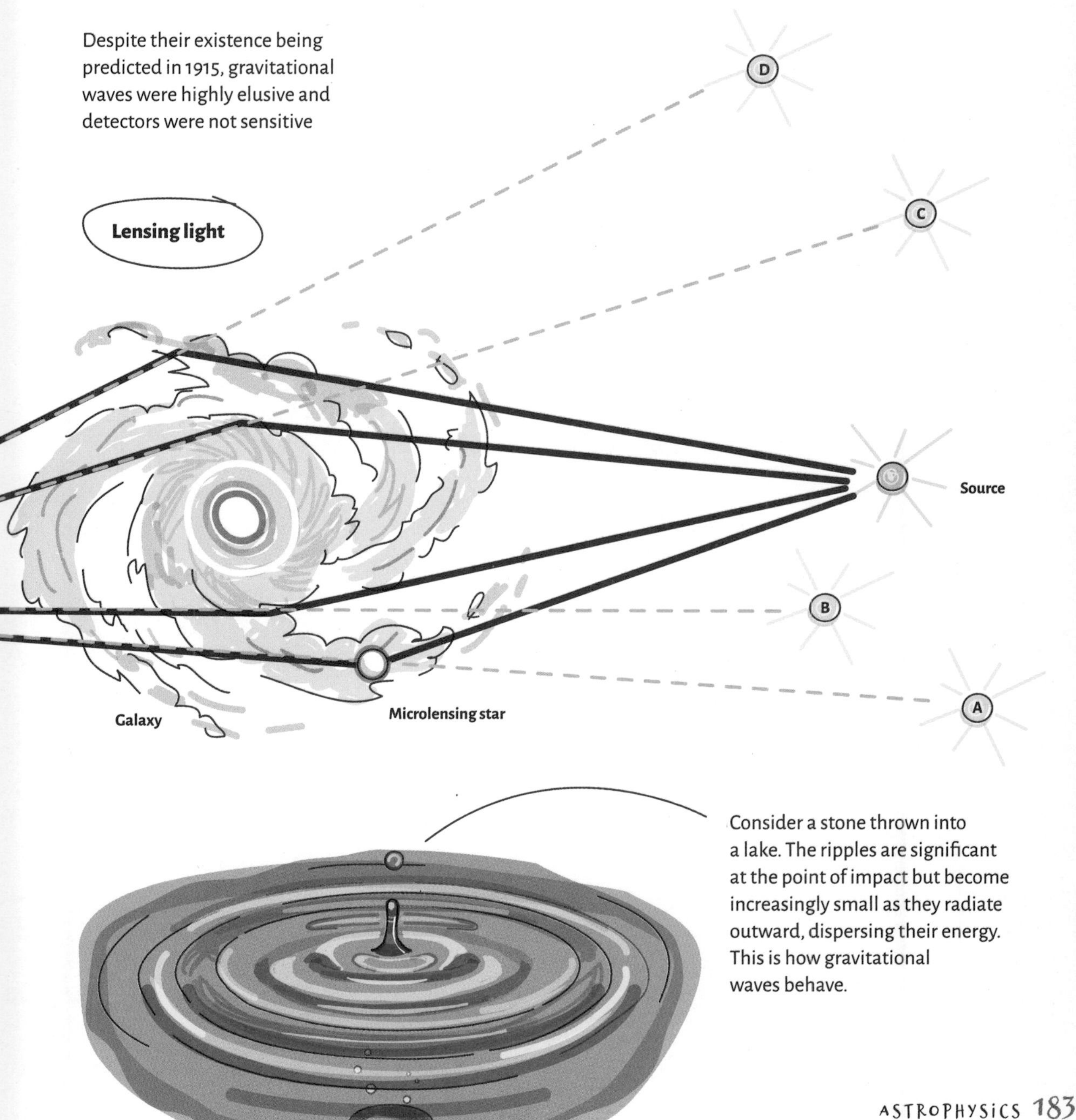

Consider a stone thrown into a lake. The ripples are significant at the point of impact but become increasingly small as they radiate outward, dispersing their energy. This is how gravitational waves behave.

BLACK HOLES

Exotic and mysterious, black holes have a profound impact on the space and objects that surround them. They are formed from the demise of supermassive stars, and it is thought that they reside in the middle of most galaxies, perhaps binding them together or creating the focus on which they formed.

What is a black hole?

All masses produce a gravitational field, causing an attractive force to masses close enough to its influence and distorting the region of space in close proximity to the mass. The more intense the gravitational field, the bigger its effect becomes.

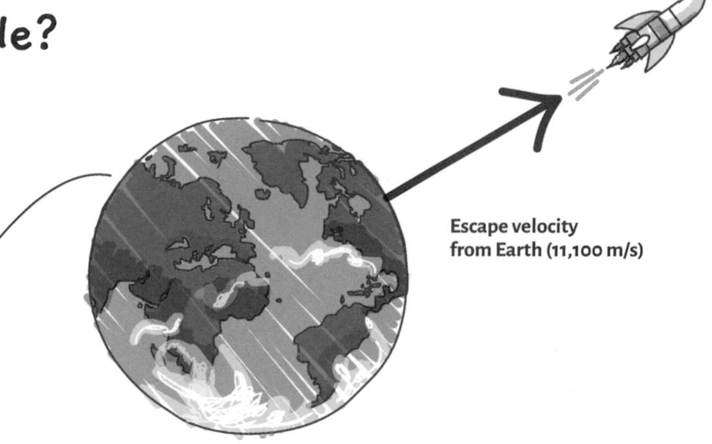

Consider throwing a ball on Earth directly upward. If enough energy is provided (a large enough velocity), the ball will escape Earth's gravity and will be free of its influence. This is known as the escape velocity and varies on different planets.

By equating the energy required to overcome Earth's gravity (gravitational potential energy of the Earth) with the kinetic energy of the ball, it can be shown that the escape velocity is given by the relationship:

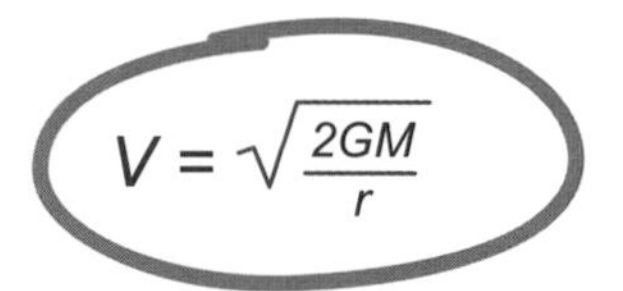

$$V = \sqrt{\frac{2GM}{r}}$$

where G is the universal gravitational constant, M is the mass of the Earth, and r is its radius.

For Earth, escape velocity is approximately 11,100 m/s. If the velocity required to escape the gravitational pull of a larger, more concentrated mass, such as a collapsed star, exceeded the speed of light ($\approx 3 \times 10^8$ m/s), then it would appear black due to no light escaping, and therefore become a black hole.

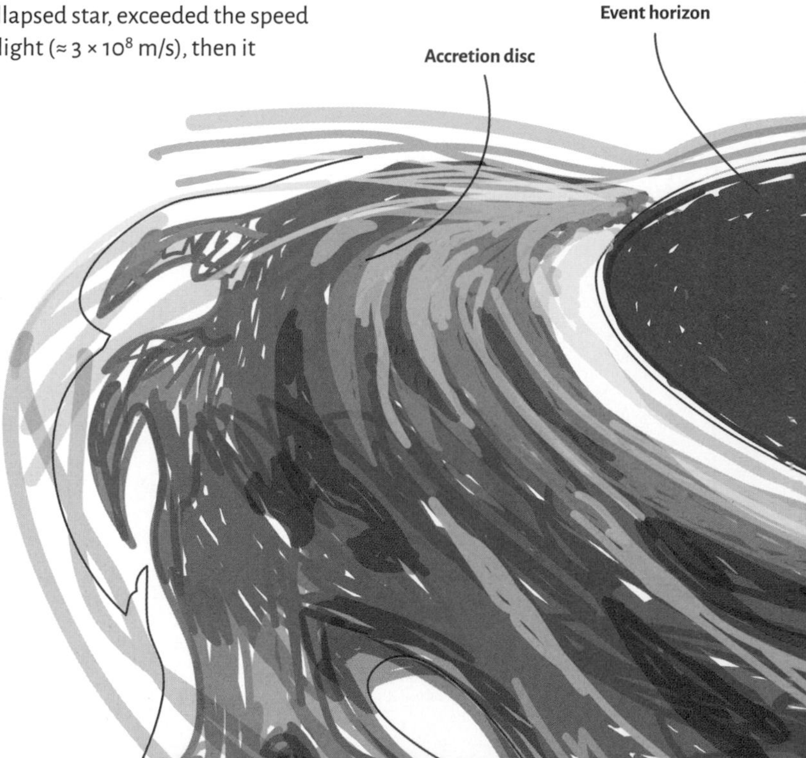

How a black hole is formed

As a very massive star reaches the end of its life cycle, its final phase is a supernova spectacular explosion that reveals a very hot dense core that is no longer fusing nuclei.

For stellar cores greater than approximately 5 solar masses, there is insufficient outward pressure to prevent complete gravitational collapse. Consequently, the remaining central mass becomes increasingly condensed until it reaches a point where the escape velocity exceeds the speed of light—and so a black hole is born.

The radius of a specific mass of star, *M*, at which this phenomenon happens is called the **Schwarzschild radius**, named for the German physicist Karl Schwarzschild and is given by the relationship:

$$R_{sch} = \frac{2GM}{c^2}$$

where *G* is the universal gravitational constant, *M*, is the mass of the stellar core, and *c* is the speed of light.

For a solar mass black hole, its radius would be a little under 3.2 km (2 miles).

The structure of a black hole

As all stars are thought to be spinning, as a star collapses, its spin speed increases because of conservation of the angular momentum (see chapter 4). When the radius reaches the critical Schwarzschild radius, light can no longer escape from the surface of the sphere formed—this is called the **event horizon**. Inside this rapidly spinning surface, the laws of physics become impossible to predict, and all information within it is lost from the universe. At the center of this spinning information sink is the **singularity**, a point of infinite density and zero volume, while very high speed streams of particles are accelerated along the axis of rotation producing **relativistic jets** that are formed outside the event horizon.

Black holes can be formed in a binary system, even if each star individually has insufficient mass to do so—the more massive star may form a black hole while its companion is stripped of its atmosphere, feeding material into the black hole, superheating it, and radiating huge quantities of energy as X-rays.

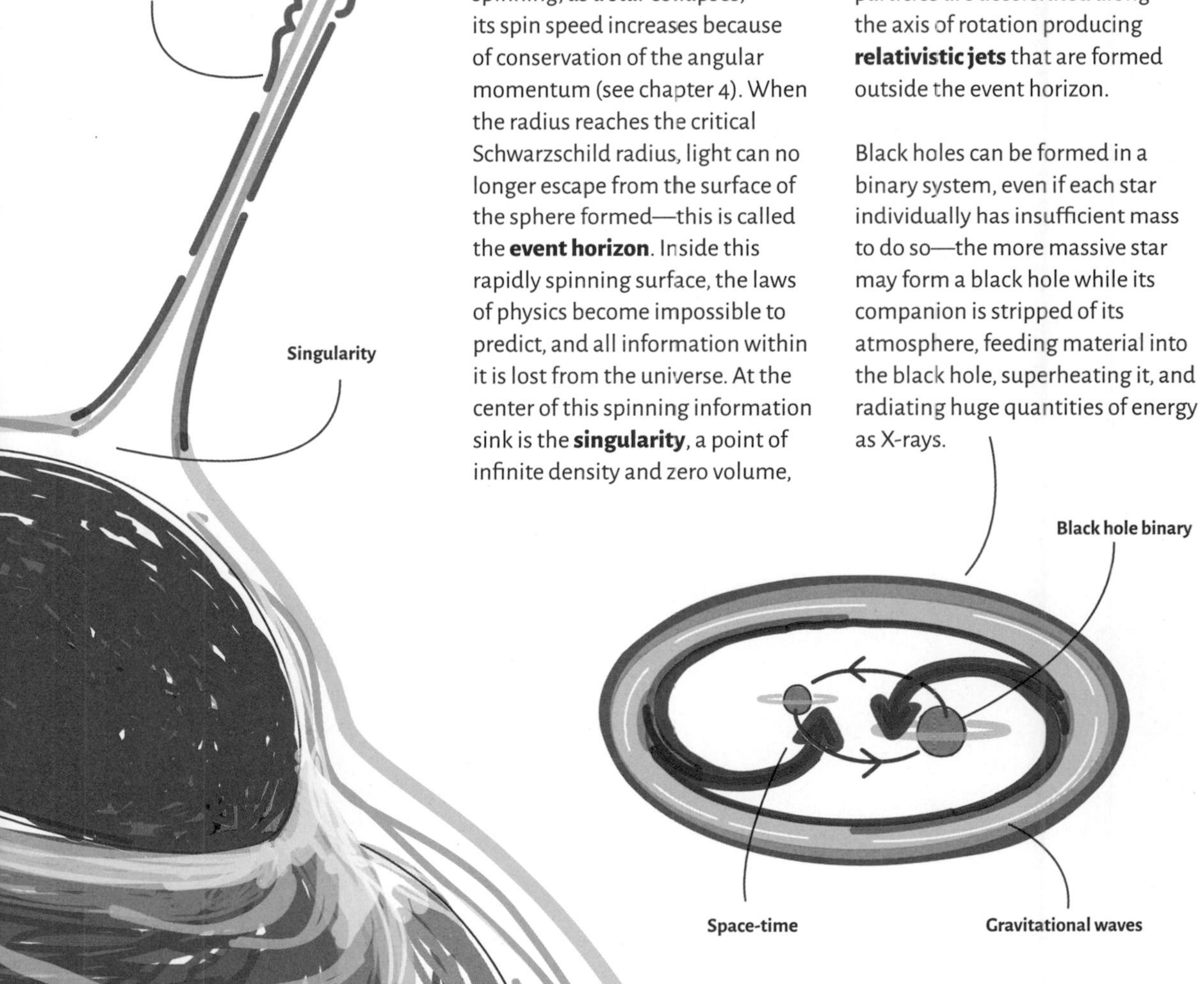

ASTROPHYSICS

STELLAR EVOLUTION

NEBULAE

Giant clouds of dust and gas where stars are made.

SMALL STARS

Small stars (-10 solar masses) become red giants, then white dwarfs.

MASSIVE STARS

Large stars (+10 solar masses) collapse, creating supernovas, then either a neutron star or a black hole.

THE HERTZSPRUNG-RUSSELL DIAGRAM

Visual categorization of stars by their temperature and brightness.

THE END OF THE UNIVERSE

GRAVITATIONAL LENSING

Changes the path of light rays, making objects appear to be elsewhere.

GRAVITATIONAL WAVES

Ripples in space-time created when massive bodies interact.

BLACK HOLES

Collapsed, condensed massive stars from which no light escapes.

CRITICAL DENSITY

A specific density at which the fate of the universe is decided.

EXPANSION

Universe expands infinitely; stars die.

THE BIG CRUNCH

Universe contracts into a singularity, and the cycle starts again.

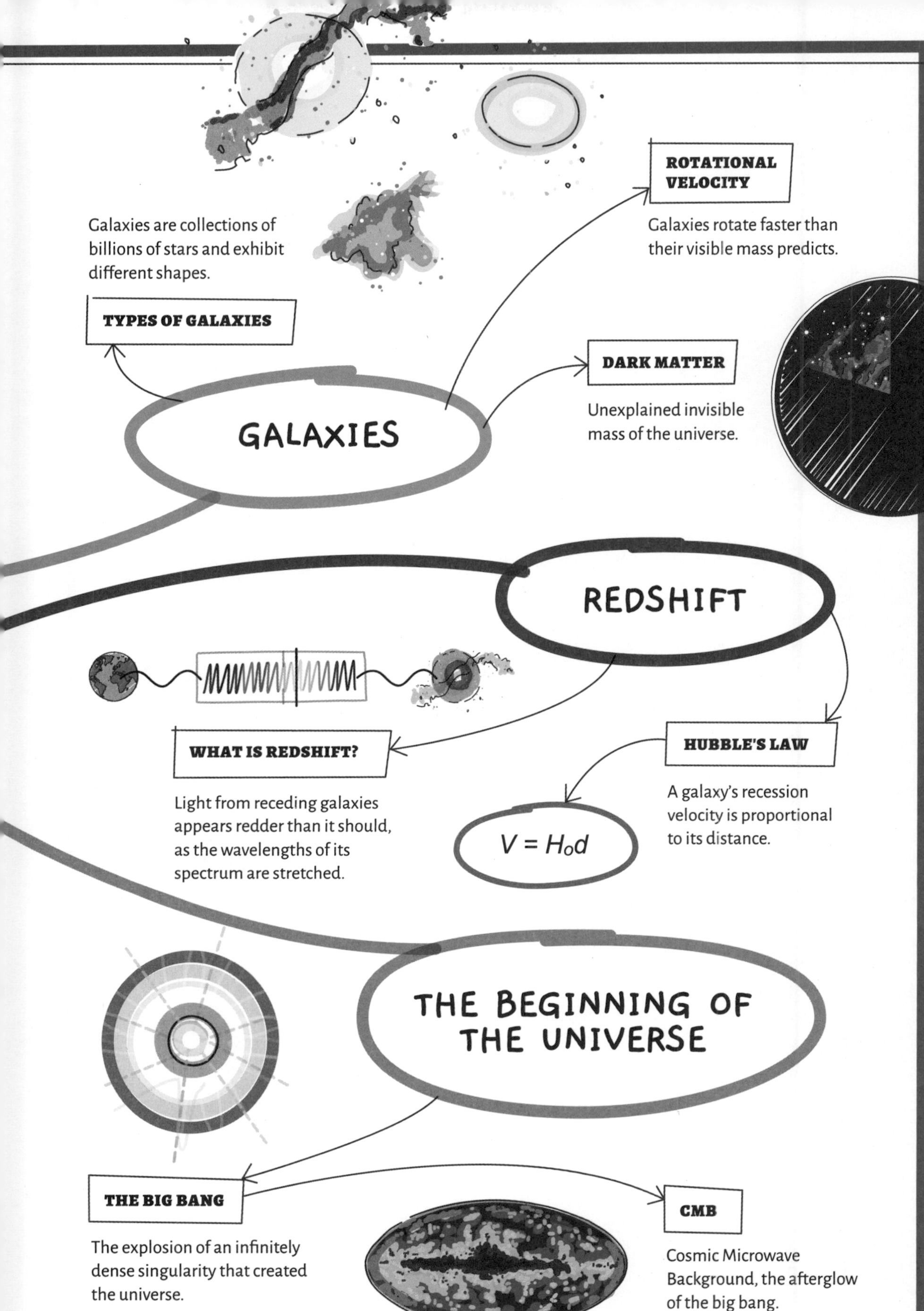
Galaxies are collections of billions of stars and exhibit different shapes.
TYPES OF GALAXIES
ROTATIONAL VELOCITY
Galaxies rotate faster than their visible mass predicts.
DARK MATTER
Unexplained invisible mass of the universe.
GALAXIES
REDSHIFT
WHAT IS REDSHIFT?
Light from receding galaxies appears redder than it should, as the wavelengths of its spectrum are stretched.
HUBBLE'S LAW
A galaxy's recession velocity is proportional to its distance.
$V = H_0 d$
THE BEGINNING OF THE UNIVERSE
THE BIG BANG
The explosion of an infinitely dense singularity that created the universe.
CMB
Cosmic Microwave Background, the afterglow of the big bang.

INDEX

A

absolute zero 129
absorption lines 158, 175
acceleration 9, 11, 15, 22–3, 24, 25, 28–9, 37
  equivalence 152, 164
  linear motion 26
  motion graphs 30–3
  oscillation 101, 103
  rotational motion 38, 40, 42, 45, 48, 49
  uniform 28, 29, 31, 33, 34–5
  see also gravity; Newton's laws of motion
aircraft 13, 41, 147, 148
air resistance 10, 13, 15, 35, 147
albedo 115
alpha particles 154, 156
alternating current (AC) 88, 97
ammeters 68, 69
Ampère, André-Marie 64
amperes (amps) 63, 64
angle of incidence 106, 107, 116–17, 126
angle of induction 88, 97
angle of reflection 106, 126
angle of refraction 107, 116–17, 127
angular momentum 51, 52, 58–9, 60, 185
angular speed 45, 49
antinodes 110
applied forces 10, 12, 24, 39
Archimedes' principle 144–5, 146, 149
ascent of projectile 35, 36
astronomical units 167
astrophysics 166–87
atomic number 155, 156
atomic structure 18–19, 82, 154–5, 158, 162, 165
auroras 81

B

balanced torques 46–7
bar magnets 20, 80, 88
batteries 68, 69, 71, 72
Bernoulli, Daniel 147
Bernoulli's principle 147, 148
beta particles 156
big bang theory 178–9, 187
big crunch 181, 186
bioluminescence 93, 97
black holes 58, 94, 124, 153, 169, 172, 177, 183–6
blue giants 169, 171
Bohr, Niels 6, 155, 158, 165
bosons 160, 161
bungee jumping 101
buoyancy 144–5, 149

C

capacitance 73, 75
capacitors 68, 69, 73, 75, 82
cars
  acceleration 29
  circular track 39, 40, 45
  displacement 27
  Doppler shift 111
  drag 147
  friction 10, 24, 40, 44, 45, 48, 53
  hybrid 89
  pistons 131, 137
  suspension 103
  wheels 39, 44, 131, 136
cathode ray tube (CRT) 83, 84
Celsius 129, 140
center of rotation 41, 58
centripetal acceleration 40, 45, 48, 49
centripetal forces 40, 42, 45, 48
chain reactions 19, 157, 165
charge
  neutrality 18
  subatomic particles 161
  transfer 63, 64, 66, 74
circular motion 39, 40–1, 48
circular orbits 39, 49
closed-circuit loops 70, 74
closed systems 50–3, 61
coherent waves 109, 125
collisions 51, 54–6, 61
color
  auroras 81
  CRT 83
  stars 171
  subatomic particles 161
  sunlight 106, 122
  thin-film interference 123
concave lenses 118, 119, 127
concave mirrors 118, 127
conductance 67
conduction 130, 133, 141
conductors 66, 73, 75, 91
conservation laws 44, 50–61, 70, 134–5, 185
constant acceleration *see* uniform acceleration
constant speed *see* uniform speed
constructive interference 109, 110, 112, 123, 124, 125, 126
contact forces 8, 10–13, 24, 38, 40, 76
control rods 19
convection 133, 141, 149
conventional current 63
convex lenses 118, 119, 127
convex mirrors 127
cooling 132–3, 141
Copernicus, Nicolaus 43
cosmic microwave background (CMB) 179, 187
Coulomb, Charles-Augustin 63
coulombs 18, 63, 64, 73, 74, 82, 157
Coulomb's law 82, 84
critical angle 117, 127
critical density 180–1, 186

D

dark energy 180
dark matter 173, 187
daughter elements 156, 165
Daytona racing 40
de Broglie, Louis 159
de Broglie wavelength 159
deceleration 9, 11, 29, 30, 32, 38
density 167, 169, 172, 178, 185
  critical 180–1, 186
  fluids 143–5, 146, 148–9
  optical 107, 116, 127
descent of projectile 35, 36
destructive interference 109, 110, 113, 125
dielectric 73
diffraction 106, 108, 113, 114, 125, 159
diodes 68, 69
displacement 34
  linear motion 26, 27, 28
  motion graphs 30–3
  oscillation 101, 103
displacement-time graphs 32–3, 37, 100
distance 27
  motion graphs 30–1, 32
  particles 27
  perpendicular 46
  range 35, 36
  torque 39, 44
distance-time graphs 22, 32
doping 162–3
Doppler shift 111, 112, 174
double-slit experiment 125, 159
drag 13, 22, 147

E

Earth 106, 151, 153, 174, 175
  escape velocity 184
  gravitational field 78–9, 85
  gravity 102
  magnetic field 81
  orbit 42
Einstein, Albert 6, 92, 151, 152–3, 158, 164, 174, 181, 182–3
Einstein ring 182
Einstein-Sitter model 181
elastic collisions 55, 56, 61
elastic potential 54
electrical circuits 68–73, 74
electrical resistance 64, 65–7, 70–1, 72, 75, 91
electric charge 60, 63, 74
  conservation 51, 52, 70
electric current 63–4, 67, 68, 74–5, 87–9
electric fields 73, 77, 80–3, 84–5, 104, 120
  see also electromagnetism
electricity 62–75, 86, 88, 90
electromagnetic induction 87–91, 96–7

electromagnetic radiation 92–3, 96–7, 115
electromagnetic spectrum 92–5, 96, 179
electromagnetic waves 98, 151, 156
electromagnetism 81, 86–97
electromagnets 21
electromotive force (emf) 70, 71, 72, 74, 90
electron gun 83
electrons 18, 63, 64, 65, 73, 81, 87, 154–6, 158–9, 161
  CRT 83
  energy shells 158, 159, 162
  semiconductors 162
electrostatic forces 14, 18, 25, 48, 76, 82, 84
elementary charge 63
elliptical galaxies 172
elliptical orbits 39, 49
emission lines 162, 175
energy
  conservation 44, 50–5, 60–1, 134–5
  grid 91, 97
  loss 90–1, 97
  quanta 158
  transfer 90–1, 97, 98, 104, 128–41
entropy 53, 138–9, 140
equilibrium 46, 47, 48
  oscillation 100–1, 103
  point 102, 99
equivalence principle 152, 164
escape velocity 184, 185
ether 151
event horizon 185
exclusion principle 159
expanding universe 174, 178, 180, 186
explosions
  momentum 57
  supernovas 94, 167, 169, 185

F

Fahrenheit 129, 140
Faraday, Michael 73, 87
Faraday's law of induction 87, 96
farads 73
fermions 160
ferromagnetic materials 80, 84
fiber optic cables 114, 117, 127
fissile neutrons 19, 157
fixed resistors 69
flat universe 180
Fleming, Sir John Ambrose 87
Fleming's left-hand rule 87, 96
flight time 35, 36
floating 144–5, 149
fluid dynamics 146
fluid flow 146, 149
fluids 142–9
  see also gases; liquids
food and energy 137
forces 8–25, 44, 76–85
  applied 10, 12, 24, 39
  contact 8, 10–13, 24, 38, 40, 76
  imbalance 9, 22, 46, 148
  magnitude 13
  noncontact forces 8, 14–21, 25, 38, 76–85
  see also electrostatic; gravitational; magnetic
free electrons 64, 65
frequency
  of light 92, 94–5, 96
  of waves 99, 105, 107, 108, 109, 111, 112
friction 10, 24, 40, 44, 45, 48, 53
fuel elements or rods 19
fuses 66, 69

G

galaxies 172–7, 178, 182, 187
gamma radiation 19, 92, 94, 96, 156, 157
gases 128–31, 134–5, 138, 140–1
  particles 55
  pressure 17, 143, 148
  see also fluids
general relativity 152–3, 164, 182
generators 88–9, 97
geocentric solar system 43
gluons 160, 161
gradient of graph 12
gravitational fields 15, 77–9, 85, 152–3
  black holes 184
  strength 79, 85
gravitational forces 14, 15, 16–17, 25, 76, 173, 180, 181, 186
gravitational lensing 153, 182
gravitational waves 153, 183, 186
graviton 17
gravity 11, 16–17, 34, 35, 43, 48, 102, 152–3, 167, 180

H

harmonics 110
heat 33, 51, 55, 61, 91
  thermistors 68, 69
  thermodynamics 128–41
  see also temperature; thermal
Heisenberg, Werner 159, 164
heliocentric solar system 43
helium 18, 51, 144, 154, 155, 156, 157, 167
hertz 94, 96, 99
Hertzsprung, Ejnar 170
Hertzsprung-Russell (HR) diagram 170–1, 186
Hooke, Robert 12
Hooke's law 12, 24, 103
Hubble, Edwin Powell 176
Hubble Deep Field (HDF) 176
Hubble's constant 176, 180
Hubble's law 176, 187
Hubble Space Telescope (HST) 166, 176, 177
Huygens, Christiaan 158
Huygens' principle 92
hybrid cars 89
hydrogen 17, 51, 155, 157, 158, 167, 168, 169
hydrostatic equilibrium 17

I

ice skaters 58, 173
ideal gas 55
incident waves 106, 107, 115, 126
  angle 106, 107, 116–17, 126
incoherent waves 109
incompressible particles 54
induction, electromagnetic 87–91, 96–7
inelastic collisions 55, 56, 61
inertia 9, 15, 151
  moment of 51, 59, 60
infrared (IR) 95, 96, 121
in phase waves 109
instantaneous velocity 45
insulators 66, 73, 75, 162
interference 109–10, 112–13, 114, 123–5, 126, 159
  constructive 109, 110, 112, 123, 124, 125, 126
  destructive 109, 110, 113, 125
  pattern 125
  thin-film 123, 126
  two-slit 125, 126
interferometer 123, 124, 126, 151, 183
internal energy 132, 134–7, 140
internal resistance 70, 71
ions 63
isotopes 64, 155, 157

J

joules 64

K

Kelvin 129, 140
Kepler, Johannes 43
kinematics 29, 150
kinetic energy 45, 53, 54, 61, 79, 103, 128–31, 136, 143, 146, 156–7
Kirchhoff's laws 70–1, 74

L

laboratory reference frame 175
lamps 69
left-hand rule 87, 96
lenses 114, 118–19, 120, 127
leptons 160
levers, principle 44, 49
lift 13, 41, 147, 148
light 86
  double-slit experiment 125, 159
  electromagnetic spectrum 92–5
  Hertzsprung-Russell diagram 170–1
  optics 114–27
  plants 137
  recession velocity 175
  redshift 174
  scattering 121, 122
  speed of 150, 151
  waves 104, 106–7, 108, 114–17, 184
  waves/particles 92, 158–9
light-dependent resistors (LDR) 68, 69
light dispersion 107
light emitting diodes (LED) 69
LIGO 124, 183
linear momentum conservation 51, 52, 56–7, 61
linear motion 26–37
liquids 128–31, 138, 140
  see also fluids
longitudinal waves 104, 113
low-pressure system 146
luminosity 170–1

M

magnetic fields 20–1, 77, 80–1, 84–5, 104, 120
 permanent 21, 80, 81
 strength 21, 80, 84
 see also electromagnetism
magnetic forces 14, 20–1, 25, 76
magnetized bodies 20
main sequence stars 168, 171
Mars 43
mass 11, 15, 16, 22, 23, 25, 77
 gravitational field 79, 85, 152, 153
 inertia 9
 momentum 51, 56, 57, 58–9, 60
 point 78
 rotational motion 45
 subatomic particles 161
massive stars 58, 169, 185, 186
mass-spring systems 103, 113
mechanical energy 88, 89, 90, 97
Michelson interferometer 123, 124, 126, 151, 183
microscopes 118
microwaves 95, 96, 179
Milky Way 172, 173
mirrors 114, 115, 118–19, 127
moderator fluid 19
momentum 23, 24, 54, 60
 angular 51, 52, 58–9, 60, 185
 conservation 51, 52, 53, 56–9, 60–1, 185
 linear 51, 52, 56–7, 61
 negative 56, 57
 positive 56
 rate of change 23
 sum of 57
monochromatic light 124, 125
moon 16–7
Morgan-Keenan (MK) system 171
morphology 172
motion
 circular 39, 40–1, 48
 equations of 34, 37
 graphs 30–3, 37
 linear 26–37
 Newton's laws 15, 22–3, 24, 28, 38, 45, 77, 103
 projectile 35
 retrograde 43
 rotational 38–49, 58–9, 173, 187
motor generators 89, 97
motors 89, 97
multistage velocity-time graphs 31
muons 161

N

nebulae 167, 168, 186
negative momentum 56, 57
negative velocity 28
net displacement 27
neutrinos 161
neutrons 18, 19, 51, 155–7, 160, 161, 162
neutron stars 58, 169
Newton, Sir Isaac 6, 9, 16, 92, 150, 158
newtons 9, 12, 15, 16, 23, 24
Newton's cradle 57
Newton's law of gravity 16, 25, 43, 78, 85, 152, 153
Newton's laws of motion 15, 22–3, 24, 28, 38, 45, 77, 103
nodes 110
noncontact forces 8, 14–21, 25, 38, 76–85
 see also electrostatic; gravitational; magnetic
nonmagnetic metals 80, 84
nonvertical normal reaction force 40
normal reaction forces 10–11, 24, 39, 40
n-type semiconductors 162
nuclear decay 156
nuclear fission 19, 157, 165
nuclear forces 14, 19, 25
nuclear fusion 93, 155, 157, 165, 167
nuclear physics 154–5, 165
nuclear power stations 19, 157
nuclear reactions 19, 93, 94, 155, 156–7, 165, 167
nucleons 19, 156
nucleus 19, 154–7, 159, 165

O

oblique collisions 54
ohms 64, 66, 67
Ohm's law 67, 75
optical density 107, 116, 127
optical telescopes 118
orbital motion 17, 39, 42–3, 49
orbital path 42
orbital period 42, 43, 49
orbital radius 43
orbital speed 42, 49
origin 27, 31–3, 36
oscillations 98, 99–105, 113, 129
out of phase waves 109

P

parabola 35
parallel circuits 72, 74
parallel electric fields 82, 84
parallel gravity fields 79, 85
partial reflection 107, 116
particles 36, 82
 collisions 54–5
 incompressible 54
 light 92, 158–9
 see also electricity
pascals 129, 143
Pauli, Wolfgang 159, 164
pendulums 113
 simple 102
Penzias, Arno 179
permanent magnetic fields 21, 80, 81
photons 81, 92–4, 158–9, 161
pistons 131, 137
pivot point 46, 47
Planck, Max 6, 158, 164
Planck's constant 159
planetary nebulae 168
plum pudding model 154
polarization 120, 126
polaroid lens 120
positive momentum 56
positive velocity 28
potential difference 63, 64, 70, 71, 73, 82, 83
potential energy 53, 54, 61, 79, 103
power cables 91, 97
power cells 68, 69
power stations 88, 90–1, 157
pressure 143–7, 148
primary coil 90
prisms 107
probability wave function 159
projectile motion 35
protons 18, 19, 155–7, 159–60, 161, 162
p-type semiconductors 162
pulling forces 10, 24
pushing forces 10, 24

Q

quanta 92, 158
quantization of light 158
quantized energies 155
quantum mechanics 159
quantum physics 150, 158–9
quantum tunneling 159
quarks 160–1, 164
quasars 177

R

radar 111
radial electric field 82
radial gravitational fields 78, 85
radiation 133, 141, 179
radioactive decay 19, 54, 156, 165
radioactive elements 19, 157
radio waves 58, 92, 95, 96, 108, 111, 179
radius of the circular path 45, 49
rainbows 107
range 35, 36
rate of change of momentum 23
reaction forces 10–11, 24, 39, 40, 47
real images 119, 127
recession velocity 175, 178
recoil 57
red giants 168, 171
redshift 174, 175, 178, 187
reflected waves 106, 110, 115, 126
reflecting telescopes 118, 127
reflection 106, 113, 114, 115–17, 121, 126–7
 angle of 106, 126
 partial 107, 116
 total internal 117, 127
refraction 106, 107, 113, 114, 116, 117, 118, 127
 angle of 107, 116–17, 127
refractive index 116, 127
relative velocity 151
relativistic jets 185
relativity 151–3, 164, 182
restoring forces 12
retrograde motion 43
rotational dynamics 45–7, 48–9
rotational motion 38–49, 58–9, 173, 187
rotational velocity 38, 40–2, 45, 49, 173, 187
rotation torque 89
Russell, Henry Norris 170
Rutherford, Ernest 154–5, 165

S

scattering experiment 154, 165
Schrödinger, Erwin 159, 164
Schwarzschild, Karl 185
Schwarzschild radius 185
secondary coil 90
semiconductors 162–3, 164
series circuits 72, 74
side force 41
silicon 162
simple harmonic oscillators 100–1, 102, 103, 113, 129
singularity 178, 181, 185, 186, 187
sinking 144–5, 149
sinusoidal 101
Sitter, Willem de 181
small stars 17, 168, 186
Snell's law 116, 117, 127
solids 128–31, 138–9, 140
sound 33, 51, 53, 55, 57, 61, 91
  waves 52, 98, 104–5, 108, 111
space-time 152–3, 164, 182
special relativity 151, 164
spectral lines 158
spectroscopy 174
spherical universe 180
spin 161
spinning cylinders 59
spinning tops 59
spiral galaxies 172
spring constant 12
spring forces 10, 12, 24
Standard Model 160–1, 164
standing waves 109, 110, 112
stars 17, 58, 167–73, 182, 185, 186
  Hertzsprung-Russell diagram 170–1
  nuclear fusion 157
static electricity 18, 63
statics 39, 45, 46, 48
step-down transformers 90, 91, 97
step-up transformers 90–1, 97
sun 153, 167, 168–9, 171, 173, 182
sunlight 106, 107, 115, 120–2, 132–3, 137
supergiants 171
supernovas 94, 167, 169, 185
SUVAT equations 34

T

tau 161
telescopes 118, 166, 174
televisions 83
temperature
  electrical resistance 65
  fluids 143, 146
  gradient 133, 141
  Hertzsprung-Russell diagram 170–1
  light sources 93, 97
  see also heat; thermal
tension 10, 11, 41
tesla 21, 80, 84
thermal energy 130, 134, 141
thermal expansion 131, 141
thermal neutrons 19
thermistors 68, 69
thermodynamics 128–41
  laws of 134–9, 140
thin-film interference 123, 126
Thompson, Sir John 154
thought experiments 153
3D glasses 114, 120
thrust 13, 22, 144–5
tidal forces 17
time 36, 164
  linear motion 26
  motion graphs 30–3, 34
  oscillations 99, 103
  relativity 151–3
  rotational motion 45, 49
  waves 105, 112
torque 39, 44, 46–7, 49, 89
total internal reflection 117, 127
total magnitude of the displacement 27
trajectory 35, 36
transformers 90, 91, 97
transverse waves 104, 110, 113
traveling waves 104–5, 113
turning moment 46, 48, 89
two-slit interference 125, 126

U

ultraviolet (UV) light 94, 96, 121
uncertainty principle 159
uniform acceleration 28, 29, 31, 33, 34–5
uniform electric field 83
uniform speed 32
uniform velocity 30
universal gravitational constant 16, 43, 78, 82, 184
universe
  beginnings 176–7, 187
  end of 180–1, 186
  expanding 178, 180, 181, 186

V

valence bonding 162
valence electrons 162
variable resistors 68, 69
vectors 13, 24, 27, 28, 29, 36, 40, 42, 56
velocity 9, 22–3, 28–9, 79
  Bernoulli's principle 147
  collisions 55
  escape 184, 185
  instantaneous 45
  linear motion 26
  momentum 51, 56–7
  motion graphs 30–3, 34
  oscillation 101
  recession 175, 178
  relative 151
  rotational 38, 40–2, 45, 49, 173, 187
  vertical component 28, 36
  waves 105, 111, 112, 116
velocity-time graphs 30–1, 33, 34, 37
vertical force 41
Very Large Telescope (VLT) 166
virtual images 119, 127
visible spectrum 95, 96
voltage 63–4, 67, 68, 71, 72–3, 74–5
  semiconductors 162
  transformers 90, 91
voltmeters 68, 69
volts 64, 75

W

water
  fluid flow 146
  pressure 144–5, 149
  temperature 129, 130, 132–3
  waves 98, 104, 108
wavelength 94, 96, 99, 105, 107, 108, 109, 110–11, 112, 116, 123, 125, 159
  de Broglie 159
  recession velocity 175
wave mechanics 159
waves 98–113, 114–17
  amplitude 99, 112
  frequency 99, 112
  interference 105
  light 92, 158–9
  particle duality 92, 158–9
  peak 99, 109
  properties 98, 106–8, 113
  speed 105, 113
  trough 99, 109
wave shadow 108
weather systems 146
weight 11, 13, 15, 16, 22, 25, 145
wheels 39, 44, 131, 136
white dwarf 168, 171
Wilson, Robert 179
work done 44, 134–7, 140

X

X-rays 94, 95, 96, 121, 185

Y

Young, Thomas 125

Z

zero velocity 32, 33, 57

Acknowledgments

I would like to extend my heartfelt thanks to a number of individuals who have provided invaluable support and guidance throughout the writing of this book. First, my thanks go to Lindsey Johns, who introduced me to the project back in November 2019, providing me with this exciting opportunity. I offer my immense gratitude to Kate Duffy for her professionalism and patience when the first chapters were painstakingly drafted and redrafted with her careful guidance. Thanks also go to Sarah Skeate for her beautiful illustrations, Kathy Steeden for her editing skills, Mike LeBihan Studio for the polished design, and Viv Croot, who has provided so much valuable advice and good humor while subtly cracking the whip. Finally, and not least, I wish to thank my wife, Victoria, and my boys, William and Jason, for their patience and support throughout, giving me determination and fortitude when the task seemed overwhelming.